CORN Publication Series 11

ISBN 978-2-503-51779-7
D/2009/0095/14

Exploring the food chain

Food production and food processing
in western Europe, 1850-1990

Edited by Yves Segers, Jan Bieleman
& Erik Buyst

CONTENTS

LIST OF CONTRIBUTORS

JAN BIELEMAN   Rural History Group, Wageningen University, The Netherlands

PAUL BRASSLEY   School of Geography, University of Plymouth, United Kingdom

ERIK BUYST   Workshop in Quantitative Economic History, University of Leuven, Belgium

JEAN-MICHEL CHEVET   INRA-ALISS, Ivry Cedex, France

TED COLLINS   Dept. of Agricultural and Food Economics, University of Reading, United Kingdom

MARIA DE WAELE   STAM-Stadsmuseum Ghent, Belgium

ALAIN DROUARD   Directeur de recherche au CNRS, Paris, France

WIM LEFEBVRE   Workshop in Quantitative Economic History, University of Leuven, Belgium

CARIN MARTIIN   Swedish University of Agricultural Sciences, Uppsala and Stockholm University, Sweden

FLEMING JUST   Institute of History and Civilization, University of Southern Denmark, Esbjerg, Denmark

PETER A. KOOLMEES   Dept. of Veterinary Public Health, Utrecht University, The Netherlands

PETER LUMMEL   Open Air Museum, Domain Dahlem, Berlin, Germany

MOGENS R. NISSEN   Institute of History and Civilization, University of Southern Denmark, Esbjerg, Denmark

YVES SEGERS   Interfaculty Centre for Agrarian History, University of Leuven and Hogeschool-Universiteit Brussel (HUB), Belgium

HANS JURGEN TEUTEBERG   Department of History, University of Münster, Germany

LIST OF FIGURES

LIST OF TABLES

Preface

Food production and the available range of foods in Europe have changed dramatically over the past two centuries as a result of many different factors. Population growth and urbanisation during the nineteenth century led to a quest for higher yields and improved productivity. Partly thanks to the introduction of new crops and cultivation techniques, broader scientific knowledge and the increased input of technology, farmers succeeded in meeting and even exceeding the increased demand for food. The shortages from previous centuries were converted into surpluses, particularly after the Second World War. This was associated with an (initially cautious) increase in scale, commercialisation, capital intensification and specialisation of the sector.

Population growth was not the only driving force; increasing purchasing power also played a part. Initially, the urban elite boosted the demand for luxury foods and therefore for innovation. Starting in the second half of the nineteenth century, the purchasing power of broad sections of the Western European population increased. They no longer ate only basic foods, such as potatoes and bread, but also broadened their range of food. Meat, fish, dairy products (milk, butter and cheese), vegetables and fruit were increasingly within their grasp, as well as more expensive foods or products from the emerging modern food industry, such as chocolate, canned food, sweets, etc. Together with the increased number of town-dwellers and industrial workers, who had to buy most of their food from the commercial circuit, these developments gave a huge boost to the modernisation of food production and processing.

The processing of foods therefore underwent far-reaching modernisation from the second half of the nineteenth century onwards. Until then, the food industry had been a traditional sector, mainly restricted to a few traditional activities, with many small-scale enterprises: breweries, gin distilleries, mills, sugar refineries, etc. The links between agriculture and food processing were very tight. However, as in the primary sector, an increase in purchasing power and in the (urban) population, together with technical innovations, led to the establishment and development of modern margarine factories, (cooperative) dairy factories or creameries, industrial slaughterhouses, chocolate and biscuit factories, canneries, etc. A lot of these new companies immediately used steam power or even gas and electricity and introduced machinery into the production process. By the end of the twentieth century, food processing had been transformed into a modern, often high-tech industry, dominated by a few large enterprises in every subsector, offering a wide range of products. In only a few decades these industries were transformed from an important complement to primary agricultural production on the farms to an all-embracing industrial business.

The food industry is an important link in the food system, which is understood to mean a chain of activities from farm to household. Recent studies show the significant importance of this sector in the economic and agricultural development of the nineteenth century. At the end of the twentieth century this situation had not changed. On the contrary, today the food industry in Europe is one of the most important economic sectors in terms of turnover and employment.

I.   An evolving food chain

A key theme of this book is the place of the primary and processing sectors in the food chain. Since the middle of the nineteenth century, the food chain has changed dramatically, as have the various links in the chain and their mutual interdependence. Three basic processes can be distinguished. Firstly, *lengthening* of the chain. The number of intermediate links in the food system increased substantially. Around the mid-nineteenth century, agriculture on the European continent was aimed in particular at self-sufficiency and was not very commercially oriented. Agricultural products were processed chiefly on the farm itself. Farmers' wives were responsible for processing milk into butter and cheese, which they themselves took to the weekly market in the neighbouring town. The first dairy factories did not appear until the end of the century, sometimes in the form of cooperatives. They usually operated on a relatively small scale and within a regional or national context. Over time, produce was no longer marketed by the producers themselves. Gradually, (market) traders, cooperative sales associations and modern shops took over these tasks. Moreover, the geographical distance between the various links also increased. With the development of a more effective transport network, new means of transport and better communication techniques, the geographical distance between the beginning and the end of the chain increased. Around 1850, farmers delivered their animals for slaughter to butchers or to the slaughterhouse in the neighbouring town. Production and certainly processing of meat usually took place in the immediate vicinity of the large consumption centres. The arrival of modern steamships meant that frozen meat could be shipped to Europe from the United States.

A second evolution can be described as a process of *differentiation*. The various links or phases in the chain have themselves become much more complex over the past two centuries. A good example in this context are of course developments on the farm. Farmers not only had to possess more technological know-how, but they also had to comply with increasing numbers of rules regarding hygiene and health (often prescribed by the government). One of the consequences was that farmers became increasingly dependent on third parties, which brings us to the final development.

Thirdly, we also observe a process of *narrowing* in the food chain. The interdependence of the various links continued to increase. Growing system interrelation emerged. The activities in the chain became increasingly geared to one another. The shift in the processing of farming products into food to the food factories of course meant that both sides had to reach firm agreements. Dairy farmers had to ensure that they delivered their milk to the dairy factory under ideal hygienic conditions.

In parallel to this narrowing, a *change of power* also seems to be taking place in the chain. In the mid-nineteenth century farmers and market gardeners themselves largely determined what they produced and how. Of course, even then they also took into account the wishes of the market and the consumers, but their autonomy seems to have been much greater. At that time, the agricultural producers were the ones who largely shaped the operation of the chain or attempted to steer it through their own initiatives, such as cooperatives. Nowadays, the farmer has perhaps become the weakest link in the entire chain. In some sectors, through the system of contract farming, farmers are bound to the requirements, needs and desires of the food industry or of supermarkets.

II. A collection of papers

Despite the economic importance of the modern food industry (and the critical relationship with the farmer) relatively little research has so far been conducted into the history of food processing and the alimentary industry in Western Europe. What little has been performed, was achieved primarily within a national context. Agricultural and economic historians traditionally focussed on land and output, agricultural policy, agro-systems, agricultural science and education, etc. Consumption historians confined themselves chiefly to mapping food consumption, involving a quantitative approach and more recent cultural themes such as eating out, the rise of restaurants, the role played by cooks and cookbooks in the diffusion of eating culture, food and hygiene, etc. In particular, the extremes of the food chain were studied. Many of the intermediate links in the chain remained largely underexposed, such as food processing or the food industry. For this reason, this new Corn publication is important; it fills a gap in international historiography. The book also fits in well with Volume 7 in the Corn Publication Series: "Land, shops and kitchens".

The essays in this book are the result of a conference held in Leuven (Belgium) in November 2003, organised by the Interfaculty Centre for Agrarian History (K.U.Leuven), the Workshop in Quantitative Economic History (K.U.Leuven) and the Rural History Group of Wageningen University (The Netherlands). Agricultural historians and food historians came together with the intention of gaining a better understanding of the food chain, particularly the development and interdependence or interaction between the first two links: food production and food processing or farming and the food industry, against the background of modernisation, globalisation and expansion. The papers in this book cover virtually all Western European and Scandinavian countries and concern not only farmers, but also butchers, *chocolatiers*, fishermen and viticulturists. The period between 1880 and the Second World War is examined in detail, although some authors discuss a period of almost 50 years. In most cases, one product or product group is the focal point. A surprising amount of attention is devoted to meat, dairy, fish and wine. Focusing on one product makes it possible to discuss various links in the chain in greater detail. In addition, some topics take a more prominent place in several contributions. We summarise these briefly below; a thorough evaluation is given at the end of this book (see Brassley in this volume).

The first topic is the increasing role of technology. New scientific and technical knowledge brought about far-reaching changes in the production process, on the farm and in the factory. The introduction of new conservation techniques formed the basis for the modern canning industry. New fishing techniques and the introduction of steam traction on fishing vessels substantially changed the fishing sector around 1900 (see the contributions by Drouard and Teuteberg). It is striking that new techniques did not always lead immediately to the elimination of the traditional methods. In many cases, they continued to exist alongside one another and complemented one another. On small farms and dairies, people were not always keen to alter tried-and-trusted practices, or they simply did not have the necessary capital (see Segers and Lefebvre and Martiin).

The second topic is the emergence of cooperative enterprises. Through these partnerships, farmers and market gardeners attempted to obtain better prices for their products. From the end of the nineteenth century onwards, farmers began to set up purchasing, credit and production cooperatives in many countries (see Chevet and Bieleman). Through these partnerships, not only did they gain greater economic power, but they also acquired increasing political influence and, when necessary, they could attempt to influence agricultural policy at national and, later, even European level (see Just and Nissen).

The third topic is the increasing focus on health, hygiene and quality, not only by the government but also among consumers. The industrialisation and chemicalisation of food production at the end of the nineteenth century led to suspicion among consumers about the quality, origin, composition and health of some foods. Some new foods were composed of elements from widely varying raw materials and ingredients. Their agrarian origins were often no longer evident or were unclear, which of course made forgery and fraud possible more than previously and in other forms (see De Waele and Koolmees).

III. Acknowledgements

Various people and institutions contributed to the organisation and smooth running of the conference and to the publication of this volume in the Corn Publication Series. Firstly, we wish to thank the authors for their enthusiasm and expertise in their approach to this topic. The conference was also well served by commentators on the individual papers. Therefore, we wish to extend our sincere thanks to the many referees: Paul Brassley, Adel den Hartog, Pim Kooij, Janken Myrdal, Mark Overton, Peter Solar, Peter Priester, Peter Scholliers, Margreet van der Burg, Eric Vanhaute and Leen Van Molle. We thank Wim Lefebvre for his help with the practical preparation and organisation of the conference. Roeland Hermans and Bieke Verhoelst were responsible for the technical editing of the present publication. Alexis Vermeylen was in charge of design. The organisation of the conference and the publication of this volume were made possible thanks to the financial support from the Flemish Fund for Scientific Research (FWO-Vlaanderen), the CORN Research Community, the University of Leuven and Brepols Publishers.

Yves Segers
Jan Bieleman
Erik Buyst

1 The emergence of mechanised dairying in the northern Netherlands, and particularly in the provinces of Drenthe and Friesland

Jan BIELEMAN, Wageningen University

This paper looks at the shift in dairy production from farms to (cooperative) creameries in two, geographically quite different provinces in the northern part of the Netherlands. The focus is on the emergence of mechanised dairying in the province of Drenthe which I examine along with developments in the adjacent province of Friesland. This is usually depicted as the country's first and foremost dairying province, although in this it had to compete with the provinces of South- and North-Holland. By contrast, farming in Drenthe was for a long time described as backward and stagnant. Only recently has it become clear that farmers in the open-field farming regions were also affected by international economic developments which pushed them into adapting their farming system and technology. The emergence of mechanised dairying around 1890 had drastic effects on the farming system in Friesland, but even more so in the sandy parts of Drenthe. In this province especially it initiated a wide complex of changes, including the promotion of small farming which had far-reaching social consequences.

I.   Introduction

During the nineteenth century, the daily menu of many Europeans gradually changed as a slowly increasing prosperity allowed them to spend a growing part of their income on more 'luxury' agricultural products like meat, sugar, vegetables, fruit, and – of course – dairy products. Dutch farmers benefited from this gradual change in the mass consumption of these products, encouraged to do so by the rapid developments in transport systems with the advent of the steam engine and by the liberal winds that had changed trade in the wake of Britain's repeal of the Corn Laws in 1846. The abolition of import restrictions meant that important products became relatively cheaper, in turn stimulating demand.

In the eighteenth century, livestock farmers in Friesland, the province next to Drenthe, had already begun to profit from a growing demand for butter from Britain. From time immemorial in this province, the production of butter had been an important aspect of the livestock farming system alongside stockbreeding. However, during the eighteenth century, the processing of milk into butter became the mainstay of the farmers' activities. This shift in production aims can be seen from the amount of butter supplied to the market in the town of Sneek, one of the main Friesian butter markets in those days – if not the main one. The supply had increased from less than 30,000 vierendelen ($\approx$ 1.2 million kg) around 1760 to a long-term peak of 52,000-54,000 vierendelen ($\approx$ 2.1-2.2 million kg) in the 1820s, most of which was shipped to Britain (Figure 1). At the same time, supplies in other Friesian markets, like those in Leeuwarden and Harlingen had also grown.

Figure 1.1 Market supply of butter in Sneek (Friesland) in vierendelen (á 40 kg), 1711 - 1877 (annual figures and 10-years moving avarage)

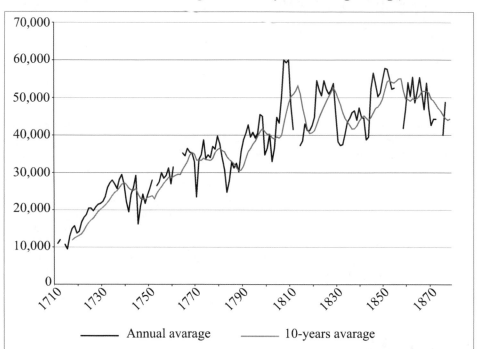

Source: After Faber, 1972: 598 (graph IV.15) (with special thanks to Prof. Faber (Bennekom) who kindly made the underlying data available); *Verslagen van de landbouw*, 1855-1877.

For a long time, regions in the southern parts of the province of Holland – the Rijnland region (around the city of Leiden) and Delfland region (around the city of Delft) – had been the most important butter-producing regions in the Dutch Republic. However, they were surpassed when Friesian farmers began to develop their dairying activities, turning Friesland into the butter-producing region par excellence of the time. From 1780 onwards, Friesian butter was no longer being exported to Britain through the port of Amsterdam but directly from the Friesian port of Harlingen. Even during the Napoleonic era and in spite of the Emperor's economic policy towards Britain, Dutch dairy farmers (i.e. mainly Friesian farmers) managed to retain almost one-quarter of the British butter import market, and in 1815 almost three quarters of Dutch butter exports went through Harlingen (Bos, 1978: 232). This meant, however, that the price Friesian farmers got for their butter was, to a large extent, decided on the London market.

At the same time that Friesian farmers in the typical pasture areas in the lower lying parts of the province were turning to the British market, farmers in the eastern, sandy part of this province, the Wouden area, were gradually converting from arable farming to livestock and dairy production. As the area under cultivation was reduced, the number of cattle increased considerably (Faber, 1972: 202-203). Around 1800 it was reported from the directly adjacent area to the north of Overijssel (the area around the town of Steenwijk) that for some time most of the winter sown rye was being fed to the animals, because farmers 'had begun to take advantage of the benefits cattle brought them compared to arable farming. For the same reason they had also started to cultivate turnips and spurry for animal fodder'.[1]

Thus, it appears that a kind of 'butter frontier' formed by butter-producing farming systems slowly moved from the holocene, green and wealthy pasture areas of lower-lying Friesland to the east, up onto the pleistocene, sandy parts of the hinterland, i.e. Drenthe.

II.   The emergence of a new farming system

Physically, the landscape in the province of Drenthe (2,622 km^2; 82,800 inhabitants in 1850) consists of a large boulder clay plateau with hardly any relief ('The Drenthe Plateau'). It is covered by a blanket of sand and partly by vast peat bogs. Along its fringes, the physical situation allowed farmers to have more and better grasslands. From the eighteenth century onwards, farmers specialised in livestock farming as the number of their cattle increased. Nevertheless, farming on the boulder clay plateau was dominated by open-field farming, and the arable complexes together with the villages lay like small islands in a virtually endless sea of rough moorland which was used as a commons. Ever since the Late Middle Ages, and even more after 1650, the cultivation of rye had become the mainstay of the farming system. Cattle played only a subordinate role as the sale of rye had become the farmer's first and foremost source of income. By contrast with developments in the lower fringe areas, the number of cattle in the open-field villages dwindled significantly according as sheep became the farmers' most important producers of manure (Bieleman, 1987).

---

[1] Translated from a quotation in Slicher van Bath, 1957: 570.

Table 1.1 **Number (in %) of dairy farmers according to their stock of dairy cows in the two main dairying districts[2] and the Wouden district in Friesland and in the province of Drenthe, 1910**

	1-2 cows	3-5 cows	6-10 cows	11-20 cows	≥21 cows	Total - index	Total absolute
Friesland: 2 dairying districts	11	12	18	21	38	100	5.119
Friesland: the Wouden district	41	27	16	11	5	100	9.676
Drenthe	45	34	17	4	0	100	13.549

Source: National Archive in The Hague, Archief Directie van Landbouw, Landbouw-economische aangelegenheden 1813-1945, inv.no. 48.

From the results of the first national cattle-census, held in 1910 (Table 1.1), it appears that in Drenthe, as well as in the adjacent sandy Wouden district of Friesland, small holdings dominated, as respectively 79 per cent and 68 per cent of all keepers of dairy cows had no more than five cows. In the two dairying districts in Friesland herds were much larger, as there 77 per cent of all dairy farmers had at least six dairy cows. In Drenthe, this was also, if not largely, the result of the developments that had been taking place ever since the mid- nineteenth century. However, earlier, at the beginning of that century, the situation was, comparatively speaking, hardly different: 5,440 farmers held an average of seven to eight head of cattle in 1800 (Bieleman, 1987: 336-342).

With the change in the international economy around 1850, farming in Drenthe changed too. Benefiting from the opportunities offered by the introduction of steamships and railways, and spurred by the advent of free trade throughout Europe after 1846, Drenthian farmers turned to foreign markets for livestock products (Bieleman, 1996).

The 'price revolution' that went along with this, or more specifically the differentiated price development for livestock products as compared with that for arable products (Figure 1.2) caused a shift in farming that proved to be both structural and influential.

---

[2] These two districts held 28 per cent of all Friesian dairy farmers; the Wouden district 53 per cent.

Figure 1.2   The price-index of rye and butter in Drenthe, 1815-1910 (1812/1821 = 100)

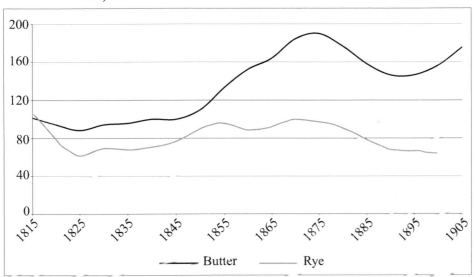

Source: After Bieleman, 1987: 171 (graph 3.5).

As the price of rye remained fairly stable and the price of butter increased by 90%, farmers transformed their farming system, which had been mainly oriented towards selling rye, into one directed at selling butter and pigs, both products with a much higher added value. Up till then livestock had served the arable part of the farming system, but from now on the arable served the livestock. In short, in order to produce more high-grade products, the former cash crop, rye[3], now became a fodder crop. Buttermilk together with ground rye and potatoes was used for feeding calves and fattening pigs and piglets. From then on, dairying coupled with cattle breeding and pig feeding became the pivot of the new farming system and a new source of income.

In 1845, the editor of the annual Report on Farming in the Netherlands wrote that in Drenthe 'some husbandmen in the province were increasingly applying themselves each year to the production of butter which is sent to Holland' (Verslag van de Landbouw, 1845: 81). At the general meeting of the Agricultural Society in Drenthe in 1877, one of the speakers remarked that 'compared to the situation about 25 to 30 years ago when Drenthian farmers' income came largely from rye cultivation, now the main source of wealth is their livestock. The 'korenesch' [= the open-field] is being used to maintain their families and their livestock'.[4] Soon, farmers also began to buy increasing quantities of additional, imported fodder stuffs like grains and concentrates like oil cakes. A notebook from a farm in the hamlet of Ten Arlo (in the municipality of Zuidwolde) indicates how this process affected individual farms: between 1835 and 1870 the revenues from dairy-

---

[3] And to a lesser extent buckwheat and potatoes also.
[4] Translated from a quotation in Homan, 1947: 102-103.

ing on this farm increased from 25 per cent to 54 per cent of the total income, while at the same time the share of the grain (rye) sales fell from 10 per cent to 6 per cent and those of cattle from 43 per cent to 32 per cent. In the same period expenditure on the purchase of animal feed rose from 2.5 per cent of the total costs in 1835 to 44 per cent in 1870 (Bieleman, 1987: 386). The labour for this new set of activities on the farm was made possible by mobilising the women. From then on, the processing of milk became very much the domain of the farmer's wife, possibly helped by her daughter(s) or maid. Only very rarely did men become involved in the actual dairying process, a hard, time-consuming and labour intensive activity (Compare: Van der Burg, 2002).

Besides the one in Meppel, butter markets were also established in other Drenthian towns in this period, e.g. in Assen (the capital of the province), in Coevorden (a seventeenth century fortress town in the southeast) and in Hoogeveen (a town that flourished thanks to its peat trade). And as the supplies of butter in these markets increased, so the supply of rye decreased. In Assen, for instance, the annual supply of rye decreased from almost 15,000 hl (or 1 050 metric tons[5]) in the years 1851-1860 to only 1,100 hl in 1876-85 (Bieleman, 1987: 728-729, Appendix 5.9). In time, Meppel was to become a major outlet for farmers selling butter they had produced themselves. By the beginning of the nineteenth century the butter trade in Meppel was already important, as more and more livestock farmers from the low lying, grassy southwestern corner of the province sold their butter there. It was reported then that:'from time immemorial, in the autumn during the months of September, October and November, important barrel-butter markets are held, which are visited by buyers from Texel (an island in the north of the province of Holland), Friesland and Overijssel. These markets are often thriving as many wagons loaded with butter can hardly find a place inside the town and have to be parked outside'.[6]

Around 1810 the supply of barrel butter was as much as 2,300 to 2,400 vierendelen (c. 40 kg) annually, while at the same time a 'rather considerable' slab of loose, unwrapped butter was brought in, which was usually sold in rolls of 1½ pond (=735 gram), the so-called 'Meppeler kluiten' (a 'kluit' is a ball of butter). Compare this with the last decades of the eighteenth century when supplies of Friesian butter on the Sneek market amounted to as much as 40 000 vierendelen or 1.6 million kg annually (Faber, 1972: 598, graph IV.15). This slab butter must have matched Friesian butter in quality, since Friesian merchants bought it, rewrapped it, labelled it with a Friesian mark and then sold it as Friesian butter. Most of the butter supplied to the Meppel market was eventually shipped to Amsterdam to be subsequently exported, particularly to Britain (Bieleman, 1987: 382-386).

---

[5] Based on a weight of 70 kg per hl.
[6] Translated from a quotation in Bieleman, 1987: 383.

Figure 1.3   Supply of butter on the market at Meppel in kg, 1810-1889

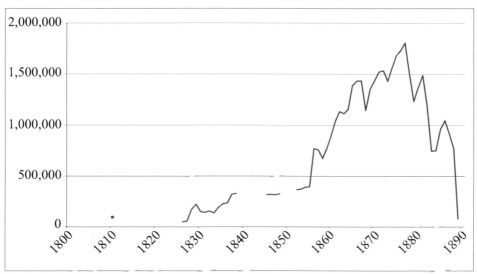

Source: After Bieleman, 1987: paragraph V.3.6 , pp. 343-345.

From the early nineteenth century, supplies in Meppel grew. In 1820 the local weigh house weighed a total of 133,800 kg of butter and in the years 1826-35 quantities had increased to an average of 151,150 kg. By 1838 the turnover had reached 329,000 kg (Figure 1.3). At this time, most of the butter would be brought in during the months of September and October (Figure 1.4), when most farmers had accumulated enough butter to make it worth their while to take it to the market. In those two months almost half of the total annual amount was brought in (Alstorphius Grevelink, 1840: 92).

Figure 1.4   Average monthly supply of butter on the market at Meppel in kg, 1826-1828 and 1836-1838

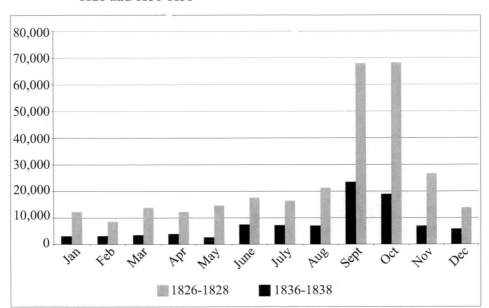

Source: Alstorphius Grevelink, 1840: 92-93.

Market supplies reached their peak during the 1870s, when 1,537,400 kg was sold annually. In the absolute peak year, 1877, this was as much as 1,806,200 kg (Figure 1.3). So, compared to the situation around 1830, supplies had increased more than tenfold in less than five decades. In those years, the Meppel market outranked the one in Zwolle and, together with Leeuwarden and Sneek, was one of the three major butter markets in the Netherlands. In 1877, the amount of butter supplied to the Meppel market was only just a little below that for Sneek. About three-quarters of a century earlier, market supplies in that Friesian town had been about twenty times higher than those in Meppel.

In the southwest of the province, farmers usually brought their butter to the market themselves – and particularly to the market in Meppel. There it was sold to specialised butter merchants, who then sold it on through further trading agencies to be eventually shipped to Britain. The fattened pigs were purchased by travelling buyers all year and sent to Amsterdam or to Germany, especially to Hamburg.

Further east, into the hinterland, farmers usually sold their butter in 'kluiten', or balls, to itinerant buyers who went from farm to farm. In fact, farmers bartered it for all kinds of commodities like animal feedstuffs and groceries. In other cases farmers purchased animal feedstuffs and oil cakes from local shopkeepers on credit that was then deducted from the revenues from the sale of butter delivered to them after the summer. So, in effect, in the interior of the province, the whole butter trade was a barter trade. In particular, farmers without enough financial elbowroom could do nothing but buy the animal feedstuffs and concentrates they needed on credit. They were therefore condemned to barter with the butter traders. Most small villages had at least one butter trader ('boterdrager' , literally a butter carrier), whereas the larger, central villages had usually two or even three. As they were completely dependent on this barter trade, farmers were forced to agree on prices below the actual market price.

Some better-off farmers could afford to sell their butter directly to the market, like that in Meppel. However, even there bartering was a common practice. Farmers who bought the most feeding stuffs and oil cakes got the best prices for the butter they 'sold' to the merchant. Usually, before making an offer for the farmers' butter, the merchant would ask how much in products the farmer would buy in return, and after that, a price was agreed (See for instance: Uitkomsten, 1890: 12).

Concerning the methods of butter-making: in virtually the whole of Drenthe, including the larger farms, the most common technique of processing milk into butter at that time was milk churning. Only in some farms in the southwest, where stocks were larger than elsewhere in the province, was cream churning the practice (Rapport, 1926). In 1914, in a series of articles based on interviews with several elderly people, Ten Heuvel depicted the way farmers worked (Ten Heuvel, 1914, 277-284; Tiesing, 1927: 410-417). According to these informants, there were generally two milkings per day, and as stocks were small and milk yields low, milk from different milkings was collected and stored in casks. Usually two of these casks would be filled with several days' milk waiting to ripen, i.e. to become thick and sour. To speed up the process, the farmer's wife would pour a kettle of hot, boiling water into the cask that was to be churned the next day; when ready, the sour, thick milk was poured into the churn and possibly some hot water was added to get the right temperature as the churning began.

In most farms the milk was churned by hand in a dash churn or plunger churn. Others had a churn that was driven through a lever device, called a 'winde'. According to data provided by the agricultural publisher Tiesing in 1927, these hand-driven churns contained eighteen to 25 litres of milk and were 1.25 to 1.50 metres high. The diameter at the top of a churn like this was 60 to 70 centimetres and at the bottom 75 to 80 centimetres (Tiesing, 1927: 410-417).The churns were made of pinewood and the disc of the agitator was made of oak. This usually had eight to ten holes and a diameter of about 1½ cm. Churns with a lever device could hold about 30 to 45 litres of milk. Sometimes, on the larger farms, a dog or a horse turned the churn, which was again somewhat larger.

Experts in those days estimated that the (sour) milk churning method produced one kilogram of butter from 39 litres of milk. Using the system of cream churning (and skimming milk) 44 litres of milk were needed for one kilogram of cream butter (Staring, 1862: 1087-1090).[7] So in fact milk churning was more effective than cream churning.

---

[7] Ten Heuvel wrote that in the early twentieth century with the help of separators creameries were estimated to produce 1 kg of butter out of 28 litres of milk, see Ten Heuvel, 1914: 283. This figure is confirmed by data on production in the first generation of creameries in Drenthe given by the Verslag van de Landbouw over 1898. These creameries, which made butter with the help of both manually powered and steam powered separators, produced 1 kg butter from 28 to 29 kg milk.

However, another reason for using a milk churning system instead of cream churning must also have been the fact that the latter demanded much more attention and care as it was more inclined to fail. Milk churning was much less risky too, especially under the poor hygienic conditions then pertaining in farms in the sandy regions like Drenthe. The considerable disadvantage of milk churning was that the butter produced did not keep as long as cream butter.

III. A sales crisis

Around the mid-nineteenth century the British market still seemed to offer endless possibilities (Bieleman, 1996: 25-30). Experts at the time depicted it as 'a bottomless well that can never be filled' ('een bodemlooze put die nooit gevuld kan worden') (Hengeveld, 1865: vol. 2, 114). However, as the British demand for butter grew (tripling between 1851-55 and 1866-70), it also opened the door to all kinds of tampering. Traders, for instance, forged butter by mixing it with water, potato starch, gum and other substances (Croesen, 1931: 39-40). They also mixed high-quality butter from renowned livestock regions with cheaper inferior butter that was produced under the poorer conditions of the sandy parts of the country. Although this was not an entirely new problem – complaints had been heard about this sort of fraud as early as the seventeenth century – it now became a serious threat to the export position of Dutch dairy products, as other countries were entering the British market with increasing quantities of high-quality products (Bos, 1978: 235-236). Even more serious at this time was the competition from margarine and the way its producers, already in the business as intermediary butter traders, added it to farm butter for export. Butter merchants from the province of Brabant had taken up the production of margarine in the early 1870s (Hoffmann, 1969; Wilson, 1954; Verbeek, 1992; Bakker, 1991; Bakker, 1992). And although the share of Dutch butter exports still amounted to 45 per cent of British imports in 1885, this was largely due to the rising sales of margarine.[8]

As more and more butter of doubtful quality was shipped to Britain to compete with high-quality products from elsewhere, problems grew and the price of Dutch farm butter decreased. Between the 1840s and the years 1871-80, the price of butter on the Meppel market increased from 0.69 guilders per kg to 1.31 guilders per kg (a faster rise than on the Leeuwarden butter market[9]) (Hylkema, 1913[3]: 620-621), subsequently to drop to 1.01 guilders per kg in 1886/95 (Bieleman, 1987: 706, Appendix 3.3). Although the problems were already noticeable in the mid-1860s, the turning-point in price development came about 1876. From then on problems became more serious year by year.

As these problems increased and became increasingly more evident there were many discussions at all levels about whether Dutch dairy farming needed a different approach to the organisation of its production and its marketing systems (Van der Burg, 2002: 105-

---

[8] Until 1886 export statistics did not distinguish between butter and margarine. Bakker, 1991: 21, Table 2; Bakker, 1992: 108, Table 4.2.
[9] Although the Meppel price level was lower than that on the Leeuwarden market, prices rose there from 0.87 gld/kg to 1,44 gld/kg, only to fall to 1,14 gld/kg. Hylkema, 1913[3]: 620-621.

117).[10] In particular during the 1870s, dairying and the improvement of dairying became a top item on the agendas of all kinds of meetings and conferences of the agricultural and intellectual vanguard.

IV. Discussions

One important, national platform for the debates was the Dutch Agricultural Confer- ence. This Nederlandsch Landhuishoudkundig Congres had been established in 1846 and was meant to organise large national conferences on an annual basis and in different parts of the country (Van der Poel and R.J.C. Wessels, 1953). In the early 1850s discussions were already being held on the desirability of establishing collective dairy processing units, based on the idea of the Swiss 'fruitières'. However, the conclusion was that the very character of dairying in the Netherlands made this approach unsuitable and would not be popular amongst farmers. For some time experts were in favour of an approach based on improving dairying methods on the farms themselves.

However, after 1871, discussions became more intense, and there was a shift in favour of large scale, factory processing. Gradually it became clear that processing milk into butter and cheese on a factory scale was preferable to producing them on the farm (Van der Burg, 2002: 137-139). Comparisons were made with the situation in countries like England, Switzerland and Denmark. Many became convinced that mechanised production processes would allow the quality of the products to be guarded effectively. The Govern- ment's annual report on agriculture in 1878 argued that such a method would produce the best quality butter, which because of its quality and homogeneity could easily be sold directly to the export trade (Verslag van de Landbouw, 1878: 36-37).

In Friesland, then the foremost dairying province, livestock farmers had turned to the export trade in the second half of the eighteenth century. Farms there were usually far from small and holdings with 20-25 or more head of dairy cattle were quite common (Faber, 1972: 211). Here, the issue of improving dairying methods was discussed within the framework of the Friesian Farmers' Society (Friesche Maatschappij van Landbouw) which had been established in 1852 and which from the mid-1870s on became the forum for initiating discussions. The main issue was the way dairying methods in Sweden, Denmark and Schleswig-Holstein were improving and, in particular, the way in which the so-called Swartz system of skimming milk contributed to that.

According to this system – introduced by Johann Gustav Swartz in 1864 – milk to be skimmed was poured into containers of ca. 50 cm high, made out of tinned sheet-steel with an oblate, oval crosscut. These containers could contain 20 or 30 to 40 litres of milk. Once filled, they were placed into cisterns with very cold (running) water, or preferably ice water which allowed milk to be cooled down to about 4-6°C. If conditions were favourable, milk fat formed a top layer in the container within about 12 to 24 hours. This layer could

---

[10] Van der Burg discusses in particular the gender aspects of agricultural modernisation (and dairy- ing in particular) in this period.

be taken off as cream, using a specially shaped spoon. An essential feature of the system, also known as the 'ice-method', was that the cream was churned while still sweet, which had the advantage that the butter was of a superb quality and kept well for a long time.

Ever since its introduction, the Swartz system had spread to many places in Scandinavia. In the Netherlands it became known as the 'Danish system', although, in a broader sense this expression actually included everything in which the Danish dairy farmers were then ahead of the Friesians, and the adjective 'Danish' therefore often also stood for 'new'. Probably the first time the Swartz system was demonstrated in the Netherlands was on the occasion of a large agricultural exhibition in Apeldoorn (province of Gelderland), in 1878 (Van der Poel, 1967: 174).

In Friesland discussions resulted in the setting up of a commission in 1878 which went to Denmark to study dairying in that country. The commission's conclusions emphasised that the improvement of milk processing on the farm was preferable to establishing collective creameries (Geluk, 1967: 44; Bakker, 1991: 27). During the following years discussions continued on whether or not the Society should try to persuade farmers to adopt the Swartz system.

Meanwhile, from the early 1880s on, some Friesian farmers equipped their farms with 'Danish' milking cellars (the old ones were reconstructed), that is to say the Swartz system for skimming milk. Adopting the Danish method for dairying, however, also meant the purchase of other, related pieces of equipment; such as a 'Danish' (i.e. Holstein-ian) churn and a butter worker (the latter was a piece of equipment to make the freshly churned mass of butter into a homogeneous product and to remove the milk remains). Indeed, some farmers even went as far as to build ice cellars in their farmyards so as to have ice at their disposal during the summer. And although positive experiences were reported by some farmers adopting the Scandinavian method, these were more or less isolated individual initiatives. Hardly anybody was familiar with the right methods of working and most of them applied the Danish system only partially which meant that results were poor (Spahr van der Hoek, 1952: vol. 1, 525-534). Those who 'went Danish' seldom purchased all the necessary equipment that the method required. Sometimes they did not buy the special Swartz containers but stuck to their copper setting pans and placed them in cold water. Indeed, it was very difficult to go Danish 'all the way', if only because it was very difficult to have ice always available. Therefore, these Friesian farmers adopted the Swartz system by using water from a well or pumped it up from the ground water, which meant it was less cold. Without either the proper equipment or ice-cold water, it not only took considerably more time for the cream to separate but the outcome was also not satisfactory.

In practice, most Friesian farmers indeed stuck to their old methods of dairying. For them there was, in fact, little other choice. And as experts and leaders of opinion were not convinced of the idea that collective dairying would work in Friesland, things stayed largely as they were. Nevertheless, although most Friesian farmers were against collec-tive creameries, as was the Friesian Farmers' Society, it was that clear something had to be done. Then, in 1879, opportunity knocked when De Laval presented his improved separator.

In Drenthe, as in other provinces at that time, a farmers' society had been established in 1844, the Drents Landbouw Genootschap or DLG for short (Van der Zeijden, 1994: 33-58). In time, the society proved to be an important forum where the current problems in regional agriculture were discussed and solutions suggested. The organisation also undertook various initiatives for improving regional farming in various fields, not least in the field of dairying. In 1852 it organised a contest on the question 'What is the best way to make butter' (Rapport, 1926; Ten Heuvel, 1915: 139-149).

In the 1870s discussions on how to improve the quality of regional butter became more frequent and intense. In the spring of 1878 the Society invited an agricultural teacher from Limburg to give a number of lectures on the Swartz system of processing milk. In meetings during 1879 and 1880 the 'Danish method of milk processing' was put on the agenda, alongside lectures on other subjects in the field of dairying and on the way this was practised in the other provinces as well as abroad.

In 1881, the Swartz system was indeed introduced 'successfully' on a dairy farm in the southwest of Drenthe. In the same year, it was reported from this area that the use of butter workers (an implement for cleaning the butter by kneading the milk remains out of it), was becoming more and more general. In 1885, the regional chamber of commerce reported 'that many livestock farmers are trying to improve their dairying methods in all kinds of ways, and were indeed having some success' (Bieleman, 1987: 389).

In 1887 the board of the DLG installed a commission which had to look into the way creameries in the neighbouring province of Friesland were operating. It had to investigate and report on whether mechanised processing of milk would offer a solution to the problems in Drenthe as well. In the course of the same year, the commission reported that mechanised processing in Friesland indeed was very successful, and recommended it as a solution for Drenthe dairy farmers as well. However, it was not as yet clear whether this should be put into practice in a few large-scale creameries, or in a larger number of small ones. Yet, many farmers still had doubts as to whether the (fresh) skimmed milk that was to be received back from the creameries was suitable for feeding to their calves and pigs. They feared that this skimmed milk would not be as good as the sour buttermilk from their own farms, and that pig fattening in particular (the other cornerstone of their business) would no longer be possible (Rapport, 1926).

V.   The first steam-powered creameries

Meanwhile in 1879, a private creamery had already been established in the Friesian village of Veenwouden (about 15 km east of the capital Leeuwarden). Its founder, a private entrepreneur called Bokma de Boer, produced butter and skimmed-milk cheese. The latter was shaped like a brick and since it was intended for the German market was given the German name 'Backsteinkäse'. As a private entrepreneur Bokma de Boer had great difficulty in obtaining enough milk at a reasonable price, probably because of the usual opposition of the farmers (Van 't Hull and Ockers, 1992). A year later, in 1880, a milk plant was established in the town of Leeuwarden, which used a separator to skim the milk mechanically, an event that is usually taken to be the beginning of continuous milk processing in the Netherlands (Geluk, 1967: 44-45). Soon there were more private factories, and by 1885 there were five in Friesland alone (Spahr van der Hoek, 1952: vol. 1, 537-547).

However, it was not until 1886 that the first (steam-powered) cooperative creamery was established. By then 23 dairy farmers in the Friesian village of Warga (some 6 km southeast of Leeuwarden), with between them 715 dairy cows (i.e. an average of 31 per farm), had had the courage to join together in setting up a cooperative, despite the opposition and scornful remarks of their neighbours (Wiersma, 1959: 67-72).[11] Their immediate motive was a conflict with the management of a private creamery in Leeuwarden about their deliveries. The construction of the Warga creamery needed an investment of 35,000 guilders, and the money was raised by taking a debenture stock against 4 per cent (Spahr van der Hoek, 1952: vol. 1, 544-545; Geluk, 1967: 49-50).

Soon the number of (cooperative) creameries increased. By 1890 there were already 45, of which sixteen were explicitly cooperatives, and in 1898 there were 113 of which 66 (60%) were cooperatives (Spahr van der Hoek, 1952: vol. 1, 555). Statistics for the cooperatives show that those with the largest capacity were, indeed, found in the southwestern, 'green' quadrant of the province, the livestock farming area with its grassy pastures on peat and clay-on-peat soils. In that year, 60 per cent of these cooperatives processed more than two million litres, with about a quarter (24%) processing even as much as 4 million litres or more. The largest creameries were those in the villages of Giekerk (the municipality of Tietjerksteradeel) and Weidum (the municipality of Baarderadeel), which processed 7,591,000 and 7,580,000 litres respectively (Verslag van de Landbouw, 1898: vol. 2, 250-258).[12] The smallest factories were in the Wouden district, the sandy part of Friesland adjacent to Drenthe, where conditions were like those in Drenthe. In 1898, the 113 Friesian creameries together produced 41.5 per cent of the factory-produced butter in the Netherlands (Spahr van der Hoek, 1952: vol. 1, 560-561).

---

[11] Although the Warga cooperative creamery is usually depicted as the place where cooperative milk processing in the Netherlands began, it should be stated here that before the fall in dairy prices in the province of Noord-Holland, some small-scale cooperative cheese creameries had already been established. Bieleman, 1992: 296.

[12] Compare also Spahr van der Hoek 1952: vol. 1, 548-549.

In Drenthe, however, the situation was quite different. Here, the first cooperative creamery began in January 1889. A year before, in January 1888, the Central Board of the DLG, hoping to get things moving, had organised a meeting in De Wijk, a municipality in the grassy, livestock farming region in the southwestern-most part of the province. De Wijk and its surrounding region was known for its large and wealthy farms, which had been orientated towards dairying more and more since the second half of the eighteenth century (Bieleman, 1987: 304-397). Although there were still some doubters, an enterprise like this was more feasible here than in any other part of the province. Once more an excursion was made to Friesland to look at a private creamery in the village of Elslo.

In the course of that year farmers in the De Wijk region indeed expressed enthusiasm for the idea. Some of the required capital had already been raised and a piece of land had been purchased. In June 1888, the first Cooperative Steam Creamery in Drenthe was established in the hamlet of Rogat in the municipality of De Wijk. The Rogat Creamery began production at the end of January 1889. The total cost of the enterprise (including the architect's fee and course salaries for two employees) amounted to 19,326 guilders, that is to say roughly 55 per cent of what the Warga creamery had cost. In its first year – that is from 21 January 1889 until 1 May 1890 – it processed 1,118,600 litres of milk. Soon more farmers joined the cooperative and in 1900-1904 the average annual quantity processed was 3,3750,000 litres.

Figure 1.5 The blueprint of the Rogat creamery

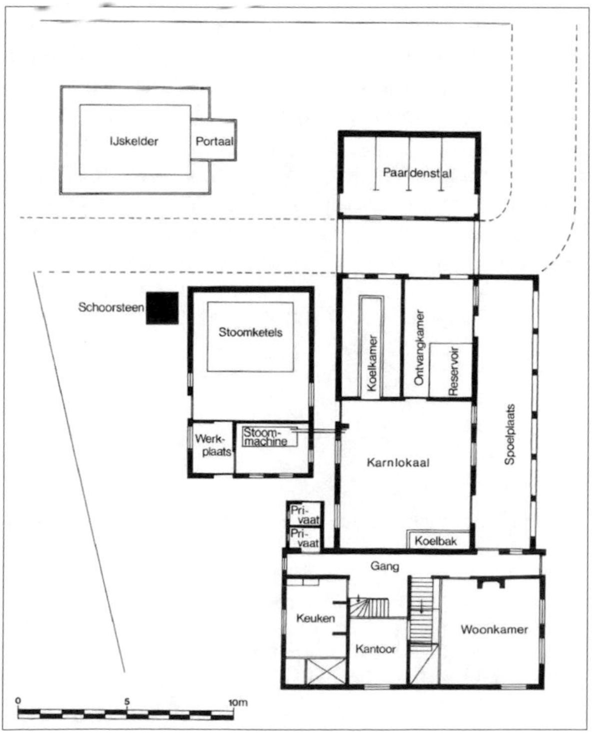

Although good working separators were available after 1879, some of the first creameries were, in fact, basically designed to work on the basis of the Swartz milk skimming system. This was the case in Friesland[13] and also in Drenthe. From the blueprint of the Rogat creamery kept in the municipal archives of De Wijk, it appears that this creamery was designed to use the Swartz skimming system (Bieleman, 1987: 390).[14] Photographs of personnel and implements in other, often smaller creameries also indicate that indeed the Swartz system was used in many other places during these early days.

In the autumn of 1889 a second steam creamery was established in Drenthe in the municipality of Dalen, in the southeastern, grassy fringe area of the Drenthe Plateau. During the first years, milk was brought to the creamery by the farmers themselves. In fact, this creamery began as a limited company but was transformed into a cooperative ten years later in 1909. A third steam-driven creamery began in May 1890. From the outset, a corn mill was connected to it, using power from the steam engine to drive it, but it was not very successful and was sold to a private entrepreneur in 1897. In 1909 it became a cooperative. Two more steam-driven creameries were set up in 1892, and in the following year seven cooperatives were established, all equipped with a steam engine. Many other village communities followed this trend, setting up steam-driven creameries. From the very start these all had a corn mill connected to them. Later, others were extended to include a mill. Although the number of windmills had increased rapidly with the emergence of the mixed farming system (Bieleman, 1987: 426- 428 and 775, ref. 490; Ten Heuvel, 1915: 39, ref. 3)[15], after 1850 many central villages also established steam-driven, cooperative corn mills where farmers could have their rye and oats ground to feed to their animals. In fact, the combination of a steam-driven creamery connected to a corn mill became a very characteristic phenomenon on the Drenthe plateau from the last decade of the nineteenth century. Eventually, in the mid-1920s, 33 of the 59 milk processing plants had a corn mill attached to them as well as a feedstuff store (Rapport, 1926).

According to a survey published in the annual Verslag van de Landbouw for the year 1898, Drenthe already had 22 steam-driven cooperative creameries. However, compared to those in Friesland, these establishments were rather small. The largest quantity of milk that year was processed in the creamery in Hoogeveen with 3,111,100 litres of milk. The Rogat creamery came a good second with 3,086,300 litres. Of the total number of 22 steam-driven creameries only seven – or one-third – processed 2,000,000 litres or more (Verslag van de Landbouw, 1898: 258-265).

---

[13] This was, for instance, the case in the first private creamery established by the entrepreneur Bokma De Boer, in the Friesian village Veenwouden, in 1879. Spahr van der Hoek, 1952: vol.1, 547.

[14] The ground plan of the Rogat creamery shows a water cistern for holding the typical Swartz oval/cylindrical containers (in ice cooled water) and an insulated ice house where ice cut during the winter from the adjacent canal could be stored. Instead of using ice the milk could also be cooled down by using cold ground water that was pumped up with a Norton pump.

[15] In Drenthe, the number of windmills had doubled since the early nineteenth century. In the year 1857 alone, for instance, seven new ones had been built.

Accidents did happen. In 1895, a separator in the creamery at Roden, in the north of the province exploded, killing two employees. The factory had to stop for some time until a new separator could be delivered by the purveyor (Rapport, 1926).

VI. Manual power in creamery dairying

Before the cooperative, steam-powered creamery in Rogat actually even began, some had indeed supported the idea. However, they also argued that creameries like this in fact would be much more needed in those parts of the province where butter prices were even lower and the butter quality even poorer. Even in 1892, three years after the Rogat creamery had started, the sigh could be heard …'that the establishing of cooperative creameries is much desired now, especially considering the difficulty of stopping the bartering of butter for animal feedstuff' (Van Stuijvenberg, 1949: 101).

Two years earlier, in January 1890, the dairy adviser to the farmers' association in the neighbouring province of Overijssel, Van Weydom Claterbos, had been invited by the DLG to give a number of lessons on the theme of dairying. He continued to do this in many other villages in Drenthe in the following years. Women in large numbers from all over the region now came to attend these meetings also, sometimes in their hundreds. The annual report of the DLG stated: 'Interest in improved dairying in this province has generally been aroused by the lectures and courses on this matter, which are followed by an acceptance of improved methods, both on farms, and in creameries' (Rapport, 1926). On these occasions Claterbos did much to encourage the system of hand-powered creameries that was in use in several villages in the north of the province of Limburg in the south of the country, a method he had come to know while visiting the province.

There in Limburg, in the village of Tungelroy, close to the Belgian border, the local schoolteacher had taken the initiative in establishing a small creamery similar to those in Belgium. Farmers joined together to process their milk with hand-driven equipment, a separator and a churn (Verheij, 1917?). The separator was bought in the Belgian town of Remicourt, the hometown of Jules Mélotte who shortly before had invented a new type of separator that functioned more lightly and quickly than other available models. It was therefore much easier to handle and was less liable to wear and tear. The Mélotte separator, which had been put on the market in 1890, appeared to be an immediate success (Van Dijck and Van Molle, 2001: 118-119).

The Tungelroy experiment was soon copied in other villages in the region, and soon dozens of these small creameries were established there. The main advantage of this system, which before long was to become known as the 'Limburg system', was the fact that it was cheap. The cost of setting up the first Tungelroy creamery came to only 600 guilders, as the rent of the small building was 50 guilders.

Another example of the simple and low profile these small creameries had was that in the village of Groesbeek called Eigen hulp ('Self help'). There every farmer had to operate the separator for himself, so that the labour costs were kept as low as possible

(Geluk, 1967: 50, 77-78; Dekker, 1996). A government report from 1910-12 states: 'A minimal, small compartment can serve as a creamery. The inventory consists of a Mélotte hand separator, a churn, a butter worker, a heating pot as preheater, a scales, some small equipment for testing milk and a number of milk cans' (Overzicht, 1912: 284).

After Claterbos had advocated the use of the 'Limburg system' in Drenthe on several occasions, the first hand-powered separator was installed in a farm in the village of Ekehaar in central Drenthe in 1894. In the same year, and again at the instigation of Claterbos, a delegation from the small village of Erm (in the municipality of Sleen) went to Brabant and Limburg to look at three hand-driven creameries in the villages of Beugen and Meijel (about 160 and 200 km respectively from Erm). This initiative led to the first hand-driven creamery on the 'Limburg system' being established in Erm. It started working in May 1894 and initially functioned with a Mélotte separator, which was replaced after some time by an Alfa-Laval.

The Erm initiative was followed by a group of farmers in the east-Drenthe village of Borger where a hand-driven creamery began to be used in October 1894. It needed an investment of 1,325 guilders and during the first summer it employed three workers besides a manager. Initially it had only two separators but got a third one in 1897. In March 1898, farmers also began to be paid according to the fat content of their milk rather than only for the quantity delivered. Milk control had been introduced in Friesland in 1894 as a means of selecting dairy cows for breeding purposes. Now this also proved to be a way of preventing tampering with the milk supplies.[16] Eventually, in 1908-09, the hand-driven creamery at Borger was converted to steam, and in 1916 a 'large' steam-driven corn mill was built onto it.

In 1895, one year after these small-scale creameries had taken off, about 39 per cent of all butter in Drenthe was produced in them, the rest still on farms. In Friesland this share was already 43 per cent by then. Another year later, the volume of creamery-produced butter exceeded the volume of farm butter. Seven years later, in 1903, 83 per cent of all butter in Drenthe was produced in creameries (in Friesland this was as much as 86%) while just before World War I, virtually all the milk was processed in the creameries. Only 2 per cent of creamery-processed butter was made in non-cooperative creameries (Figure 1.6) (Croesen, 1931: 190-191, table 1.2 and 1.3).

---

[16] According to Spahr van der Hoek in Friesland the milk control in creameries started from 1904 onwards! In 1894 milk control on fat content had started as a means for selecting dairy cows for breeding purposes. Bieleman, 2002: 135; Spahr van der Hoek: 1952: vol 2, 256.

Figure 1.6   Percentage of all butter produced in creameries in Friesland and Drenthe, 1895, 1900, 1903 and 1912

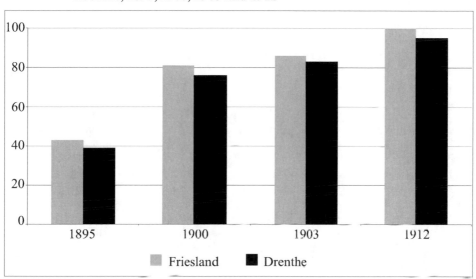

Source: *Verslagen van de landbouw*, concerning years.

There is no doubt that the arrival of the separator was a decisive factor in the rapid success of the creameries once they were built, especially after the Mélotte model became popular. Many of the creameries had two separators, one of which generally served as a back-up. According to data on cooperative creameries in the Verslag van de Landbouw for the year 1898 (and almost all of the creameries in Drenthe were cooperatives – see below), 23 out of 40 hand-driven creameries had two separators, fifteen had one and two had three. Five of the 22 steam-driven creameries also had only one separator, eleven had two and six had three of this brand new piece of technology (Verslag van de Landbouw, 1898: 258-265). It is a clear example of the small-scale character of these hand-powered creameries in those days.

VII.   A 'kneading centre'

In most of the small, hand-driven creameries, the butter prepared was not ready for trading. Usually, it was packed as a semi-finished product, i.e. in graniform state, into transport barrels and then sent to a sales agent in Sneek (Friesland). This meant that the net weight of the butter could not be measured at the creamery but was done by the buyer, a possible source of conflict between the two parties involved. For the same reason, the sales of dairy products were still very much a matter of concern, as tampering indeed was still quite common (Geluk, 1967: 144-148). Undeniably, the small hand-driven creameries often had difficulty in disposing of their butter and some therefore suggested that it would be more sensible to send the butter to be auctioned to the town of Zutphen (some

70 km south of Meppel, in the province of Gelderland). This auction had been set up by the GOZ, the union of dairy cooperatives in Overijssel and Gelderland.

For the same reason a similar auction was established in Meppel in 1900. In its early years, some 400,000 to 500,000 kg of butter were supplied (less than a third of the supplies during the peak years of the market in the early 1870s). After that, however, supplies dropped. The Meppel butter auction lasted until 1914, when the war put an end to it and it was never resurrected afterwards (Rapport, 1926).

In addition to all this, there was still too little notion of the importance of cleanliness in the butter-making process, and the quality of the end product left much to be desired. The milk was produced by cattle that were still kept under very poor hygienic conditions in potstallen (pit stables), where they stood in their own muck (a mixture of manure and heather sods). The lack of a steam boiler made it impossible to pasteurise the cream (and so the butter was more perishable), the skimmed milk or the buttermilk. Moreover, the fact that the last two were given back, un-pasteurised, to the farmers as animal feed increased the risk of all kinds of diseases, such as tuberculosis, spreading amongst the stocks of all the farmers using the creamery. It was clear that the hand-driven creameries still did not offer an efficient solution to the problems farmers in the sandy regions had in finding a better link to the (international) market.

One idea put forward to solve the problem was to centralise the final processing and salting of the butter and its sales. In the summer of 1896 this idea had already been put forward at a meeting of the DLG. At first, the associated larger, steam-driven creameries were very much against the idea, fearing that such a centralised institution would encourage the emergence of even more hand-driven creameries with all the associated problems. For the same reason they also opposed the idea that the development of a pasteurising device could help overcome the problems connected with the manual processing of the milk.

In spite of the apparent disadvantages of the small, hand-driven creameries however, a 'Cooperative Society for Central Kneading and Exports' was established in Assen, the provincial capital in 1901. Thirteen hand-driven creameries in the surrounding area immediately joined it (Rapport, 1926). The kneading centre became effective in June 1902. The aims of the society were the efficient kneading, packing and exporting of butter, and the bringing together of different smaller batches with larger ones to facilitate sales. Soon additional butter was bought from non-allied creameries to increase the turnover. This was not always profitable because many creameries had also begun to finish their own butter. In the end, a lack of work and bad management lead to the closure of the kneading centre in 1908. It proved to have been a well-intentioned but impractical effort to get better sales for the products of small cooperatives. The future proved to lie in larger steam-driven creameries.

VIII.   The shift to steam

In 1903 – the year the ministry published its report on 'butter production and butter control in the Netherlands' (Bieleman, 1904). – there were 88 cooperative creameries in Drenthe. Of these, 50 were hand-driven, the rest steam-driven. As well as these cooperatives, there were another nine 'speculatieve' or private hand-driven creameries and also five private steam-driven creameries (Figure 1.7).

Figure 1.7 Cooperative and private creameries in Drenthe, 1903

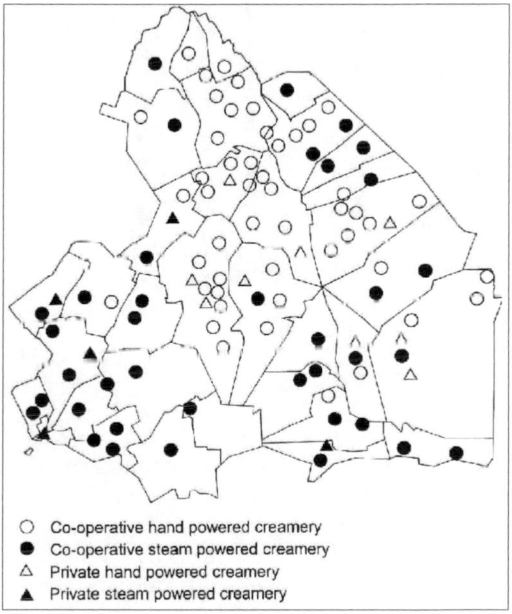

○   Co-operative hand powered creamery
●   Co-operative steam powered creamery
△   Private hand powered creamery
▲   Private steam powered creamery

Source: Bieleman, 1903.

Most of the steam-driven creameries (cooperatives or not) were to be found in the grassier, fringe regions of the Drenthe plateau. There, dairy stocks were on the average much larger than those in the open-field villages in the central part of the province. Yet, the distinction between 'cooperative' and 'private' creameries was less sharp than both terms might suggest. Many of the 'private' creameries were, in fact, set up and owned by farmers and corporations only in name.

The names of the cooperatives that were set up in this period reflect something of the 'economic spirit' of the times: 'De Vooruitgang' (Progress), 'Vooruitgang Zij Ons Doel' (Let Progress be our Aim), 'De Hoop' (Hope), 'Onderling Belang' (Mutual Interest), 'Ons Belang' (Our Interest), 'Gemeenschappelijk Belang' (Collective Interest), 'Algemeen Belang' (General Interest), 'Ons Voordeel' (Our Advantage), 'De Goede Verwachting' (Good Expectations), 'De Eendracht' (Concord), 'De Volharding' (Perseverance), 'De Drie Eenheid' (The Trinity), 'Concordia',and 'Excelsior' (Verslag van de landbouw, 1898: vol. 2, 258-265).

Figure 1.8   Number of cooperative creameries in Drenthe, 1895-1940

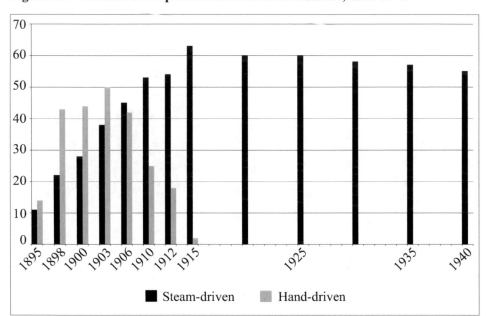

Source: *Verslagen van de landbouw*, concerning years.

After 1903, the number of hand-driven creameries decreased as one after another they disappeared (Figure 1.8). From then on and stimulated by an improving economy, production was concentrated in a smaller number of steam-driven establishments. In the year 1898 for the first time a hand-driven creamery was changed over to a steam-driven one, and after that others followed. In some cases, the management simply installed a steam boiler and connected machinery. Elsewhere, completely new plants were built. An important push towards the disappearance of the hand-driven creameries came with the closing down of the Cooperative Kneading Centre in Assen in 1906. However, it was not until 1918 that the last one finally disappeared. The importance attached to the village community having its own creamery meant that many of the hand-driven ones continued to exist for longer than was advisable.

Of course, it was important to consider whether a shift from hand to steam power was economically realistic for a small creamery. In a letter sent to a newspaper in 1897, one expert argued that a daily total of 2,000 litres of milk had to be processed for the change over to steam to be economical (Schaafsma, 1897).[17] Given a lactation period of 300 to 320 days per year, this meant an annual figure of 600,000 to 640,000 litres. In 1898 only one quarter of the hand-driven creameries met that standard (Verslag van de landbouw,

---

[17] J. Schaafsma sent a letter to the section 'Landbouw en Veeteelt' ('Agriculture') of the Provinciale en Drentsche Courant of October 7th 1897. The author, Schaafsma, was the manager of a hand-driven creamery in the village of Noordwolde in the adjacent part of Friesland.

1898: vol. 2, 258-265).[18] Yet, in 1899, a report by a commission led by the State Dairy Adviser concluded that a hand-driven creamery processing 400,000 to 500,000 litres per year could successfully make the shift. And the number of creameries reaching this total was increasing (Rapport, 1926).

A shift from hand-power to steam made creameries also much more efficient. A hand-powered creamery, using the traditional method of cream churning, could produce 1 kg of butter from the cream of 44 litres of milk (see above). Production figures from the Coöp. Zuivelfabriek en Korenmalerij 'Borger' (Cooperative Creamery and Corn Mill 'Borger') in the village of Borger in the eastern part of the province, show that when the hand-driven separators were used (it had three of these) it needed 30 litres of milk to produce 1 kg of butter (1896-1908). By the early 1930s, after a steam engine had been installed in 1908 and the hand-powered separators had been discarded, only 25 litres of milk were needed to produce 1 kg of butter (Zestig jaar Coöperatieve Zuivelfabriek en Korenmalerij 'Borger', 1954: 16-17).

As early as 1896 the volume of butter produced in creameries exceeded the volume still produced on farms. In 1903, 580,000 kg of butter (16%) was produced on farms and 2,950,000 in creameries (84%) (Verslag van de landbouw,1903: 86-87). Seven years later, in 1910, virtually all the butter was produced in creameries, with only 5 per cent still being made on farms (Table 2), nearly all of which were steam-powered creameries by then. In that year, only 316,864 kg out of a total amount of 4,018,410 kg of butter, or 8 per cent was made in manually-powered creameries. Yet, the share of Drenthe's cooperative creameries in the total of national creamery-produced butter was modest. In 1910 they produced 9 per cent of all creamery-produced butter in the Netherlands which amounted to as much as 46 million kg, whereas the Friesian creameries then produced 33 per cent of it (Verslag van de Landbouw, 1910: 60-61).

Table 1.2   Butter production in kg on farms and in creameries in Drenthe, 1895 and 1910

	Farms		Creameries	
	Abs.	in %	Abs.	in %
1895	1,533,200	61	978,954	39
1910	194,600	5	3,823,810	95

Source: Croesen, 1931: 191 Table III; Verslag van de landbouw, 1910: 60-61.

By about 1910 the volume of butter produced annually by one cow was about 79 kg. According to several sources this was twice as much or even more than during the early nineteenth century. In terms of money, productivity was four times as high by 1910 (Bieleman, 1987: 381-382 and 397). In 1906 it was reckoned that for a small farm (with

---

[18] That is to say 9 of the 36 hand-driven creameries that were active throughout all of 1898.

3 ha of land and three dairy cows during the summer) in central Drenthe, the sales of milk to the local creamery contributed as much as one third of the total revenues of the farm (Table 1.3) (Schetsen, 1912: 366).

In 1940 there were 55 cooperative creameries left in Drenthe, as well as one that was not a cooperative (Verslag van de landbouw, 1940: 203).

Table 1.3   **Revenues and expenditures (in guilders) on a small farm in central Drenthe in 1906**[19]

Revenues	Abs.	%	Expenditures	Abs.	%
Pigs piglets	458.00	49	Fertilisers	50.00	22
New-born calves	22.00	2	Animal feed	175.00	78
Milk (8 090 litres)	303.37	32			
Eggs and chicks	94.50	10			
Potatoes (40 hl)	60.00	7			
Total	937.87	100		225.00	100

Source: Schetsen van het landbouwbedrijf, 1912: 366.

IX. A sea of skimmed milk

Meanwhile, the amount of milk processed by all creameries together and the total production of butter had increased enormously. Between 1903 and 1925, for instance, the volume of butter they produced had tripled from 2,059,000 to 6,177,400 kg (Rapport, 1926). The quantity of processed milk delivered by all the dairy farmers in Drenthe to their cooperatives had increased from 100 million kg in 1900 to 175 million in 1925 and to 305 million in 1940 (Figure 1.9).[20] At that time, the bulk of the milk deliveries to the creameries was between April and August, as in these months the total amount was above the yearly average.[21]

---

[19] The farm had 1 ha of arable land and 2 ha grassland with three dairy cows during the summer and two during the winter

[20] In 1895 all cooperative creameries taken together had processed 29.9 million kg. Verslag van de landbouw, 1895; Boijenga, 1963: 16.

[21] In 1952, for instance, in the peak month of May the deliveries were 121 per cent higher than those in November, the month when the least milk was delivered. Centraal Bureau voor Statistiek, Zuivelstatistiek, 1953.

Figure 1.9 Supply of milk in all creameries taken together in Drenthe in millions of kg, 1900-1940

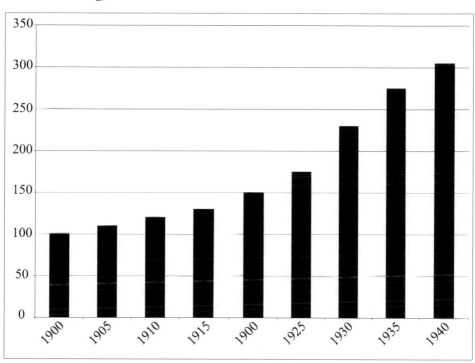

Source: Boijenga, 1963: 16.

The causes of this vast increase were multiple. At first, of course, it was due to the increase in the number of dairy cows, a direct consequence of the appearance of the cooperatives and the incentive they gave to small farming. Between the mid-1880s and 1939 their numbers increased from 39,700 to 101,300, a growth of 155 per cent (Verslag van de Landbouw). Then there was the (genetic) improvement in the dairy cattle itself. In the slipstream of cooperative milk processing, local breeding societies and milk recording societies were established in many places, and these did much to improve cattle. They usually were allied to the cooperative creameries (Ten Heuvel, 1916, 14-18). In the mid-1920s there were 43 breeding and milk recording societies, with 1,134 members and almost 8,700 cows under their control, i.e. about 16 per cent of the whole dairy cattle stock (Rapport, 1926). Meanwhile, the fat content of the milk had become an important criterion in stockbreeding: 'Breed butter cows!' was the motto (Bieleman, 2000: 135-136). However, what probably contributed most to the increasing productivity per animal was the improvement in nutrition: more and better quality hay and, above all, an increasing use of concentrates.

In order to process the growing pool of skimmed milk, cheesemaking departments were added to some of the creameries. In the mid-1920s Drenthe still had as many as

59 milk processing plants (46 driven by steam; the rest by other kinds of mechanical power) of which fourteen had begun to make cheese. Most of them were to be found in the southwestern part of the province. One of the plants made milk powder, another made condensed milk and a third made both (Rapport, 1926). The addition of these new branches gave the cooperatives a much broader basis.

But as the flow of milk kept increasing, it became ever more difficult to handle the skimmed milk in a cost-effective manner, since turning it into cheese, when feasible, also had economic limitations. During the 1930s the problem worsened. In the early 1930s the creameries received 200 million kg of milk, of which 150 million kg was returned to the farms as skimmed milk (Boijenga, 1963). And as long as this stream of milk kept growing, the number of pigs decreased, because of production regulations the government introduced to tackle the economic crisis after 1931. At the same time regulations in the field of calf-rearing also limited the number of outlets for skimmed milk. Farmers could simply not absorb the return of any more skimmed milk, and there were already signs that they were feeding too much of it to their stock. In some cases, cans with skimmed milk were simply emptied out into ditches.

In order to use the growing volume of skimmed milk economically, 21 cooperatives in Drenthe decided to establish the 'Drentse Ondermelk Organisatie' (Drenthe Skimmed Milk Organisation), or DOMO, on 5 March 1938 (Boijenga, 1952; Boijenga, 1963). The idea was that the skimmed milk could be made more valuable if more efficient use was made of the processing capacity of the joint cooperatives – initially 15, soon to be joined by another 7. During the early years the functioning of the DOMO organisation was indeed limited to the transportation and reselling of the skimmed milk to creameries which were still able to process it economically.[22]

X.   Factory dairying and its social consequences

In retrospect, it is clear that the socio-economic significance of factory dairying involved much more than simply producing butter on a large, collective scale. In fact, factory dairying appears to have been a companion, if not the motor, of a true revolution in farming, and an important factor in the emancipation of the small peasantry. This had a great effect on rural life in sandy regions like Drenthe.

Once a cooperative creamery in a village had been established, it became possible to make one or two milking cows economically productive as their milk was sold to the creamery. At the same time, that is to say from 1890 onwards, pig fattening developed in new ways as the British market became the outlet for a more lightly fattened type of pig (up to about 50 kg), the 'Londense biggen' (London piglets). This type of pig fattening proved to be well suited to small farms in particular, the short period of fattening ensuring

---

[22] After World War II it was decided to transform the DOMO into a central establishment for the processing of milk. At that time, 49 of the 52 cooperatives in Drenthe were allied to it. It became the largest cooperative dairy organisation in the Netherlands and became known as the supplier of drinking milk to the American army in Germany.

a quick return on the money invested. Also, as dairy activities on the farms ceased or at least decreased, female labour was now deployed in poultry keeping. The production of eggs became a booming business after 1900, and especially during the 1920s. For many farmers, the establishment of a cooperative creamery was also a springboard to other forms of cooperative entrepreneurship, like cooperative buying associations, cooperative credit banks and breeding associations.

Meanwhile, and as part of a new and energetic policy towards the agricultural sector, the national government began appointing extension officers, one in every province. Soon also dairy advisers, cattle breeding advisers and other specialised consultants were appointed by the state. Together, these officers played a crucial role in channelling the recent findings of the growing body of scientific agricultural research, and they acquainted practising farmers with all the innovations available.

This wide complex of changes – of which the cooperative creameries were more or less the pivot – strongly favoured small farming and after 1890 the numbers of small farms in the sandy regions increased very rapidly, as these changes made possible a rapidly growing rural population in the region. Between 1890 and 1930 the number of holdings in Drenthe with 1-5 ha of land increased from about 4,000 to almost 7,800 (Verslag van de landbouw, 1890/91: 192; Rural History Group, Agricultural Statistics Collection). In just a few decades the sandy regions became the domain of small farming, as numerically small holdings outstripped the traditional, larger farms. The results in terms of the number of dairy cattle per farm can be seen from the membership figures for the cooperative 'Gieten-Bonnen', in the village of Gieten in the northeastern part of the province (Figure 1.10) at the end of the 1930s (this village can be seen as representative of the central parts of Drenthe). Of the 197 members in this cooperative, 101 (51%) had 1-4 dairy cows and only seventeen members (9%) had ten or more (Stoetman, 1986: 109, Table 34b and c).

Figure 1.10 Number of members of the cooperative creamery Gieten/Bonnen according to their numbers of dairy cows, in 1937/38

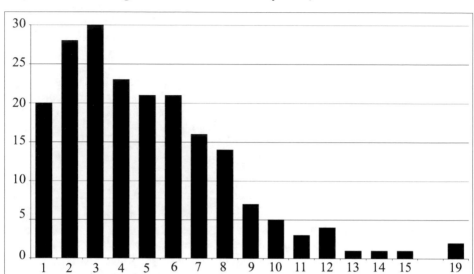

Source: Stoetman, 1986: Table 34b and c.

In hindsight, the rapid way in which creamery dairying came about shows the true audacity of farmers who together with others made investments in a time of uncertainty – after all it was the time of the 'agricultural depression'. Moreover, they had to deal with the fact that henceforth their milk had to be given into the hands of others, and they had to wait to see what products were made from it; products that until then had been made on their own farms. In the future they had to deliver the milk from their cows to be processed collectively, without having any direct control over the process which now was delegated to the personnel of the cooperative. In doing so, the farmers often had to stand up to an understandably negative response from their own wives. Butter-making had been the domain of the farmer's wife and one of her basic skills, and the shift of dairying activities from farm to creamery deprived her of this labour and the status connected to it. Then there was the innovation introduced by modern, 'steel technology': the centrifugal skimming of milk. It should be realised that at that time most farmers, especially in regions like Drenthe, were still used to working mainly with wooden implements.

Seen in this light, and taking into account later developments, the Swartz system as well as the small-scale hand-driven creameries of 'the Limburg system' should not be censured as being less advanced methods or technologies. The Swartz system was not designed for Dutch conditions at all and the level of the farmers' theoretical knowledge needed to work with it successfully was inadequate. The small hand-driven creameries were, in fact, a practical solution, given the overall situation in most villages in Drenthe at that time and constituted an important intermediate phase before the larger – and much more expensive – steam-driven creameries took over.

However, there was much to be gained and this was especially important for smallholders: farmers now got cash into their own hands which they could spend freely. They were no longer condemned to the detested system of barter trade. Or, as the annual agricultural report of the municipality of Vries for the year 1898 testified: 'With the founding of 7 cooperative creameries, dairy farming has entered a completely new era, making farmers prosper. Not only has dairying become more profitable, but the, in many respects pernicious, barter trade of butter for goods has been killed. The farmer has become more free and makes his purchases wherever he finds is best for him' (Verslag van de landbouw, 1898: vol. 2, 258-265).

XI. Postscript

Initially everybody had great confidence in the changes mechanised dairying (and all the innovations connected to it) brought to the rural economy and rural society in a province like Drenthe. Very few realised that these improvements had in fact made the farmers entirely dependent on the sales of the products (butter, pork and eggs) on foreign markets, as well as on the supply of 'raw materials' like fodder, concentrates and fertilisers from abroad. In the 1920s, however, scarcely more than one generation after cooperative creameries had effected a true emancipation for small farming, the problems connected to this became apparent. Developments begun in 1850 and that resulted after the 1880s in a growing disproportion between the available area of land and the number of people dependent on it. Because of the worsening economic situation, it became clear then that the number of people making their living on a relatively small area of land was, in fact, much too large. The emergence of a large group of 'mini-farms' had its reverse side in a far too-low income per head or – to put it in other economic terms – a far too-low labour productivity. One can say that the problem that became manifest in the 1920s was, to a large extent, if not completely, the result of the success of mechanised creaming and the other innovations connected with it. Soon it was to become known as the 'Kleine-boeren Vraagstuk', or the 'small farmers' problem', and it was this problem that eventually was to dominate the agricultural political agenda after the 1950s.

Bibliography

Manuscript sources

Rijksarchief in Drenthe (Assen), Archief van de Rijkslandbouwleraar/Rijkslandbouwconsulent voor Drenthe (1892-1936), aanvullingen inv.no. 86.

Rapport omtrent omvang en toestand der zuivelbereiding en de fabriekmatige zuivelbereiding in de provincie Drenthe (1926).

Rural History Group, Department of Social Sciences – Wageningen.

University Agricultural Statistics Collection, Wageningen.

Printed and secondary sources

Alstorphius Grevelink, P.W. (1840) *Statistiek van de Provincie Drenthe, voornamelijk uit het oogpunt van Nijverheid en Volkswelvaart; met opgave der hoofd-middelen te opbeuring van dat gewest*, Assen.

Bakker, M.C.S. (1991) *Boterbereiding in de late negentiende eeuw*, Zutphen.

Bakker, M.C.S. (1992) 'Boter', Lintsen, H.W. et al. (eds.) *Geschiedenis van de techniek in Nederland: de wording van de moderne samenleving 1800-1890. Techniek en modernisering, landbouw en voeding,* Zutphen, vol. 1, pp. 103-133.

Bieleman, G.J. (1904) *Boterproductie en botercontrole in Nederland,* 's-Gravenhage (Verslagen en Mededeelingen van de Afdeling Landbouw van het Department van Waterstaat, Handel en Nijverheid, nr. 1).

Bieleman, J. (1987) *Boeren op het Drentse zand 1600-1910. Een nieuwe visie op de 'oud' landbouw,* Wageningen (AAG Bijdragen 29).

Bieleman, J. (1992) *Geschiedenis van de landbouw in Nederland 1500-1950. Veranderingen en verscheidenheid,* Amsterdam and Meppel.

Bieleman, J. (1996) 'Dutch agriculture 1850-1925 - Responding to changing markets', *Jahrbuch für Wirtschaftsgeschichte,* 1996, 1, pp. 11-25.

Bieleman, J. (2000) 'Dieren en gewassen in een veranderende landbouw. De georganiseerde veeverbetering', Lintsen, H.W. et al. (eds.) *Techniek in Nederland in de twintigste eeuw, deel III, Landbouw en voeding,* Zutphen, pp. 130-153.

Bos, R.W.J.M. (1978) *De Brits-Nederlandse handel en scheepvaart 1870-1914; een analyse van machtsafbrokkeling op een markt,* S.l.

Boijenga, Tj. (1952) *De start van de Domo,* Assen.

Boijenga, Tj. (1963) *Vijfentwintig jaar Domo. 1938-1963,* S.l.

Burg, M. van der (2002) 'Geen tweede boer'. *Gender, landbouwmodernisering en onderwijs aan plattelandsvrouwen in Nederland, 1863-1968,* Wageningen (A.A.G. Bijdragen 41).

Centraal Bureau voor Statistiek (1952) *Zuivelstatistiek 1952,* Utrecht.

Croesen, V.R.IJ. (1931) *De geschiedenis van de ontwikkeling van de Nederlandsche zuivelbereiding in het laatst van de negentiende en het begin van de twintigste eeuw*, 's-Gravenhage.

Dekker, J.C. (1996) *Zuivelcoöperaties op de zandgronden in Noord-Brabant en Limburg. Overleven door samenwerking en modernisering. Een mentaliteitsstudie*, Middelburg.

Dijck, M. Van and Molle, L. Van (2001) 'Arme en rijke boeren. Landbouw en platteland in een wijde cirkel rond de hoofdstad (1880-1950)', De Maeyer, J. and Heyrman, P. (eds.) *Geuren en kleuren. Een sociale en economische geschiedenis van Vlaams-Brabant, 19de en 20ste eeuw*, Leuven, pp. 103-135.

Faber, J.A. (1972) *Drie eeuwen Friesland. Economische en sociale vernieuwingen van 1500 tot 1800*, Wageningen (A.A.G. Bijdragen 17).

Geluk, J.A. (assisted by C.F. Roosenschoon) (1967) *Zuivelcoöperatie in Nederland. Ontstaan en ontwikkeling tot omstreeks 1930*, 's-Gravenhage.

Hengeveld, G.J. (1865) *Het rundvee, zijne verschillende soorten, rassen en veredeling*, 2 Vols., Haarlem.

Heuvel, A. ten (1914/1915/1916) 'De ontwikkeling van den landbouw in Drente', *Cultura* 26 (1914) pp. 145-157, pp. 181-193, pp. 277-284, pp. 315-327; 27 (1915) pp. 37-46, pp. 139-149, pp. 321-328; 28 (1916) pp. 14-18.

Hoffmann, W.G. (1969) 'De ontwikkeling van de margarineindustrie', Stuijvenberg, J.H. van, *Honderd jaar margarine 1869-1969*, 's-Gravenhage, pp. 1-28.

Homan, J., (ed.) (1947) *Gedenkboek uitgegeven ter gelegenheid van het 100-jarig bestaan van het Genootschap ter bevordering van de Landbouw in Drenthe*, Meppel.

Hull E. van 't and Ockers, B. (1992) 'Freia in kort bestek. De geschiedenis van de bouw en inventaris', *Stoomzuivelfabriek Freia en de ontwikkeling in de Nederlandse zuivelindustrie, 1850-1970*, Arnhem, pp. 17-32.

Hylkema, H.B. (1913[3]) *Leerboek der zuivelbereiding*, Leeuwarden.

Poel, J.M.G. van der (1967) *Honderd jaar landbouwmechanisatie in Nederland*, Wageningen.

Poel, J.M.G. van der and Wesssels, R.J.C. (1953) *De verslagen van het Nederlandsch Landhuishoudkundig Congres 1846-1953*, S.l.

Schaafsma, J, 'Letter to the section 'Landbouw en Veeteelt', *Provinciale en Drentsche Courant*, 7th of October 1897.

Schetsen van het Landbouwbedrijf in Nederland, 's-Gravenhage, 1912.

Slicher van Bath, B.H. (1957) *Een samenleving onder spanning. Geschiedenis van het platteland in Overijssel*, Assen.

Spahr van der Hoek, J.J. (assisted by O. Postma) (1952) *Geschiedenis van de Friese landbouw*, 2 Vols., Drachten.

Staring, W.C.H. (1862) *Huisboek voor den landman in Nederland*, Haarlem.

Stoetman, W. (1986) *De zuivel in Drenthe. Een onderzoek naar de achtergronden en ontwikkelingen van de coöperatieve zuivelfabrieken in Drenthe, 1880-1939*. Unpublished essay, Assen.

Stuijvenberg, J.H. van (1949) *Centraal Bureau. Een coöperatief krachtenveld in de Neder- landse landbouw 1899-1949*, Rotterdam.

Tiesing, H. (1927) 'Zuivelproductie voor de totstandkoming van de boterfabrieken in Drenthe', *Landbouwkundig Tijdschrift*, 39, pp. 410-417.

Uitkomsten van het onderzoek naar den toestand van de landbouw in Nederland, ingesteld door de Landbouwcommissie, benoemd bij Koninklijk besluit van 18 September 1886, no. 28 (1890), 's-Gravenhage.

Verbeek, N.W.H. (1992) 'Margarine', Lintsen H.W. et al. (eds.) *Geschiedenis van de techniek in Nederland: de wording van de moderne samenleving 1800-1890. Techniek en modernisering, landbouw en voeding*, Zutphen, vol 1, pp. 135-169.

Verheij, W. (1917) *Het 25-jarig bestaan der eerste Coöperatieve Boterfabriek in Limburg te Tumgelroy*, S.l.

Verslagen van (over) de(n) landbouw in Nederland (Annual governmental report on the situation in Dutch agriculture), over various years.

Verslagen van het Nederlandsch Landhuishoudkundig Congres, 1846-

Wiersma, J.P. (1959) *Erf en wereld*, Drachten.

Wilson, C. (1954/1970) *Geschiedenis van Unilever: een beeld van economische groei en maatschappelijke verandering*, 2 Vols., Den Haag.

Zeijden, Hans van der (1994) 'De eerste halve eeuw', Bieleman, J. et.al., (eds.) *Boeren- landschap in beweging; anderhalve eeuw boerenbedrijf in Drenthe en het Drents Land- bouw Genootschap*, Groningen, pp. 33-58.

Zestig jaar Coöperatieve Zuivelfabriek en Korenmalerij 'Borger' te Borger (1954) Borger.

2 Industrialising the countryside.
The cooperative dairy industry in Belgium during the interwar period

Yves SEGERS, University of Leuven and Hogeschool-Universiteit Brussel, Belgium
Wim LEFEBVRE, University of Leuven, Belgium

I.   Introduction

Business historians mostly study the development of large and important companies or of prominent business leaders. Small businesses generally are ignored as sources of information are scarce and of little use or do not go far enough back in time (Coppejans-Desmedt, 1998; Buntinx, 2001). This lack of attention is deplorable, not only because we know lamentably little about a number of sectors and branches of industry, but most importantly because some of them are of great economic importance. The food industry is one such 'forgotten' sector. The number of publications that deal specifically with the development of the modern food industry in Belgium over the past two centuries can be counted on the fingers of one hand (Landuyt, 1984; Sas, 1999; Devos, 1999).

In the present contribution we study the development of the cooperative dairy industry during the interwar years on the outskirts of Brussels and in the Brabant periphery. During this period, the Belgian dairy industry consisted of numerous small units, which to a large extent operated locally but whose importance cannot be underestimated. During the 1920s and 1930s, the dairy sector accounted for some 25 per cent of the total value of farm production. By comparison, the production of meat represented barely 12 per cent to 13 per cent . According to contemporary estimates, the total value of dairy produce was 2.5 to 3 thousand million BEF more than the output of the Belgian coal or steel industry (De Winter and Tambuyzer, 1956: 422).

Up to the present the limited Belgian (and also international) historiography rates the dairy sector during the interwar period quite positively. It points out the trend towards larger steam-driven dairies and the introduction of new techniques for the production of butter and the processing of milk. There is a lot of praise for the efforts of the government and of the sector itself to overcome the economic difficulties of the 1930s (De Baere, 1973). Nevertheless, does this view truly correspond to reality? After all, during the Second World War the German occupier ordered a fundamental reorganisation of the dairy sector, including the abolition of numerous dairies, the introduction of compulsory supply and a stress on quality and product differentiation (*Pasinomie*, 1940: 285-289 and Roger, 1944: 204-231). Was this intervention appropriate, or was it an attempt to increase control over the production of food?

In this study we are in any event joining in a wider international discussion. A great deal of attention has recently been paid to the part played by small, traditional businesses in the process of industrialisation and technological development, with some authors even assigning them a key role. Arni Sverrisson points out that smaller businesses are often very successful and can continue to operate alongside large enterprises, thanks to their ongoing adaptation of traditional knowledge to new technological possibilities. He distinguishes between the functioning of two parallel *technological complexes*, one carrying on large-scale production and the other characterised by a gradual development of small intermediate technology systems (Scranton, 1997; Sabel and Zeitlin, 1997; Sverrisson, 1993 and 2002). The idea of technological complexes implies a socio-technical approach to the processes of industrialisation. The type of industrialisation is primarily determined by technical aspects (the use of particular machines and processes of production, the organisation of work and so on), but the choice of these techniques is made by social groups of players. Sverrisson distinguishes between two different technological networks, which are characterised by a number of stereotypes, namely, town versus rural, large-scale versus small-scale, innovative versus conservative and the like. This is a question of theoretical constructions. Most technological networks fall somewhere between the two extremes (Sverrisson, 1993: 11-13 and 44-46; Sverrisson, 2002: 247-250). Our assumption is that in the Brabant (cooperative) dairy sector also, various industrial-technological networks are active, with proximity to Brussels as an important centre of consumption being a predominating factor. We pursue the ideas of Sverrisson further. First, we examine not only the processes of industrialisation in their socio-technical aspect but also their profitability and commercial organisation. Next, we look at how small-scale dairies compete with one another and with private companies (and not only if they survive).

For this contribution we consulted the very valuable files of the Dairy Consultancy of the Belgian Farmers' Union, which was formed in 1907 at a time when the number of cooperative dairies was rapidly rising. The dairy consultants gave advice to the very recently established and associated dairies on bookkeeping, business technology, personnel training and management. They visited associated dairies at least once a year and then drew up an in-depth report. The records and files that have been preserved give detailed basic information concerning the way in which rural Flanders endeavoured to join in the advancing industrialisation and technological modernisation of the 1920s and 1930s. We consulted almost 40 files of dairy factories.[1]

This paper is divided into four parts. In the first part we present a general overview of the dairy sector in Belgium and in the Brabant region during the interwar period, with special attention to cooperative initiatives. The second and third parts examine the technological and economic development of small and large cooperative dairy factories in the vicinity of Brussels and in the periphery. The fourth and last part presents an explanation for the developments, focusing in particular on the social players.

---

[1] We present a full list of these enterprises in the bibliography.

II. The dairy sector in Belgium and Brabant during the interwar years

II.1. A young sector fully engaged in the process of development

Industrial dairy processing in Belgium began at the end of the nineteenth century. Under the pressure of the *Agricultural Invasion,* numerous farmers shifted from arable farming to stock breeding and horticulture. The growth of livestock, the growing yields of milk and various technical innovations led to the establishment of the first dairies in Belgium. At the end of the 1890s, there were some 300 manual and steam dairies, including many cream-skimming sections working for a central butter factory.[2] Some two-thirds of all dairies became cooperatives. Between 1895 and 1907 their number rose from 63 to 544 units. The impact of this substantial growth must however be placed in perspective. At that time, barely 22 per cent of the milk produced was processed industrially (i.e. at a factory). Production consisted mainly of butter; with only some factories (principally private ones) concentrating on milk for consumption (Niesten, Raymaekers and Segers, 2002: 5-13).

The First World War slowed down the rapid development of the dairy industry in Belgium. The regulations introduced by the German occupier (such as compulsory deliveries of milk at fixed prices) caused many farmers to turn away from cooperative dairies during this period of shortages. Supplies from cheap farm cream-skimming producers and the existence of a lucrative black market intensified this process even more. During the war about a quarter of the dairies closed, particularly in the provinces of Hainaut, West and East Flanders, Luxemburg and the Herve region (Mommens, 1985: 30).

The problems that arose during the war years continued to play a part for some time to come. Livestock was completely decimated and its recovery needed years. Nevertheless, between 1919 and 1921 the import of German livestock (partly as war reparations) and a prohibition on the slaughter of calves ensured a 35 per cent increase in livestock. The post-war shortages resulted in high milk and butter prices, which kept many farmers from returning immediately to the cooperative dairies. The government and the Belgian Farmers' Union endeavoured to give a fresh start to the industrial production of milk and butter, but at the beginning they succeeded only partially. In comparison with foreign countries, large arrears had accumulated. In the traditional dairy countries such as The Netherlands and Denmark, industrial dairy production experienced further growth, together with all kinds of modernisation at the technological level (Dekker, 1996 and De Baere, 1973: 32-34).

Only from about 1923 onwards did industrialised milk processing again attract more interest, resulting in the establishment of new dairies. It is not easy to map the volume and structure of the Belgian dairy industry during the years 1920-1940 as the statistical material is scarce and incomplete. In 1921 there were some 500 dairies, a quarter of which

---

[2] A similar process occurred in other Western European countries; see Ó'Gráda, 1977; Henriksen, 1999 and Taylor, 1976. The number of cows grew in Belgium between 1880 and 1910 from 790,000 to 980,000. The average milk yield increased in the same period from 1,900 to 2,700 litres.

were in Limburg and a good 20 per cent in East Flanders. Brabant accounted for 17 per cent, Antwerp for 15 per cent and Luxemburg for 12 per cent. At the end of the 1920s, there were 650 milk businesses (Mommens, 1985: 41). We discern two important reasons for this increase. First, the economic upswing caused agricultural workers to change to more lucrative jobs in the secondary or service sector. After World War I, help on the farm was difficult to find and was also very costly. In many cases, dairy production on the farm was discontinued, so that members of the family could apply themselves to other activities. Second, consumers demanded more and better quality products, which gave rise to a diminishing demand for farm butter (and also lower prices).[3]

Through the Dairy Consultancy, farming organisations and in particular the Belgian Farmers' Union promoted the establishment of dairy cooperatives. In publications and circulars, they stressed the advantages of cooperation such as:
1) savings on equipment and business operating costs;
2) better quality and hence higher prices;
3) time saving;
4) thanks to the cooperative element, farmers could remain their own bosses and the influence of distributors would be eliminated;
5) a central dairy would be good for local trade and employment;
6) and at the end of the day, the farmer's wife who traditionally carried out milk processing, had more time for housekeeping, agricultural education and demonstrations.[4]

The heyday of the 1920s brought about the emergence of the first large regional dairy factories (in Limburg and the Antwerp Kempen), and moreover a deliberate changeover from manual dairies to steam dairies. This shift to an increase in scale and in professionalism was greatly encouraged by the Dairy Consultancy. Nevertheless, from the beginning of the 1930s, the new élan received a heavy blow. The economic crisis had serious effects on all sectors, as well as on companies in general. The government's deflationary policy affected internal purchasing power. For a part of the population, milk for consumption and butter became too expensive. Margarine, which more than ever revealed itself as an adequate substitute, grew into an even bigger competitor and on top of that came large-scale imports of butter from, among other places, The Netherlands, which dumped its surplus butter on the Belgian market. The Belgian government reacted by drawing up a system of quota restrictions. Existing measures restricting imports were strengthened. Traders and shopkeepers saw their turnovers fall because of disappointing sales and the socialists advocated the sale of food at affordable prices. The farmer-stockbreeders also had grounds for complaint. The low producer price of milk and the rising price of cattle feed placed their income under pressure more than at any time.[5]

The dairy sector and the government attempted to pump up stagnating consumer demand through greater product variety and the launching of promotion campaigns. For example, an attempt was made to reduce the milk lake by encouraging the production

---

[3] KADOC, Archief centrale bestuurs- en adviesorganen van de Belgische Boerenbond, nr. 7.2.3.3.
[4] KADOC, Archief centrale bestuurs- en adviesorganen van de Belgische Boerenbond, nr. 7.2.3.3.
[5] In this period a lot of 'milk strikes' were organised: cattle breeders refused to deliver their milk to the dairy factories.

of cheese, the development of new specialised by-products with greater added value, such as children's and dietary foods, milk powder for the manufacture of chocolate, condensed milk and casein. Nevertheless, the results were slight, but the changeover to the distribution of milk for consumption was more successful. Initially, the sector played on consumers through large-scale events such as milk parades, press publicity, lectures, posters, brochures and the introduction of a promotional service. Hygiene and quality were stressed in communications to the general public. In 1938 for example, government control of the production of butter, milk for consumption and (Herve) cheese was introduced, even if only on a purely voluntary basis (Niesten, Raymaekers and Segers, 2002: 76-78; Kwanten, 2001: 154-159).

At the end of the 1930s, some 550 to 650 dairies and creameries were active in Belgium, employing a total of approximately 3,400 personnel.[6] The fall in the number of companies was due to the amalgamation of many small manual dairies into larger steam dairies and the disappearance of small-scale dairy factories. Table 2.1 gives an idea of the structure of the dairy sector in 1934 and emphasises that the majority of the dairies were still small scale, both in the cooperative and private sectors. It is noticeable that creameries had almost exclusively a cooperative structure and that they represented a good 40 per cent.

Table 2.1   The Belgian dairy industry in 1934, number of factories

	< 5000 litres milk/day	> 5000 litres milk/day	Total
Cooperative sector			
Dairies	148	43	191
Creameries	80	35	115
Total	228	78	306
	(75%)	(25%)	(100%)
Private sector			
Dairies	168	44	212
Creameries	23	7	30
Total	191	51	242
	(79%)	(21%)	(100%)
Total	419	129	548
	(76%)	(24%)	(100%)

Source: Boerenbond, Documentatiedienst, DD 1195.

[6] *Economische en sociale telling van 27 februari 1937*, part 2, 202.

II.2. The cooperative dairy sector in Brabant

Contemporary specialists estimate that the cooperative sector in Belgium in about 1940 controlled some 45 per cent of industrial dairy production. Since the focus of these contributions is on the cooperative dairy industry, it is appropriate to point out briefly the specific qualities of these businesses. A nutshell definition of a cooperative is found in Nilson: 'A cooperative is a user-owned and user-controlled business, which distributes benefits on the basis of use'. The cooperative brings together persons who are economically weak and who have little negotiating power in the market, but wish to maximise the safeguarding of their interests through cooperation.

Cooperative dairies, which are generally active in the areas of both production and distribution, are based on four important principles and we will illustrate these by means of the model statutes of the Dairy Consultancy. The cooperative is first and foremost a democratic organisation. All members assembled at an annual general meeting and elected the members of the board. Every member has one vote, regardless of the size of his herd or the volume of his milk production. This board appoints the director of the dairy and all other members of its personnel. A second characteristic is the members' desire to achieve the highest possible price for the milk supplied by them. The interests of the members are more important than the profit earned by the cooperative and so the margin between purchase and sale prices is deliberately kept as low as possible. The distribution of dividends (which are determined according to the size of the quantities of milk delivered) is of rather subordinate importance.

Third, a cooperative is an open association and there is no limit to the number of members. Every one is welcome, as long as the basic regulations are obeyed. Members cannot participate in and supply other dairies or similar factories except with the permission of the board. In principle, the cooperative only buys milk from associated stockbreeders. In practice, however, during the 1920s and 1930s this rule was followed only rarely. Finally, the cooperative also wanted to play a social role by providing training and education and promoting mutual cooperation among farmers (Nilson, 1997).

The Brabant region occupied a special place in the Belgian dairy country. This is shown by table 2.2, which contains a general overview of the cooperative industry at the end of the 1920s. In comparison with other provinces, there were few dairies. Turnover was to a large extent achieved by the sale of drinking milk which in Brabant, represented no less than 45 per cent of the total turnover, while being only 22 per cent and 15 per cent in Antwerp and East Flanders respectively. The Limburg and West Flanders dairies produced butter almost exclusively. The proximity of large residential centres played a significant part in the structure of production and in Flemish Brabant, the influence of Brussels was naturally very great. Many private enterprises were established in the region, making the competition in that region greater than elsewhere. Important private enterprises included Pax in Brussels, Nutricia in Laken, Lacsoons in Rotselaar, Meert in Opwijk and Delco in Louvain.

Table 2.2   The cooperative dairy sector in Flanders (all provinces, 1928)

	Factories	Members	Cows	Milk, mill. BEF	Butter, mill. BEF	Byproducts, mill. BEF
Antwerp	43	5.760	17.266	14,2	46,8	5,9
Brabant	22	5.030	10.760	18,9	20,8	2,6
West-Flanders	3	1.526	4.105	-	15,2	-
East-Flanders	38	5.775	14.283	6,9	40,7	1,9
Limburg	46	5.363	15.007	0,2	36,5	0,1

Source: KADOC, Archief centrale bestuurs- en adviesorganen van de Belgische Boerenbond, nr. 7.2.3.3; rapport coöperatieve melksector 1928-1929.

During the interwar years, the number of Brabant enterprises which were members of the Dairy Consultancy fell appreciably (see table 2.3). The economic crisis, which accentuated the competition within the region even more, compelled both cooperative and private companies to close down their activities. The smaller enterprises in particular stopped working first, but the sector nevertheless retained a mixture of some large and many small steam diaries. In 1936, the volume of milk processed annually ranged from 200,000 litres in Nederokkerzeel to 6.7 million litres in Sint-Anna Opwijk. After Opwijk, the largest cooperative dairies in Brabant were Sint-Bernardus in Lubbeek and Sint-Victorianus in Tollembeek, with an annual 4.8 and 4.2 million litres respectively. All other dairies were on a much smaller scale and processed only 1 to 2 million litres of milk.

Table 2.3   Development of Brabant dairies, member in the Dairy Consultancy, 1918-1940

1918	33
1925	24
1927	25
1928	22
1936	16
1939	15

Source: KADOC, Archief centrale bestuurs- en adviesorganen van de Belgische Boerenbond, nr. 7.2.3.3; diverse jaarverslagen.

Figure 2.2 Map of the province Vlaams-Brabant and Brussels Capital-Region

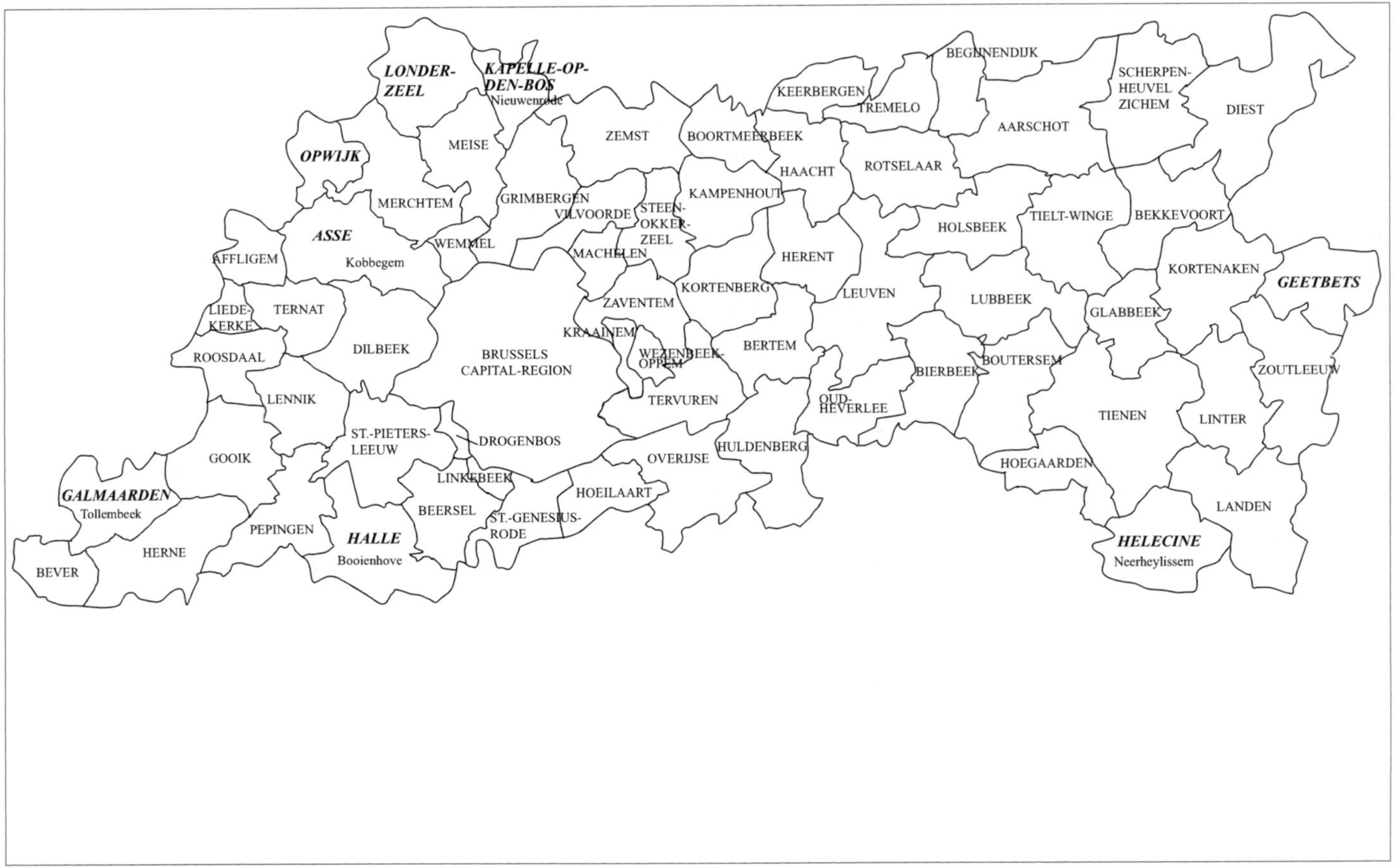

III. Technology, knowledge and hygiene

This general overview of the Belgian (cooperative) dairy sector in the Brabant region during the interwar period (albeit based on scarce information) does not include any mention of developments within individual companies. In this second part we will deal with the technical development of dairies. Are they adopting technological innovations? Does the butter and milk production process follow the rules? Are machines and buildings being maintained in line with requirements?

III.1. Technical equipment[7]

According to the dairy consultants of the Belgian Farmer's Association, the technical equipment of a dairy should be renewed every ten years. In Brabant, however, there are only two dairies (Nieuwenrode and Opwijk) that have invested in new installations. Other dairies are hardly modernising at all during the 1920s and 1930s. Generally speaking, old machines are patched up, second-hand equipment is bought and product quality suffers appreciably as a result. Then, when renewal is after all undertaken, only a few machines are replaced at the same time. This is of little use in helping dairies to move forward, because dairy processing is a process industry where various treatments meticulously follow one another and temperature and rapid processing are of crucial importance. The quality and rapidity of processing is dependent on the weakest link. The renewal of the installation ensures the best results, allowing the various stages to follow one another smoothly and ensuring unvarying quality and processing capacity.

The dairy buildings also do not always meet modern requirements. The dairy is usually accommodated in an existing building, sometimes even in a residential building which is only minimally suitable and where there is hardly any room for expansion. The reports by dairy consultants are also clear on the point that the technical equipment of most dairies is obsolete and does not function as well as it should. Some cooperative dairies in the 1930s do not even have a cooling machine!

III.2. Knowledge of the process of production

With regard to theoretical and practical knowledge of the process of producing butter and milk for consumption, most cooperative dairies are performing below par. It is striking how little insight dairies (most importantly the smaller ones) have into the best possible production processes. The butter-making cycle takes too long and in some places the milk is not being treated at all. Naturally, such lack of knowledge also has an adverse effect on the output of the dairy. The Londerzeel dairy, which after all is one of the largest businesses in the region, has been struggling for years against an excessively high loss

---

[7] This paragraph is based on several reports from the dairy consultants, i.e. KADOC, BB-M, Kobbegem, report De Brabander, June 28th 1939; Londerzeel December 7th 1944; Kobbegem report Van Esbroeck December 26th 1933; Kobbegem report Mercelis April 27th 1938, several reports Nieuwenrode 1937-1938; reports 1930-1938 Sinte-Anna Opwijk and Nederokkerzeel, report Segers, January 15th 1928.

of fat (in 1944 worth some 225,000 BEF).[8] The employees of the Dairy Consultancy try to help to put this matter to rights by visiting the dairy regularly and keeping an eye on the process of production. They give tips on the temperature and duration of the butter-making cycle, cooling, the use of pure cultures and the like. Sometimes, the consultants even show personnel how they should use the machines. They realise, however that their good advice is very rarely being followed closely. The generally inadequately trained personnel hold on to the traditional, classical methods of production.[9]

III.3. Hygiene

During the interwar years, few cooperative dairies treated hygiene as a top priority. All dairies that are members of the Dairy Consultancy regularly receive notices concerning this subject in annual technical reports. In most dairy factories, the problems are limited to inadequate cleaning of the premises. Another frequently occurring problem arises from uneven floors with the result that standing rinse water remains in hollows on the floor, providing an ideal breeding ground for bacteria. In some companies, mould forms on the walls, in Asse even in the milking sheds which are subject to inspection. Naturally, all these factors have an adverse influence on the quality of the end-product. An additional drawback in small dairies is the fact that they do not make butter every day and, particularly during the summer months, the milk delivered quickly turns sour.[10]

Dairy consultants often complain about the lack of cleanliness in dairy factories, but their complaints usually fall on deaf ears. They also try by means of lectures to make milking on the farm comply with hygiene requirements. A lack of control nevertheless prevents any appreciable amount of progress. Following the Second World War, more attention has been paid to this problem by the sector itself and by the government.[11]

At the end of the 1930s, nearly all dairies in Brabant had a pasteurisation plant available in order to render milk free from germs. Milk for consumption, cream as well as skim milk, were then all being pasteurised, the last in order to combat the spread of infectious animal diseases such as foot-and-mouth disease. Pasteurisation was important principally to convince the consumer, and was as yet not compulsory as it was in The Netherlands or Denmark.[12]

---

[8] KADOC, BB-M, Londerzeel, anonymous report 1945 and Kobbegem, report Segers, January 7th 1932.
[9] KADOC, BB-M, Kobbegem, report Segers, November 5th 1930 and Nederokkerzeel, report Segers, June 28 1929.
[10] KADOC, BB-M, Asse Sint-Isidorus, report Segers, December 9th 1929; Nederokkerzeel several reports 1929; Vollezele, report Segers, July 25th 1929.
[11] KADOC, BB-M, Londerzeel, factory circular, March 1949.
[12] KADOC, BB-M, Londerzeel, see the letter from De Troyer to dairy consultant Mercelis, January 13th 1938.

IV. Between the hammer and the anvil: economic company management

The files of the Dairy Consultancy have enabled us to carry out a limited analysis of the company results of nine cooperative dairies in Brabant. These include both smaller and larger units located within the capital's sphere of influence (Kobbegem, Londerzeel, Nieuwenrode and Opwijk) and on the periphery of the province (Geetbets, Halle-Booienhoven, Lubbeek, Neerheylissem and Tollembeek). Table 2.4 shows the clear differences between companies on the basis of average daily milk supplies during the 1930s. Only the dairies of Lubbeek, Tollembeek and Opwijk processed more than 10,000 litres of milk a day. Each of the dairy companies in Geetbets, Neerheylissem, Kobbegem and Nieuwenrode processed 2,500 to 5,000 litres a day.

Table 2.4   Average daily supply of milk (litres, 1930-1940)

Geetbets	3,601
Lubbeek	10,910
Neerheylissem	3,674
Kobbegem	3,990
Londerzeel	8,587
Nieuwenrode	3,714
Opwijk	18,233
Tollembeek	12,821
Average	8,191

Source: KADOC, BB-M, dossiers Geetbets, Lubbeek, Neerheylissem, Kobbegem, Londerzeel, Nieuwenrode, Opwijk and Tollembeek. Own calculations.

IV.1. The struggle for producers of milk

During the 1930s, there was a surplus of milk and farmers were in a weak position. The dairies were able to make demands concerning quality and were in a position to offer a low price. This did not happen, however, because of the advantages of scale, which arose from the increase in the quantity of milk supplied. The more farmers could be enlisted as suppliers, the more cheaply could the factory produce. The result is intensive competition among dairy factories. If milk supplies can be taken away from another dairy, the power of that business is thereby weakened and it is therefore a question of attracting as many producers as possible, even if one cannot process all the milk oneself. The way to do this is of course by paying higher milk prices. The competitive struggle is not equally intensive everywhere. Particularly to the northwest of Brussels in the region around Opwijk and Londerzeel, a genuine price war is being fought. In Hageland and in Haspengouw, competition is less sharp but it is felt there as well. In the region of Louvain, there is a better understanding among cooperative dairies than in the Brussels region, so prices are lower and more stable.[13]

   The prices of milk in and around the capital are also higher. The dairies of Opwijk, Nieuwenrode and Londerzeel pay the highest prices, Geetbets, Neerheylissem and Tollembeek pay appreciably less. Nevertheless, an obviously parallel evolution in milk prices shows that all dairies take their cue from the price policy of competing businesses. Between 1930 and 1934 milk prices fell extensively, but from 1935 onwards they slowly recovered, although the 1930 levels were not reached until the Second World War. Not only does the geographical position explain the level of milk prices paid, but the operation of the dairy also a part. The lower the operating cost per litre, the more can be paid to the farmers. Accordingly, the larger businesses pay better prices than the smaller ones These elements are, for example, the reason why in the 1930s in Opwijk, an average 34 per cent more was paid than in Geetbets (Figure 2.2).

Figure 2.2   Average producer price of milk (BEF per litre, 1930-1940)

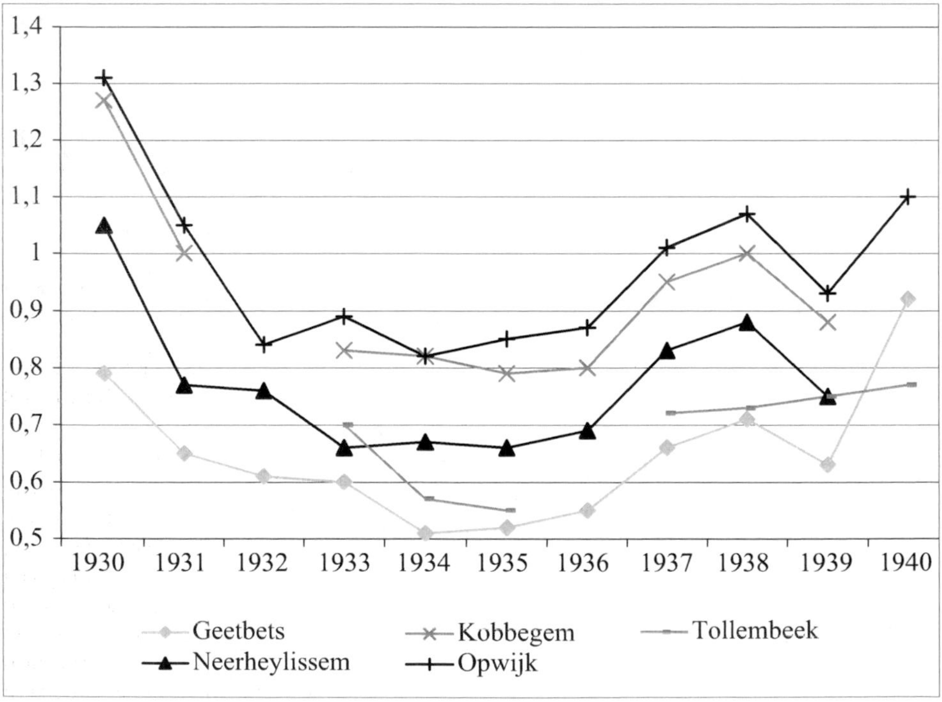

   Source: see Table 2.4.

[13] KADOC, BB-M, Londerzeel, report Delcourte May 20th 1935; Londerzeel report Van Esbroeck August 7th 1934 and report Mercelis November 13th 1939.

The competition exercised by the dairies (also the cooperatives) against one another was not always even 'clean'. Not infrequently, an attempt is made, admittedly under pressure from the farmers, to bring rivals into difficulties. Whether or not those high prices for milk are also advantageous to one's own business is of secondary interest. The margin between the purchase price and the selling price thereby often becomes extremely small. An example: the dairy of Londerzeel is forced to operate with a profit margin of 4 to 5 centimes per litre, not including the cost of transport.[14]

Carriers play a crucial part in the competitive struggle. They carry the milk on behalf of dairies from the stockbreeders in neighbouring villages. They also make the twice-weekly payments to the producers. For many farmers, the driver is their only contact with the dairy and these drivers are being regularly approached by competing dairies which try to induce them to switch their milk rounds to them. During the 1920s and1930s, numerous drivers were enticed by other cooperative or private dairies. These practices only came to en end when in the context of a re-organisation of the whole sector, the German authorities introduced compulsory milk deliveries on August 17th 1941 (Niesten, Raymaekers and Segers, 2002: 39-41).

The Dairy Consultancy of the Belgian Farmers' Union is, of course, very opposed to the competitive struggle, but is almost powerless. An attempt is being made, through the establishment of provincial dairy associations, to make the situation somewhat more normal. Among other things, these associations endeavour to reach agreements concerning the purchase (or producer) price. This nevertheless yields little result, because the large dairies such as Sint-Anna of Opwijk do not participate. Nevertheless, the dairy consultants are trying to keep the price war within limits. In place of paying high milk prices, the Belgian Farmers' Union has opted for a system of higher repayments to members at the end of the year; in that way, other cooperative dairies do not come under direct pressure. Nevertheless, the majority of stockbreeders opt for rapid results. Instead of opting for a long-term investment, the majority choose the (cooperative or private) dairy that will pay the highest price at the time. For this reason in the 1930s, few Brabant farmers had become members of a cooperative dairy, whilst the existing members remained in a cooperative, though the unfavourable rules governing leaving also playing a part in that decision. It was no accident that the only cooperative whose membership rose appreciably during the interwar period was Sint-Anna in Opwijk, which paid the highest milk price (Figure 2.2).

---

[14] KADOC, BB-M, Londerzeel, report December 1942 and letter from De Troyer to the Dairy Consultancy, December 30th 1935 and Steenhuffel, Bex report October 6th 1927.

Table 2.5   Number of shares in the respective dairies

	1930	1931	1932	1933	1934	1935	1936	1937	1938	1939
Geetbets	390	390	390	390	390	390	390	390	390	390
Halle-Booienhoven	190	190	190	190	190					
Lubbeek	900	920	920	910	900	900	900	900	900	900
Neerheylissem	370	370	300	370	380	380	380	380	350	350
Kobbegem	340	340	340	340	320	300	300	250	250	250
Londerzeel	400	525	525	520	480	425	415	400	400	400
Nieuwenrode	350	350	350	350	360	370	380	390	370	370
Opwijk	1,375	1,375	1,360	1,360	1,400	1,450	1,475	1,500	1,850	1,900
Tollembeek	920	980	1,020	1,075	1,075	1,020	980			

Source: see Table 2.4.

IV.2. Problems of distribution and sales

For many dairies, the sale of processed dairy products is a difficult task. The market for butter is burdened primarily by over-production during the 1930s, which placed the dairies in a weak position. The limited keeping properties of dairy products often compel the factories to sell them at low prices. Butter dealers make skilful use of this competition and play the dairies off against one another. Moreover, this often amateurish policy ensures that managers are not properly abreast of milk prices, marketing channels and the possibilities offered by advertising (Mommens, 1985: 54-62).

Accordingly, the cooperatives themselves try to regularise the distribution of butter through auction sales. In the 1930s, Brabant dairies sent their butter to the auctions of the Belgian Farmers' Union in Hasselt, Antwerp and Brussels, or to the auctions of the National Association of Belgian Dairies in Brussels. The advantage of the auction as a system of sales is the fact that the power of the butter merchants is thereby weakened and passes into the hands of the dairies. Using the system of bidding, dealers do not have to wait too long to purchase a good parcel of butter. The dairies play the wholesalers off against one another (Mommens, 1985: 54-62, 129-134; Niesten, Raymaekers and Segers, 2002: 11-12; Van Molle, 1990: 222, 290).

The dairy consultants of the Belgian Farmers' Union try to sell the finished butter at one of its own three butter auctions. All member dairies have access to these auctions, subject to their satisfying two conditions: butter has to be supplied regularly and must be of a satisfactory quality. First and foremost, its water content is taken into account. The butter must not contain more than 18 per cent water and this is where many dairies failed as they produce butter of too low or too irregular quality. Many other dairy factories find the cost of auctions too high and transport can also cause problems, certainly for dairies in outlying villages. Still other dairies are reluctant to supply, fearing the compulsory inspection of butter at auctions. During the interwar period, the adulteration of butter was, after all, a widespread phenomenon.[15]

In the 1930s, the sale of milk to consumers was still poorly organised. Private businesses in the Brussels area supplying milk for consumption dealt with independent milk traders or with large milk merchants. Only a few cooperative dairies engaged in retail themselves. Sint-Bernardus of Lubbeek sold most of its milk in Louvain and in 1939 employed for that purpose five traders, who were paid by the litre. From 1932 onwards, The Christian Employees' Syndicate distributed milk and cream in Brussels which originated in twelve dairies in Brabant and East Flanders.[16]

It is interesting to see whether the vicissitudes of distribution made themselves felt in the sales and profits of Brabant dairy cooperatives. The prices of butter and milk for consumption originating from various dairies underwent a distinctly similar development. Butter prices fell strongly at the beginning of the 1930s and reached a low point in 1935. The partial recovery during the following years was nullified by a fresh fall in 1939. At the beginning of the 1930s, the prices of milk for consumption also dived steeply, but remained constant from 1932 onwards. They then rose once again from 1937 onwards and a slight fall occurred in 1939. The analogous evolution of prices at the various dairies indicates that the businesses themselves have relatively little influence on price setting. In general, it is noted that the larger businesses receive better prices than the small ones, have much better operating equipment and can therefore supply products of a better quality (Figures 2.3, 2.4, 2.5 and 2.6).[17]

---

[15] KADOC, BB-M, Geetbets, letter from the director of the auction, November 3th 1933; Kobbegem, report Dreesen March 30th 1934 and Londerzeel, letter from De Troyer March 29th 1934; See also Mommens, 1985: 85-86.

[16] KADOC, BB-M, reports for Vlezenbeek and Lubbeek; also Mommens, 1985: 131-132.

[17] See sources mentioned in table 2.4.

Figure 2.3   Price of butter, 1930-1939 (BEF, kg)

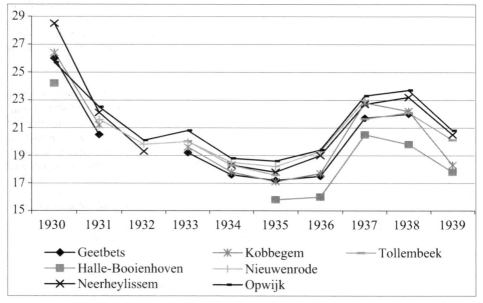

Source: see Table 2.4.

Figure 2.4   Price of milk, 1930-1939 (BEF, litre)

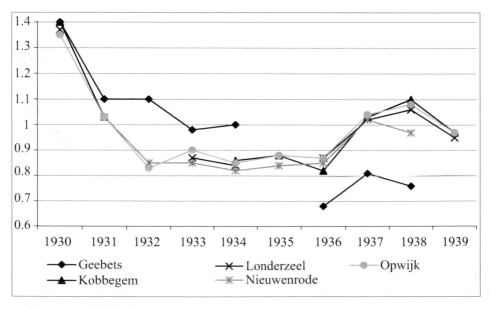

Source: see Table 2.4.

Figure 2.5   Amount of butter sold, 1930-1939 (kg)

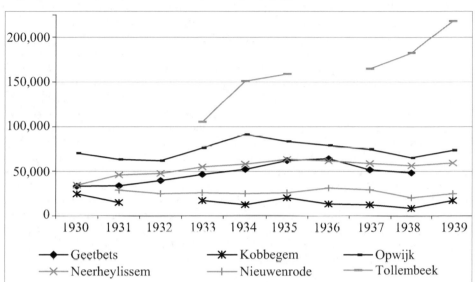

Source: see Table 2.4.

Figure 2.6   Amount of milk sold, 1930-1939 (litre)

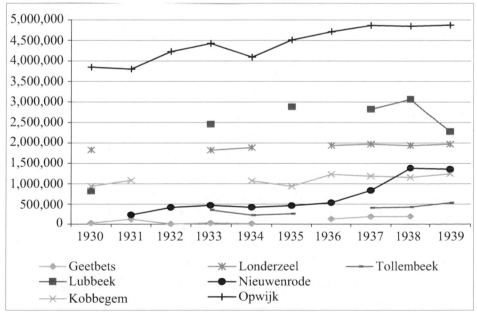

Source: see Table 2.4.

The turnover of dairies varies widely. The annual turnover at Opwijk is approximately six to eight times higher than that of its cooperative competitors. In the first half of the 1930s, the profits of dairy factories fell (under the influence of shrinking market prices), but from 1935 onwards, they began advancing. In 1939, the turnover of most factories fell once again (Figure 2.7). As far as turnover is concerned, a distinct difference can also be drawn between businesses in the immediate surroundings of Brussels (Londerzeel, Kobbegem, Nieuwenrode and Opwijk) and the dairies in the eastern part of the province which are further away from the capital (Geetbets, Halle-Booienhoven, Lubbeek, Neerheylissem and Tollembeek). Those belonging to the first group derive two-thirds of their turnover from the sale of milk for consumption, one-quarter from the sale of butter and derive over 10 per cent of their profit from secondary products such as cream, skim milk and buttermilk. On the other hand, further away from Brussels there is strong emphasis on butter and secondary products are distributed to a lesser extent. At first glance, this appears to be contradictory; in the production of milk for consumption no skim milk remains behind whilst in the production of butter it does. Nevertheless, the smaller factories send a large quantity of the skim milk back to the farmers for the fattening of calves. The larger businesses have recognised that the upgrading of by-products is essential for efficient industrial strategy. They use skim milk for many purposes, including the production of low-fat condensed milk and milk powder (Figures 2.8 and 2.9).

Figure 2.7   Total annual turnover (BEF)

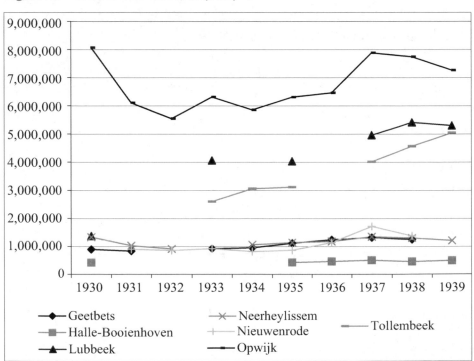

Source: see Table 2.4.

Figure 2.8   Production structure of dairies near Brussels, 1930-1940 (annual turnover)

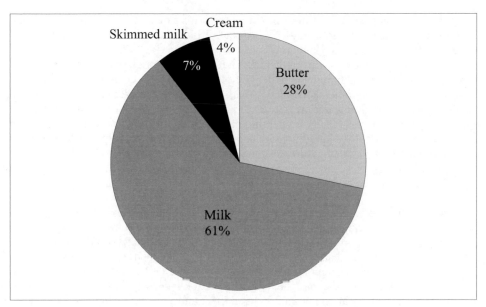

Source: see Table 2.4.

Figure 2.9   Production structure of dairies in the Brabant periphery (annual turnover)

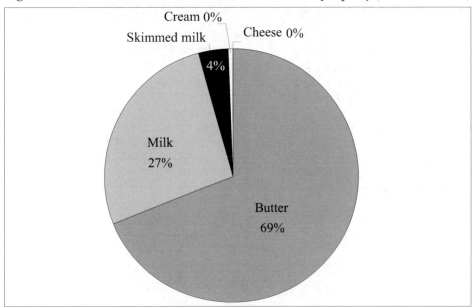

Source: see Table 2.4.

IV.3. Profits of the cooperative dairy factory

The big question remains what profits do cooperative dairy factories earn. Do rural communities succeed in establishing viable, dynamic businesses, or do they remain small-scale, traditional businesses which only just manage to survive? Three crucial points of economic management were distinguished. The keeping of operating costs per litre of processed milk low is an important challenge for a dairy. Moreover, dairy processing is a seasonal industry; in summer dairies receive appreciably more milk than in winter and a factory can work profitably only if it uses the largest possible part of its capacity during the winter months. Finally, dairy processing also regularly calls for new investments in machines and for adequate liquidity to be able to pay suppliers throughout the year. Do the dairy businesses in Brabant possess an adequate capital base to provide for this?

IV.3.1. Operating costs

There is a striking relationship between the size of an enterprise and its related operating cost. Generally speaking, the larger factories have a lower per litre operating cost; in Opwijk, for example, it is five to eight times less than in the case of its competitors. Profits are influenced by the fact that Opwijk and Neerheylissem let the farmers themselves pay for the cost of transporting their milk to the dairy, whereas in the case of other dairies, it is, next to the payment for the milk, the largest item of expense. Nevertheless, the profits of Sint-Anna in Opwijk are remarkable; this dairy factory processes milk substantially more cheaply than other cooperative dairies (Table 2.6, Figure 2.10).

Table 2.6   Structure of the operating costs, 1930-1939 (%)

	1	2	3	4	5	6	7	8	9
Milk	75.82	89.41	80.95	91.19	89.93	87.19	85.75	96.48	81.04
Transport	10.65	1.11	7.61		3.19	5.1	5.67		8.91
Wages	5.74	4.03	3.36	4.22	2.24	2.86	2.94	1.28	2.83
Fuel	1.53	1.07	0.96	1.26	0.83	1.12	1.22	0.59	0.85
Other	6.26	4.38	7.12	3.33	3.81	3.73	4.42	1.65	6.5
Total	100	100	100	100	100	100	100	100	100

Source: see Table 2.4.

1) Geetbets, 2) Halle-booienhoven, 3) Lubbeek, 4) Neerheylissem, 5) Kobbegem, 6) Londerzeel, 7) Nieuwenrode, 8) Opwijk, 9) Tollembeek

Figure 2.10 Total production cost per litre (BEF)

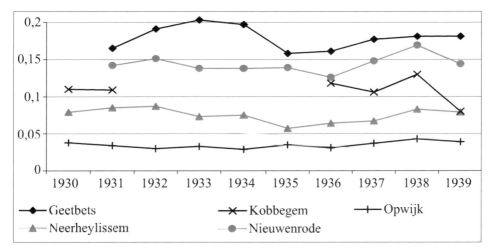

Source: see Table 2.4.

IV.3.2. Utilisation of output during winter months

Figure 2.11 illustrates the relationship between the lowest and the highest milk deliveries. If we assume that at the time of the highest milk deliveries the dairy is working at virtually its full capacity, these figures give a picture of the capacity used in winter. Once again we find that the size of a dairy is the decisive factor. During the worst period of the year, Opwijk still uses almost 80 per cent of its capacity, Nieuwenrode slightly more than three-quarters, Neerheylissem and Londerzeel work at respectively 60 per cent and 70 per cent of their maximum outputs. During the winter, the dairies of Tollembeek and Geetbets work at even less than 40 per cent of their capacity (Figure 2.11).

Figure 2.11 Lowest and highest milk deliveries, in % (1930-1940)

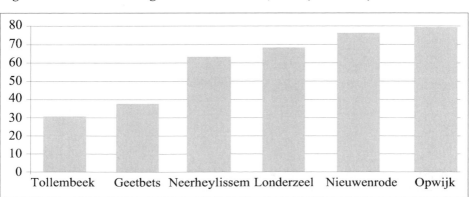

Source: see Table 2.4.

IV.3.3. Lack of a firm capital base

The cooperative dairies invest too little, a frequently recurring problem among the cooperatives in general, because the farmer-members are anxious to earn the highest possible profit at the end of the year and have very little interest in a plan of investment. Too few farmers realise that it is precisely the latter strategy which can make or keep a dairy profitable in the longer term. This attitude is of course not good for the quality of technical equipment and the maintenance of the dairy. The lack of investment is partly dictated by the limited subscribed capital of the cooperative dairy factories. Every farmer who becomes a member of the dairy must buy a share for every cow that he owns and the total of this payment is the amount invested by the farmer in the company. Nevertheless, the subscribed capital of most cooperative dairies is pitifully low and only amounts to a few thousand BEF, with the exception of the dairies in Lubbeek, Tollembeek and Opwijk.

The interpretation of profit and loss figures for the dairies under study presents problems: the final figures have more than once been tampered with, for example in order to hoodwink the tax authorities, or in order not to have to submit poor results to members at the general meeting. As the profit and loss figures in the annual accounts show, few dairies succeed in earning a profit several years in succession and the profits made are extremely modest (rarely more than 25,000 BEF a year). There is accordingly also no basis for a policy of investment and modernisation. The question moreover arises as to how the cooperative dairies manage to stay in business. They do so very frequently because they borrow money from local savings and loan associations, or from the Middenkredietkas (MKK) of the Belgian Farmers' Union which in 1935 was converted into the Centrale Kas voor Landbouwkrediet (CKL). Many dairies were kept open thanks to various overdraft facilities at the MKK.[18]

V. Technological complexes: social and psychological determinants

The global picture of technical and economic business management in the Brabant dairy cooperatives shows few positive features. The maintenance of machinery and the knowledge of production processes leave much to be desired. Hygiene receives only a limited amount of attention. An analysis of the annual operating profits of various cooperative dairies moreover shows that there is no attempt to influence the prices of processed products, that operating costs are far too high and that dairies lack the knowledge of how to recruit farmers as suppliers by paying them lucrative milk prices, despite (or perhaps because of) the price war which the cooperative dairy factories wage on one another. Moreover, during the winter months, most businesses use only a limited part of their capacity and have an inadequate capital base for investment.

---

[18] See for instance KADOC, BB-M, Tollembeek, report Missoul of CKL on December 7th 1937 and anonymous report from CKL 1942; also anonymous report from the Dairy Consultancy on the financial situation in the dairy factories located in Nieuwenrode (August 11th 1939 and May 16th 1945) and Lubbeek, September 10th 1937.

How is it that during the 1920s and 1930s, so few dairies (and certainly not those on the outskirts) opted for new and modern technology? It was certainly not only a question of financial means. With a few exceptions, there was never, even during the economically favourable years, any heavy investment in technology, knowledge or product development. A number of external factors aside, we are of the opinion that the explanation for this cautious policy must rather be sought in factors that are primarily socio-cultural and psychological in nature. Rural communities often adopt a waiting attitude towards every kind of innovation, including technology and methods of production. The managers and personnel of the smaller dairies located in the Brabant periphery have long held on to traditional techniques and only slowly are they adopting a more professional manner of working. The existence of a limited amount of competition (and the absence of private dairies) renders this possible.

A lack of know-how strikes one at first glance. The majority of cooperative dairies do not have trained personnel. Only the manager has an agricultural engineering diploma, and usually also additional dairy training; nevertheless, he has little time to devote to technical matters at the dairy, as he has to cope with the pressures of administration, a task for which in any event he has not been trained. In the majority of dairies, bookkeeping is naturally in a state of disarray with both interim and final balance sheets often being drawn up by the dairy consultants. In addition, the managers of small dairies usually have another principal occupation (in farming), so that the day-to-day responsibilities are in the hands of the foreman.

The part played by the director is, nevertheless, of crucial importance. He must know about the technical and economic working of the factory and, what is perhaps even more important, he must enjoy the confidence of the villagers and farmers and be of impeccable character. In the case of vacancies, the management often allows itself to be influenced by village politics or by family (or political) motives, with competence only taking second place. A connection with the Dairy Consultancy is therefore not always an advantage. Some managers have an extra income, 'profits', which the board is not informed about and which threaten to come to light when the dairy consultants come to examine the books. Nevertheless, the dairy consultants often overlook less serious forms of accountancy fraud or even encourage them (in order to escape taxes)![19]

A second decisive element, which may briefly be mentioned, is the mentality of farmers who are also owners of dairies. They want to get the highest possible price for their milk and also earn a high profit at the end of the year. Few members are interested in (lower) prices for milk which would give the dairy a reasonable profit margin, or in investing in new machines and maintenance. Because of the direct participation of members in the dairy at the general meeting, they have a big impact on its concrete functioning and it is not surprising then that in many businesses the board and the manager acquired more power during the 1930s to the disadvantage of the general meeting.

---

[19] KADOC, BB-M, Geetbets, confidential report Michiels August 21th 1942 and also reports for Glabbeek (1929), Sinte-Anna Opwijk (1927), Kobbegem (1930).

The third element is the sector in general, which during the interwar years was too little aware of the common interest. The ideological character of cooperative dairies puts them in a position that is diametrically opposed to that of private players, but cooperating dairy factories concluded hardly any agreements, as is clear from the fierce competitive struggle. An interest-coordinating association only came into existence in 1938 in the form of the *Algemeen Verbond der Coöperatieve Zuivelfabrieken* (General Association of Cooperative Dairy Factories). The amalgamation of neighbouring cooperating dairies, as the dairy consultants of the Belgian Farmers' Union here and there recommended, was firmly rejected by the board and the members of these factories. The stockbreeders held on to their own village dairy. Only after the Second World War did an inter-cooperative collaboration on a larger scale come into existence (De Baere, 1973: 130-132).

Nevertheless, apart from the social and psychological characteristics of the rural communities themselves, external factors also played a part, the foremost being the government's inefficient crisis policy. Nevertheless, the government did not sit idly by and, generally speaking, its policy during this period very creditably concentrated on stimulating the marketing possibilities for domestic dairy production. Nevertheless, the Belgian government paid too little attention to the ailing production structure of the dairy sector itself. Moreover, the farmers' organisations tried to curb every intervention that would be adverse to the smaller (manual) dairy. The reorganisation of the sector carried out from 1940 onwards by the German occupier (compulsory pasteurisation, abolition of many small dairies, one dairy per production area, the obligation to supply and the like) became more than justified. Through this drastic redrawing of the dairying landscape, a completely new sector came into existence after the Second World War. The remaining dairies were somewhat larger in scale and diversified their production to embrace milk powder, condensed milk, cheese and yoghurt. Moreover, the general expansion of the labour- and capital-intensive production of bottled milk gave rise to a number of inter-cooperatives. Factories that had been closed by the German authorities did not restart after the war and in five years had lost all contact with the rest of the sector (De Baere, 1973: 39-42; Mommens, 1985).

A second external factor over which the dairies themselves have no influence is the proximity of other technological networks. It is clear from the present article that there is a difference at all levels between cooperative dairies at the edge of Brussels and those in the periphery. The dairies in the immediate vicinity of the capital, generally speaking, operate on a larger scale, are almost exclusively devoted to the production of milk for consumption and have more control over their sales (they endeavour to upgrade their by-products, to diversify their product range and to obtain higher prices). The cooperative Sint-Anna in Opwijk is most successful in doing this. Efficiency in technical and economic business management in Flemish Brabant cooperative dairies appears to be related to the proximity of Brussels, where the presence of a big urban technological network prompts dairies to develop a more rational and economic management. This network resulted from the stiff competition with other cooperatives and especially with private dairies (in the hands of city enterprises) and from the presence of a large market and a well-developed transportation infrastructure.

Bibliography

Manuscript sources

KADOC, K.U.Leuven

Archief Boerenbond, Hoofdbestuur:
-nr. 7.2.3.3. Dossier zuivelconsulentschap, 1908-1950.
Archief Boerenbond, Coöperatiediensten, Melkerijen (BB-M):
- Asse, Sint-Isidorus
- Assent
- Averbode, Sint-Gerlacus
- Baal
- Duisburg
- Eppegem
- Erps-Kwerps, Melkerij van den Boerenbond
- Erps-Kwerps, Sint-Amands
- Erps-Kwerps, Sint-Jozef
- Geetbets, Sint-Paulus
- Glabbeek, Sint-Brigida
- Goetsenhoven
- Haacht
- Halle-Booienhoven, Sint-Barthelomeus
- Hever-Schiplaken, Sint-Benedictus
- Kessel-Lo
- Kobbegem, Sint-Gauchericus
- Kortenaken
- Londerzeel, Melkerij van Londerzeel
- Lubbeek: Sint-Bernardus
- Melkwezer
- Merchtem
- Molenstede
- Nederokkerzeel
- Neerhespen: Sint-Mauritius
- Neerheylissem, Saint-Sulpice
- Nieuwenrode, Sint-Guido
- Nieuwrode
- Oetingen, Sint-Antonius van Padua
- Oplinter, Sint-Genoveva
- Opwijk, Sint-Anna
- Orsmaal
- Steenhuffel, Sinte-Genoveva
- Tollembeek,Sint-Victorianus
- Vilvoorde
- Vollezele, Melkerij van Vollezele
- Zoutleeuw

73

Boerenbond, Leuven
Documentatiedienst, several dairy reports.

Printed and secondary sources

BUNTINX, J. (2001) *Gids van bedrijfsarchieven in Vlaams-Brabant*, Brussel.

BUYST, E., GOOSSENS, M. and VAN MOLLE, L. (2002) *Cera 1892-1998. De kracht van coöperatieve solidariteit*, Leuven.

COPPEJANS-DESMEDT, H. (1998) *Bedrijfsarchieven toegankelijk voor het publiek in België. Supplement op de gids van de bedrijfsarchieven bewaard in de openbare depots van België*, Brussel.

DE BAERE, J. (1973) *Een eeuw Belgische zuivelgeschiedenis*, Leuven.

DE WINTER A. and TAMBUYZER, C. (1956) 'De evolutie van de coöperatieve zui-velindustrie in België', *Landbouwtijdschrift* IX, pp. 410-439.

DEKKER, J. (1996) *Zuivelcoöperaties op de zandgronden in Noord-Brabant en Limburg, 1892-1950. Overleven door samenwerking en modernisering. Een mentaliteitsstudie*, Middelburg.

DEVOS, G. (1999) 'The first shall be the last: recent developments in Belgian business history: a first introduction', *European yearbook of business history*, 2, pp. 21-34.

GRÁDA, C. Ó (1977) 'The beginnings of the Irish creamery system, 1880-1914', *Economic History Review*, 30, pp. 284-305.

HENRIKSEN, I. (1999) 'Avoiding Lock-in: Cooperative Creameries in Denmark, 1882-1903', *European Review of Economic History*, 3, pp. 57-78.

KWANTEN, G. (2001) *Edmond-August De Schryver, 1898-1991. Politieke bibliografie van een gentleman-staatsman*, Leuven.

LANDUYT, G. (1984) 'De voedingsnijverheid', BAETENS R. (ed.) *Industriële revoluties in de provincie Antwerpen*, Antwerpen, pp. 87-103.

MOMMENS, T. E. (1985) *Politieke, institutionele en economische componenten in de ontwikkeling van de Belgische zuivelsector tijdens het Interbellum*. Licentiaatsverhande-ling, Leuven.

NIESTEN, E., RAYMAEKERS J. and SEGERS, Y. (2002) *Vrijwaar U van namaaksels! De Belgische zuivel in de voorbije twee eeuwen*, Leuven (CAG Cahier 2).

NILSON, J. (1997) 'Financing Agricultural Co-operatives under Changing Member At-tributes and Market Conditions', TABARY, P., HASLE NIELSEN, H. and VAN DIJCK, G. (eds.), *Proceedings of the European Seminar on Adapting Farmers' Cooperatives to Changes of Policies and Market Powers in the EU*, Brussels.

ROGER, C. (1944) *La politique de l'alimentation en Belgique*, Brussels.

SABEL, C. and ZEITLIN, J. (1985) 'Historical Alternatives to Mass Production: Politics, Markets and Technology in Nineteenth-Century Industrialization', *Past and Present*, 108, pp. 133-176.

SABEL, C. and ZEITLIN, J. (eds.) (1997) *World of Possibilities. Flexibility and Mass Production in Western Industrialization*, Cambridge.

SAS, B. (1999) 'Aandacht voor een 'verwaarloosde' regionale nijverheid. De ontwikkeling van de voedings- en genotmiddelenindustrie in de provincie Antwerpen in de 20ste eeuw: geschiedenis en geregistreerde bedrijfsarchieven', *Bijdragen tot de Geschiedenis*, 82, pp. 95-122.

SEGERS, Y. (2003) *Economische groei en levensstandaard. De ontwikkeling van de particuliere consumptie en het voedselverbruik in België, 1800-1913,* Leuven.

SVERRISSON, A. (1993) *Evolutionary Technical Change and Flexible Mechanization*, Lund.

SVERRISSON, A. (2002) 'Small Boats and Large Ships. Social Continuity and Technical Change in the Icelandic Fisheries, 1800-1960', *Technology and Culture*, 43, pp. 227-253.

TAYLOR, D. (1976) 'The English Dairy Industry, 1860-1913', *Economic History Review*, 29, pp. 585-601.

TRACY, M. (1993) *Food and Agriculture in a Market Economy. An Introduction to Theory, Practice and Policy*, La Hutte.

VAN DIJCK, M. and VAN MOLLE, L. (2001) 'Arme en rijke boeren. Landbouw en platteland in een wijde cirkel rond de hoofdstad (1880-1950)', DE MAEYER, J, and HEYRMAN, P. (eds.), *Geuren en kleuren. Een sociaal-economische geschiedenis van Vlaams-Brabant, 19de-20ste eeuw,* Leuven, pp. 103-135.

VAN MOLLE, L. (1986) 'Innovation technologique et changement social: le cas de l'agriculture belge, 19e et 20e siècles', KURGAN-VAN HENTENRYK, G. and STENGERS, J. (eds.), *L'innovation technologique; facteur de changement (XIXe-Xxe siècles)*, Brussel, pp. 153-184.

VAN MOLLE, L. (1990) *Ieder voor allen: de Belgische Boerenbond, 1890-1990*, Leuven.

VANNOPPEN, H. (1984) *Van twee kerkdorpen tot stadsgewestgemeente: Erps-Kwerps 1776-1976: 200 jaar demografische, sociaal-economische, politiek-administratieve en kerkelijk-culturele ontwikkeling in Brabant*. PhD thesis Leuven.

VANNOPPEN, H. (1993) 'Van de koe in de wei tot de moderne melkerij. 100 jaar melkerijgeschiedenis in Vlaanderen. Evolutie in Midden-Brabant', *Ons Heem*, 4, pp. 180-201.

VANNOPPEN, H. (1993) 'De geschiedenis van de Melkerij Vitalac in Veltem-Beisem', *Ons Heem*, 4, pp. 249-264.

3   A Swedish success story, supported by smallholders.
The increase in milk production, 1866-1913

Carin MARTIIN, Stockholm University and Swedish University of Agricultural Sciences (Uppsala)

I.   Introduction

The second half of the nineteenth and the beginning of the twentieth centuries was a dynamic period in Sweden, involving comprehensive changes in agriculture parallel to the rapidly ongoing urbanisation and industrialisation. Dairy expansion played an important role, as did the earlier link in this food chain, the development of livestock husbandry. The animals and their keepers have, however, often been neglected or regarded as extras in the background, while an increase in the amount of milk sent to the expanding dairies has been taken for granted.

Previous Swedish research in the field is limited. Some authors have focused on the dairy sector from a large-scale, quantitative perspective, mainly referring to official national statistics and contemporary literature (Niskanen, 1995 and Staffansson, 1995). Some attention has been paid to more detailed cattle farming, and this has contributed to a better understanding of farming in practice (Szabó, 1970; Björnhag and Myrdal, 1994; Myrdal, 1994; Algers, 1998; Mårtensson and Svala, 1998; Wiktorsson, 1998; Östman, 2000; Morell, 2001; Israelsson, 2002). The increased production of milk is usually attributed to better feeding and imports of high-yielding breeds. This perspective is in line with the view commonly held in the twentieth century that sees Swedish cattle farming as moving from destitution to success (Ingers, 1956 and Szabó, 1967, 1970 and 1986).

I.1.   Accounts from an estate and its surroundings

This study takes a more detailed approach. It concentrates on the herd and dairy in the estate of Krusenberg and, to some extent, on small herds in the surrounding areas. Four aspects of change in cattle farming due to expanding dairy production were distinguished. Although the aspects were in part integrated, they principally represented quantitative tasks such as litres of milk and number of cows; efficiency with regard to cattle life cycles and nutrition needs; animal welfare with reference to calf mortality and infectious diseases; and, fourth, attitudes towards livestock.

The study was mainly based on accounts from the Swedish estate of Krusenberg, which was expected to run its large herd on a commercial basis. As the accounts also gave some information about small herds in the surrounding areas, it was to some extent, possible to compare what kind of effects the dairy expansion may have had both on a large herd with about 80 cows, and on herds with one or two animals. The records were very comprehensive and informative, and covered the livestock between 1865 and 1913 and the estate dairy from 1873 to 1913. To some extent, records concerning work and economy were also accessible. For example, "The Cattle Journal" provided detailed information on names, colours, calving, milk yield, diseases, etc., which was recorded for more than three hundred animals, whose lives could be followed from generation to generation. The first cow noted in the records was born in 1852 and the last one in 1906. Complete information from birth to slaughter or sale was provided for cows born between 1866 and 1898. The archived records were clean and neat, written by the older or younger lord of the manor.[1] This indicates that the accounts were rewritten and that the owners were interested in livestock administration and spent much time doing this.

The accounts from the estate were interpreted with the support of probate inventories, contemporary and modern literature as well as answers to ethnological questionnaires.[2] The probate inventories provided detailed and sometimes shocking insights into the living conditions of poor people and their animals. The ethnological source, a detailed collection of memories mainly referring to the final decades of the nineteenth century, gave an everyday perspective on cattle husbandry in earlier times. Together, the accounts, probate inventories and ethnological questionnaires provided information at a more detailed level than has been available in earlier Swedish research.

I.2. The Swedish countryside and traditional cattle husbandry

Before the changes are discussed further, the subject requires an introductory presentation on the prevailing general conditions and characteristics. In 1870, more than 70 per cent of the Swedish population was associated with agriculture (Morell, 2001: 15). A typical parish could be built around one or a few estates and a number of farms, surrounded by small solitary units inhabited by people such as crofters, elderly people, craftsmen and the extremely poor. Cattle, especially cows, were kept by a majority of the inhabitants in the rural community, both rich and poor. The latter only had one or two cows while large herds varied from about twenty to one hundred cows.[3] The fact that almost everybody had access to dairy cattle implied a limited market for dairy products. Long distances and poor communications in Sweden also contributed to cattle farming being principally intended for use in kind. As a general background, traditional Swedish cattle farming could briefly be described by the following characteristics:

1. Cows were multifunctional, producing milk, meat, calves, heifers, draught animals, manure, skin, etc. They also had other uses such as providing capital and insurance against financial disaster – the owner of a cow had a possession that kept him out of total penury.

---

[1] In 1890, the son of the baron started as an inspector, and in 1892 he inherited the estate. Examples of handwriting from the old and young barons show that they both filled in the "Cattle Journal".
[2] Probate inventories 1878-1882 for the hundred of Ärlinghundra, county of Uppland and the hundred of Oppunda, county of Södermanland. The answers to the questionnaires (NM 60) were collected by the Nordic Museum from the 1920s to the 1950s.
[3] Israelsson, study of probate inventories from the hundred of Oppunda, county of Södermanland 1878-1882, (manuscript).

2. Swedish husbandry was based on domestic "country cattle".

3. Due to the harsh climate, cattle were kept according to two different systems during the year: in wintertime they were housed and fed with stored fodder; in the summertime they grazed outside. This required different strategies, access to different kinds of resources, adjustment to two systems, etc.

4. Strategies for feeding were aimed at keeping costs to a minimum rather than producing high milk yields. This meant that livestock was expected to survive on by-products from the fields, mainly straw, some hay and on resources gathered from the outlying areas.

5. The supply of nutrients given to the animals was unevenly distributed over the year, but was only slightly varied according to the different needs during the lactation. This undoubtedly resulted in under-nourishment and malnutrition.

6. Cattle were kept in dark and damp byres, often without almost any light or air, with the intention of keeping out the cold winter climate for fear of animal diseases.

7. Husbandry was to a great extent part of the female sphere, managed close to the household.

8. Intensive working days, including many feedings per day and time-consuming feed preparation, were common strategies intended to compensate for the limited material resources.

II. National perspective: the official success story

Traditional cattle farming had, at least since the eighteenth century, been criticised by the upper classes, who argued in favour of importing breeds and improvements in feeding and housing in order to increase milk yields. In the 1840s, after decades of discussion, the Swedish Government decided to make official investments in the import of breeds. The imported breeds were supposed to be gradually distributed nationally and to bring wealth to the whole country.[4] This step, associated with earlier physiocratic movements, occurred before the real beginning of urbanisation and may not have been realistic until some decades later when the demand for dairy products rose due to the rising urban population.

From the 1860s to the beginning of the 1900s, the estimated total Swedish milk production increased threefold and butter exports fivefold (Niskanen, 1995: 21; Staffansson, 1995: 80). As the export of butter was one of the most profitable Swedish exports, livestock became a national concern, and formerly negative attitudes towards husbandry turned into

---

[4] See J. Th. Nathorst, 1844, representing The Royal Swedish Academy of Agriculture. Nathorst accused common people of having destroyed the domestic cattle, praised English breeds and recommended official investments in the import of livestock.

enthusiasm for new strategies. The expanding market stimulated the establishment of a large number of dairies, partly replacing the former handling of milk within households. Dairy expansion started at the estates, which processed either the estate's production only or also milk from the surrounding areas. Dairies owned by businessmen came to dominate during an intermediate period, but were later gradually replaced by co-operative dairies owned by the local farmers (Niskanen, 1995: 38-41).

At the same time, arable farming changed. The cultivation of grass and clover increased both the quantity and quality of hay and the use of root crops for cattle spread rapidly. Other feeding improvements were the use of imported oil cakes and purchased bran. Furthermore, new spacious and rationally planned cowsheds with large windows were built. Obviously, at the national level, all this appears to have been a success. In order to examine important changes at the level of the herd and to find out how dairy expansion may have affected cattle husbandry in a local community, the text will hereafter refer to the estate of Krusenberg and the surrounding community.

III. Local perspective: the estate of Krusenberg

Krusenberg in the small parish of Alsike, is situated beside Lake Mälaren, about 15 kilometres from the university town of Uppsala and 50 kilometres north of Stockholm. Lists from parish catechetical meetings reported a total of 77 units by the 1890s, including twelve homesteads and estates and a large number of small units housing crofters, soldiers, shoemakers, fishermen, railway workers, elderly people and the extremely poor.

According to the statistics, the estate of Krusenberg had 313 hectares of arable land, 20 hectares of pasture, 7 hectares of land for fruit and vegetable cultivation, 600 hectares of forest and 260 hectares of other land. In 1896, about 40 per cent of the arable land was used for grain, and nearly as much for grass and clover. The rest of the arable land was used for legumes, potatoes and root crops. The livestock consisted of 40 horses, 12 oxen, 2 bulls, 120 cows, 38 heifers, 118 sheep and lambs, 44 pigs and 230 poultry.[5]

Furthermore, fifteen dependent crofts were noted. The number was in accordance with the records from the estate, which reported fourteen to sixteen crofts between 1865 and 1890. However, there was a large discrepancy between the number of cows noted in the statistics and the number found in the estate's records. At the end of December 1896, the time of the year when the livestock was counted for the statistics, the estate had only 77 cows instead of the official 120. A study of work journals, financial accounts, etc. from the estate, reveal that fourteen of sixteen crofts had cows of their own. To this could be added a number of other kinds of small units where cows also were kept, for example, those belonging to the fisherman, the carpenter or elderly people. Hence, it was obvious that the "missing" 43 cows belonged to the people living in the small units on the estate who kept one, two or three cows of their own. An explanation of the misleading statistics was found in a comment on the official statistics, which stated that it would be almost

---

[5] Primary statistics from the Agricultural Society, County of Stockholm, 1896.

impossible to treat all the small units separately (BiSOS, N, 1866). It was obviously agreed to disregard the fact that smallholders had livestock of their own, maybe because of their supposed unimportant role in the official economy and/or that they were considered a disgrace to society. Consequently, there might be a risk of overestimating the size of larger herds and neglecting small ones, thus leading to possible misinterpretations of the effects of the dairy expansion.

III.1. The dairy at Krusenberg

The establishment and development of a dairy in 1873, which produced butter for the market, was assumed to be a driving force for change on the estate, both structural and in the detailed management of the animals.

A modest start was made in the 1860s. In 1860, the cash income from the cowshed was only SEK 727, or 7 per cent of the total cash income from arable and cattle farming, which indicated livestock production was exclusively aimed at use in kind. In 1871 this income had increased to SEK 3,693 or 20 per cent of the total cash income. As the dairy was established in 1873-1874, the first year's income from the dairy was SEK 4,709. The highest cash income was recorded for 1906-1907 when the dairy generated SEK 12,917. After this, the dairy's income declined slightly until 1912-1913 when the dairy accounts were discontinued. According to the inventory, all the dairy equipment was scrapped in 1920, indicating that the estate dairy was closed down in the 1910s. This is in line with a study by Kirsti Niskanen, which describes a decline in manor dairies, while co-operatives were expanding at this time (Niskanen, 1995: 41).

III.1.1. Quantities of milk

The earliest available information on quantity indicates that about 32,000 litres were produced in 1865, a volume that doubled up until the start of the newly built dairy (see table 3.1). This, together with the increased income, indicates that the quantities of milk started to increase prior to the establishment of the new dairy. The fact that the amount of milk processed at the dairy corresponded to the quantity produced by the herd at the estate meant that the estate did not let any milk bypass the dairy, and that no noticeable quantities of milk were bought from other herds.

Table 3.1   Total milk production, Krusenberg, 1866-1870 to 1906-1910

Period	Litres	Index
1866-1870	36,327	100
1871-1875	60,828	167
1876-1880	67,691	186
1881-1885	73,905	203
1886-1890	106,742	294
1891-1895	129,162	356
1896-1900	152,906	421
1901-1905	139,104	323
1906-1910	149,450	411

Source: D 4:2, Krusenberg.

III.1.2.   Use of the dairy's milk

The dairy at Krusenberg concentrated on butter production. In table 3.2, the amounts of butter delivered from the manor are compared with the official national quantities. During the 1880s and 1890s, butter production at Krusenberg followed the same trend as national production. Lower butter production at Krusenberg after the turn of the century was compensated for by higher sales of fresh milk and cream, which probably resulted from the increased domestic demand for fresh dairy products.

Table 3.2   Butter production in Krusenberg and Sweden, 1886-1890 to 1911-1913

Period	Quantity of butter sold, Krusenberg, kg	Quantity of butter sold, Krusenberg, index	Quantity of butter, Sweden, index
1886-1890	2,354	100	100
1896-1900	4,007	170	162
1906-1910	3,670	156	205
1911-1913	2,590	110	198

Source: G 4 A:6, Krusenberg and Niskanen, 1995: 21.

Despite the opportunities for earning income from the market, the estate did not abandon milk as a source for use in kind. On the contrary, internal needs continued to be of great importance. One third of all milk produced at Krusenberg was consumed by the inhabitants on the estate, not least by the labourers, who were given milk as payment in kind. Moreover, milk was used for the breeding of calves and to increase pig production. In this way, the estate spread the risk of a too-heavy dependence on the dairy market, but still benefited from increased domestic demand for animal products. Hence, the expansion of the dairy at the estate entailed maintenance of old structures parallel with the development of new aims of production.

III.2. Livestock at the estate

So far, this article has focused on milk volume and commercialisation. We now turn to the physical realities behind the extended dairy production at Krusenberg. In other words, the reader will be introduced to the estate's thinking about practical facts such as lactation, calving and cattle feeding.

III.2.1. Herd size

During the second half of the nineteenth century, the number of cows on the estate increased sharply. As shown in table 3.3, the herd was gradually enlarged.

Table 3.3   Average number of cows in the herd at Krusenberg, 1866-1870 to 1896-1900, as well as a comparison with the increase in Sweden

Period	Total number of cows, Krusenberg	Total number of cows, Krusenberg, index	Total number of cows, Sweden, index
1866-1870	30	100	100
1871-1875	37	123	109
1876-1880	48	160	115
1881-1885	51	170	121
1886-1890	68	227	129
1891-1895	78	260	137
1896-1900	83	277	145

Source: D 4:2, Krusenberg and BiSOS, N.

The increase was relatively linear and showed no sudden changes, indicating that neither the introduction of the new dairy in 1873, nor the new cowshed built in 1877, was connected with any strategy for increasing herd size immediately. Nevertheless, the total increase in herd size during the following decades was interpreted as an effect of the dairy expansion. From the end of the 1890s, the herd size stabilised, probably due to the fact that the new cowshed was filled up by that time.[6]

According to table 3.3, the size of the herd at Krusenberg increased almost twice as much as at the national level during the same period. This may have been due to the estate's size and income compared with the national average, which probably made it easier for an estate to change and re-allocate resources such as labour, investments and arable farming in order to take advantage of the social and economic changes taking place in society at large.

---

[6] The cowshed built in 1877 is still maintained and has been visited by the author.

III.2.2.   Strategies for increasing herd size

It was obvious that the ability to increase herd size was essential for increased dairy production at Krusenberg. Was this achieved through internal recruitment or were cows bought from other herds? The records from Krusenberg show a total of 291 cows recruited in 1866-1905, of which 69 animals, or almost 25 per cent, had been purchased. Consequently, the estate practised a strategy where chiefly internal recruitment was combined with some purchases, demonstrated by the fact that some of the heifer-calves were sold as calves and replaced by purchased cows about two years later.

Table 3.4 shows a considerable variation in the origin of the cows purchased. Almost half them were reported as originating from other provinces or towns, "bought from Småland" etc., without any other information supplied. Nearly as many originated from the local community and were identified as having been sold by crofters, other small units or homesteads in the parish.

Table 3.4   Origin of cows purchased by Krusenberg in 1865-1908

Time	Dependant units, mainly crafters	Other local herds in the parish	Bought from towns or other provinces	Other manors	Total
Cows	19	14	34	2	69
%	28	20	49	3	100

Source: D 4:2, Krusenberg.

The purchase of cows was most intensive between 1878 and 1890, when 56 per cent of the total of 69 cows were bought. The average recruitment, about 13 per cent per year, varied between 9 and 15 per cent within the five-year periods, but no trend over time was found. Likewise, no clear explanation of the recruitment strategy could be discerned although some reasons may be suggested: saving the cost of some heifer breeding, interest in genetic variation or the fact that cattle marketing seemed to have been a part of social life (Israelsson, 1998). However, the advantage of the external cows seems to have been offset by problems with the new animals. On average, the purchased cows were kept only about 60 per cent as long in the herd, compared with the internal recruits. The reasons for this striking difference seem to have been a diffuse mixture of problems with health, fertility, etc. In addition, the spread of infectious diseases through cattle trading must also be taken into account and regarded as a cost directly related to the increased herd size and, indirectly, to the expansion of the dairy.

The best cow each year reflected a possible potential milk yield, which was achieved only by some of the animals. Figure 3.1 shows that the individual maximum differed considerably from the minimum yield, and that some yields were only one fifth of the maximum. It is worth noting that the differences between individual cows during a single year were larger than the average differences between the 1860s and 1910s, and that the best cows improved further while the worst remained at a very low level. This seems to have been a consequence of keeping individual animals in the herd, despite a number of bad lactations and/or long dry periods. The decline in minimum yield in 1901-1905 could be attributed to infectious diseases, which will be discussed later. Despite these differences in individual production, it was obvious that the total quantities of milk as well as the number of cattle increased significantly at the estate from the 1860s to the beginning of the 1900s. Did this encourage efficiency in cattle husbandry at the estate?

III.2.3.   Milk yield per year

Figure 3.1   The average annual milk yield per cow (in litres) at Krusenberg in 1866–1910, together with the mean maximum and minimum milk yield achieved by any cow each year

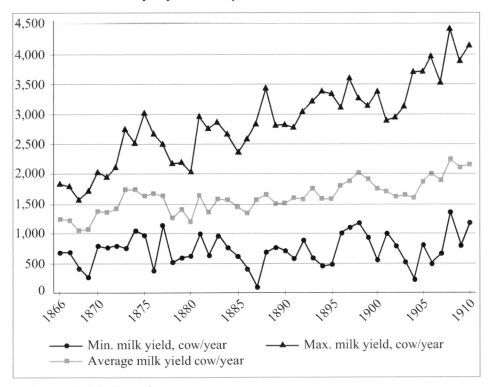

Source: D 4:2, Krusenberg.

III.3. Efficiency

The efficiency measures not only included quantitative aspects but also qualitative ones such as observations concerning growth, health and the right time for covering. The accounts made it possible to investigate two kinds of efficiency: first, whether the milk-producing part of the life cycles increased or not and second, whether the required amounts of energy and protein per litre of milk changed.

III.3.1.   Life cycles of the dairy cows

In the study, the life cycle of a cow was divided into two stages: 1) the breeding period, from birth to first calving and 2) the milk-producing period, with a number of lactating and dry periods (figure 3.2).

Figure 3.2   Main outline of the life cycle of a dairy cow

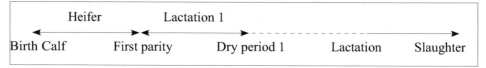

To some extent, the length of these stages and periods was flexible. According to the accounts from Krusenberg, the age of the cows at first calving varied between 21 and 54 months, while dry periods varied between one and five months. It was obvious that the proportions between the duration of breeding, lactation and dry periods as well as lifespan affected the efficiency of livestock husbandry. Table 3.5 presents the average length of the main stages in the life of the cows born at Krusenberg between 1866 and 1895.

Table 3.5   Average life cycles of cows born at Krusenberg, 1866-1870 to 1891-1895, described as age at first calving, number of calvings, age at slaughter or sale and milk production

Period	Age at first calving, months	Number of calvings	Age at slaughter or sale, years and months	Months from first calving to slaughter or sale	Milk yield per month, from first calving to slaughter or sale, litre	Total lifetime production of milk, litre
1866-1870	33	10.0	12.11	122	146	17,823
1871-1875	31	9.1	12.50	118	132	15,630
1876-1880	35	8.8	12.10	110	146	16,041
1881-1885	32	8.1	10.70	97	162	15,759
1886-1890	33	8.2	11.80	107	160	17,169
1891-1895	32	6.8	10.50	93	167	15,575

Source: D 4:2, Krusenberg.

Although there were large differences between individual cows in the herd, the average figures demonstrated only small changes in the stages of the life cycle in comparison with the huge quantitative changes in total milk production, table 3.6.

Table 3.6   Changes in some features included in life cycle efficiency. Cows born at Krusenberg, 1866-1895

	Declined, %	Unchanged	Increased, %
Age at first calving		x	
Number of calvings	- 32		
Milk yield per month, from first calving to slaughter or sale			+ 14
Milk yield per month, total time at Krusenberg			+ 10
Time from birth to slaughter or sale	- 24		
Total life production of milk	- 13		

Source: D 4:2, Krusenberg.

According to table 3.6, the breeding period remained practically unchanged while the milk yield per month improved. However, total milk production per cow at the estate declined, a consequence of the shortened time from birth to slaughter or sale. The latter was connected to a trend at Krusenberg, which increased over time, to sell old cows perhaps to small poor units instead of keeping the animals until slaughter.

The generally small changes over time differed from the common picture given in the answers to ethnological questionnaires, where the informants described large differences between old cattle farming where the milk was intended mainly for use as payment in kind, and new strategies developed for large herds with the aim of producing for the expanding dairies. The reasons for the absence of these changes at Krusenberg were not clear, but one suggestion is that the livestock were tended by people with a more or less traditional approach to the animals, and that this contributed to the continuation of some of the old strategies. As will be discussed later, animal welfare seems to have been dependent not only on modern housing or feeding, but also on tending by men and women with a feeling for cattle.

III.3.2.   Nutritional requirements

Another way of measuring any changes in efficiency during the period of dairy expansion was to study the nutrient requirements per litre of milk. Since the records from Krusenberg gave very little information about feeding, the calculations were based on energy and protein needs, and not on what the animals were really given. The standard for energy needs used was that used since the end of the twentieth century. The need for protein was reduced in accordance with recommendations in the older literature, which is considered to be more conversant with older breeds and their requirements (Spörndly (ed.) 1999) and Johnson 1955). The feeding standard applied was 5 mega joules (MJ) energy and 50 grams of digestible crude protein per litre of milk. The cows were assumed to have an average live weight of 400 kg, according to contemporary accounts from the herd at the Agricultural Institute at Ultuna, situated about 10 kilometres from Krusenberg. The total recommended supply of energy and protein for maintenance, milk production, growth and pregnancy during the average life cycle, was added and then divided by the mean total quantity of milk per cow, resulting in the needs presented in table 3.7.

Table 3.7   Energy and digestible crude protein need per litre of milk produced by cows born at Krusenberg, 1866-1870 to 1891-1895

Period	MJ/litre of milk	Protein, g	Index, MJ	Index, smb rp, g
1866-1870	15.5	118.4	100	100
1871-1875	16.5	124.5	106	105
1876-1880	15.6	120.7	101	102
1881-1885	14.5	114.3	94	97
1886-1890	14.6	113.6	94	96
1891-1895	14.2	112.2	92	95

Source: D 4:2, Krusenberg; Spörndly (ed.), 1999 and Jonsson, 1955.

The total energy and protein needs for maintenance, milk production, growth and pregnancy thus changed slightly on the estate between the 1860s and the 1890s. Long breeding periods as well as long dry-periods and/or low yield per lactation resulted in high nutrient costs per litre. Cows born in 1891-1895 needed 8 per cent less energy and 5 per cent less protein per litre of milk than did cows born at Krusenberg in 1866-1870. In 1866-1895, the total average requirements per animal were three times the energy and twice the digestible crude protein needed exclusively for the milk they produced. According to tables 3.8 and 3.9, the proportional energy and protein needs for maintenance and pregnancy diminished slightly while increasing for milk production and growth.

Table 3.8   Energy requirements for maintenance, milk production, growth and pregnancy. Percentage of cows born at Krusenberg, 1866-1870 to 1891-1895

Year	Maintenance	Milk production	Growth	Pregnancy	Total
1866-1870	60.9	32.3	3.7	3.1	100.0
1871-1875	63.0	30.3	3.6	3.1	100.0
1876-1880	60.6	32.1	4.3	3.0	100.0
1881-1885	58.3	34.4	4.2	3.1	100.0
1886-1890	58.9	34.3	4.0	2.8	100.0
1891-1895	57.8	35.1	4.4	2.7	100.0

Source: D 4:2, Krusenberg; Spörndly (ed.), 1999 and Jonsson, 1955.

Table 3.9   Protein requirements for maintenance, milk production, growth and pregnancy. Percentage of cows born at Krusenberg, 1866-1870 to 1891-1895

Year	Maintenance	Milk production	Growth	Pregnancy	Total
1866-70	40.4	42.2	11.6	5.8	100.0
1871-75	42.4	40.1	11.8	5.7	100.0
1876-80	39.7	41.4	13.4	5.5	100.0
1881-85	37.7	43.7	13.2	5.4	100.0
1886-90	38.4	44.0	12.5	5.1	100.0
1891-95	37.2	44.5	13.6	4.7	99.9

Sources: D 4:2, Krusenberg, Spörndly (ed.) 1999 and Jonsson 1955.

Consequently, the two kinds of efficiency discussed – the milk-producing share of the average life cycle and the energy and protein requirements per litre of milk – increased slightly in comparison with the large increase in total milk production and number of cows. The fact that the quantitative factors, such as total production, increased much more than the tasks more influenced by qualitative aspects indicates that there may have been a greater focus on the total quantity than on the total economy of each cow.

Accordingly, a partly passive attitude towards livestock husbandry at Krusenberg is indicated. This attitude may also reflect a degree of caution and/or a continued influence of traditional husbandry where the old structures still affected livestock production on the estate. However, the accounts reflect some modernisation with regard to feeding. Despite the sparse notations about this, some modern characteristics were revealed: no mash seems to have been used, concentrates and bran of wheat and rye were purchased and root crops and hay were cultivated. Straw, which was central in the old system of husbandry, continued to be very important. According to a feeding list for Krusenberg in 1898, the dairy cows had free access to straw, which supplemented 4-6 kg of hay per cow and day. With the use of concentrates, bran and clover, the concentration in the daily rations was increased, while the use of straw lowered the degree of concentration. This was a common feeding strategy, also found at other large farms, which presumably was regarded as offering a kind of optimum balance between yield and cash expenditures for feeding stuff.

Thus far, it has been demonstrated that the estate of Krusenberg by the time of the expansion of the dairy, revealed modern features at the same time as it maintained traditional strategies. A third aspect, which, apart from quantity and efficiency, could reflect changes caused by the expansion, was animal welfare.

III.4. Animal welfare

With reference to the answers to questionnaires, which described the dreadful health conditions involved in traditional cattle farming, it might be expected that animal welfare improved at the estate during the nineteenth century. Two indications of animal welfare will be discussed: rates of calf mortality and the occurrence of infectious diseases.

III.4.1.   Calf mortality

Calf mortality could be used as an indicator of the conditions in the herd as a whole, since if sensitive young calves survived, the conditions must to some extent have been acceptable for older cattle also. Somewhat surprisingly, the average calf mortality rate, based on 2,299 newborn calves at Krusenberg 1865-1913, was as low as 5 per cent per year, a figure regarded as good even today. Except for a peak of 10 per cent occurring between 1897 and 1908, calf mortality rates only varied marginally. The accounts did not reveal any explanations for this, although the reasons could be problems with infection and/or changes in staff – it is a well-known fact among cow-keepers that human skill, experience and interest are of great importance in maintaining good health in the stock.[7] The generally low and unchanged rates of calf mortality over a number of decades indicate that the herd was well tended throughout the period studied, and that good stockmanship to some extent may have compensated for the shortcomings of the old cowshed before the new one was taken in use in 1877.

III.4.2.   Infectious diseases

The herd at Krusenberg was struck by at least two serious infectious diseases during the period studied: contagious abortion and tuberculosis (Figure 3.3). Problems with the spread of diseases caused by imported breeds were debated in contemporary periodicals and later described in the answers to questionnaires (Landtmannen, 1898: 565 and 724). This negative aspect of cattle purchased abroad was reinforced by the accounts from Krusenberg. It is likely that the diseases were initially spread from purchased animals on the estate via the workforce and/or animals to smaller herds in the surrounding coun-tryside.

Contagious abortion caused far-reaching problems during the first years of the twentieth century. The consequences were loss of the calf, afterbirth retention and infections as well as a considerably reduced milk yield followed by a very long dry period before the eventual birth of a new calf (Vennerholm-Dahlström-Stålfors, 1920: 311-314). The records reveal that the estate decided to keep many of the infected animals, although some of the cows had two or three abortions in succession. A total of 57 contagious abortions occurred, of which 77 per cent affected the herd between 1901 and 1905. In 1902, almost every third cow had a contagious abortion. From 1907, a new period of abortions occurred, as the cow "Udda", bought from the croft Udden, brought the problem back to the estate.

---

[7] David Stone, 2003: dealing with medieval sheep, he also regards mortality as an indicator of the health status in the herd and stresses the importance of good care.

Figure 3.3   Contagious abortion and tuberculosis at the estate Krusenberg, 1881-1882 to 1911-1912

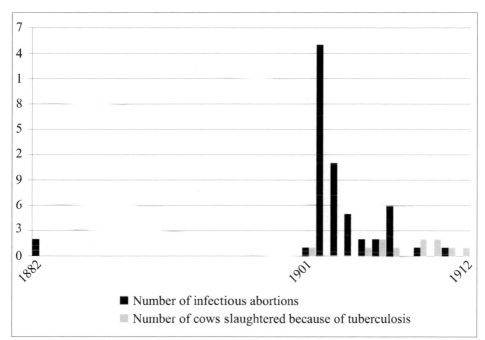

Source: D 4:2, Krusenberg.

The second infectious disease was tuberculosis, which struck both the estate herd and small herds in the surrounding areas. The first case of tuberculosis occurred in 1901. Up until 1912, a total of eleven cows were slaughtered because of tuberculosis. According to the literature, the end result was always slaughter. However, it took many years for animals infected with tuberculosis to show signs of the disease, and this furthered the spread of tuberculosis to other herds (Vennerholm-Dahlström-Stålfors, 1920: 591).

Hence, animal welfare, measured here as rates of calf mortality and infectious diseases, did not improve during the period. On the contrary, the problems with diseases seem to have been worse around the turn of the century than at any time during the period studied.

However, other factors may have developed positively over time even though it was difficult to investigate them in the records. According to the answers to questionnaires, contemporary literature and as can be seen from the still-standing building, the large new cowshed built in the latter half of the nineteenth century offered better conditions as regards light and air. This, together with better feeding, must have improved animal welfare in general.

III.5. Attitudes towards livestock

The fourth and last aspect studied, refers to attitudes. In the records from Krusenberg, attitudes were reflected in the administration of the herd. The comprehensive accounts reveal that control of livestock production was rigorous throughout the period studied, indicating that cattle administration was an integral part of the dairy process from the 1860s onwards.

The administration of the herd cannot solely be regarded as a tool for increased dairy production. Some of the information reflected other values such as names, descendants and careful descriptions of each cow, for example "Kamilla" born in 1868 who was "Yellow-brown with some white hair around the udder". Detailed written information may have made it easier to manage the herd both quantitatively and qualitatively. Moreover, it is suggested that the records were an important means of raising the status of cattle husbandry at Krusenberg. The fact that the accounts seem to have been rewritten by the lord himself demonstrated an involvement in livestock husbandry that might have been linked to the rising status of livestock production among the upper classes. Another expression of this was the number of exhibitions arranged all over the country by agricultural societies to which estate owners and modern farmers brought their best animals. Furthermore, the newly established agricultural journals wrote about successful herds and praised their owners. All this transformed cattle husbandry at estates and larger homesteads into an activity that conferred status and prestige, far from its former reputation as a necessary but disgusting activity.

IV. Local perspective: smallholdings in the neighbourhood

Questions concerning smallholdings are often difficult to answer since they are usually sparsely addressed in the archives. However, the records from Krusenberg revealed some information about smallholdings such as crofts, soldiers' crofts, carpenters' holdings and such. Of the 14-16 crofts belonging to the manor, 13 definitely and one probably had cows. In addition, a number of other smallholders also had their own cows, for example, the fisherman and the soldier. Was it possible for these groups to take advantage of the increasing demand for milk?

The first question to be asked was whether these smallholdings were able to deliver milk. Earlier, it was noted that the dairy at the estate did not buy any noticeable quantities of milk. On the other hand, it could not be proved that milk from smallholdings was not delivered to some other dairy. Still, in order to answer the question, despite this uncertainty, a comparison was made between the milk yield estimated as being possible for a smallholding, and the amount it would require for its own use. It was assumed that only surplus milk would be sold. A possible quantity for individual use was found in records from 1899, when a total of 1,460 litres of milk was paid in kind to a labourer at Krusenberg. This corresponded to the annual milk yield from one or two cows, as described in the answers to questionnaires referring to scarce conditions, and indicated that owners of one or two cows were probably not able to sell milk to any dairy and thus were unable to take advantage of the opportunities presented by the expanding market's demand for milk.

The consequences of this were that smallholders continued to use cows for their own needs, without the possibility of earning money from dairy production. This made it more or less impossible for the inhabitants of small units to utilise methods for increasing milk yields by purchasing concentrates, grass seed, glass for windows, etc. and hence kept people and cattle trapped in the old, scarce system with malnourishment and low production. The low feeding standard was exemplified by the annual quantity given to a soldier's croft at Krusenberg, which was limited to 510 kg of hay and 1020 kg of straw, thereby providing a cow with 2.8 kg of hay and 5.6 kg of straw per day during a stable period of 180 days. This feeding, realistic according to the answers to the questionnaires, resulted in a very low milk yield, far below 1000 litres per cow and year.

However, cattle husbandry at small units did not appear to have been totally isolated from the expansion of the dairy at Krusenberg. There was probably also a mutual exchange of knowledge and experience. First, smallholders could possibly have made some use of their experience of work in the large cowshed, which could be practised in the small byre at home. This was exemplified by the former labourer at the estate, C.A. Pettersson, whose experience as a stockman for the large herd was utilised at the croft Moralund when he became a crofter in 1904. Second, the manor must undoubtedly have been extremely dependent on labour, possibly often recruited from small units, and also on being able to hire extra labour among people in the neighbourhood who earned their living from a combination of self-subsistence and occasional day work. Third, calves were sold from the estate to the crofts and cows were sold from the crofts to the estate. Fourth, the large cowshed offered opportunities to earn some money from extra work. Fifth, bulls, often crossbreeds, could be used to cover cows from crofts and other small units. By using the estate's bulls, the small units gained access to genes from imported breeds, and this may have increased the potential for profit from small-scale trade in livestock. Possibly too, the ownership of crossbreeds may have implied a degree of status for the crofters.

However, it should be stressed that small units in the neighbourhood were also exposed to negative effects, such as the spread of infectious diseases from the expanding estate. The estate's progress in livestock husbandry, parallel with the difficulties in improving livestock husbandry at the small units, may have created and highlighted an increasing gap between rich and poor, and may have proved humiliating for the latter.

V.   Conclusions and discussion

The expansion of the herd at Krusenberg was the principal explanation for the large increase in dairy production at the estate. While the average milk yield per cow only rose 25 per cent during the first four decades with a commercial dairy production, the number of cows doubled. The study reveals that the expansion of the dairy did not involve a total commercialisation of the estate's milk production. On the contrary, the tradition of utilising large amounts of milk for use in kind continued to be important for the internal economy of the estate. Also in cattle husbandry, some traditional methods were applied along with new strategies. Traditional features, may have influenced the efficiency of life cycle strategies and nutrient needs, as well as the everyday care of the cows in practice. A considerable tolerance of individual discrepancies also indicates a closeness to traditional knowledge and habits.

Apart from the large increase in the number of cows, attitudes may have been one of the aspects that exhibited the most striking changes. A new kind of status came to be associated with livestock farming since it was connected to the expanding dairy sector and attracted men from the upper classes. This is in line with descriptions from that time of the transition in the image of cattle farming from an invisible occupation to a politically correct one, crucial to social and economic change.

The expanded dairy production probably not only benefited cattle husbandry on the estate at that time, but also contributed to the development of technology, arable farming, knowledge etc., all of which contributed in turn to the potential for profitable farming in the twentieth century. In contrast, the small units in the neighbourhood were very limited in their ability to take advantage of the developing commercialisation of cattle husbandry. Although it was possible to distinguish some connections between the dependent small units and the expanding dairy sector at the estate, the small units functioned in principle as a form of support for this expansion. A widening gap could probably be distinguished between the many small units, which remained within a self-sufficient system and the expanding estate which was able to take advantage of the developing urbanisation.

Bibliography

Primary sources

City archives of Stockholm:
Probate inventories 1878-1882, Ärlinghundra Hundred, County of Uppland.
The Nordiska museet:
Ethnological questionnaires, NM 60.
The regional archives in Uppsala:
Krusenberg, 1865-1913.
Ultuna, 1849-1872.
Lists from parish catechetical meetings; Alsike parish, Erlinghundra hundred, 1851-1912.
Primary statistics from the Agricultural Society, County of Stockholm, 1896.
Probate inventories 1878-1882, Oppunda hundred, Södermanland.

Printed and secondary sources

Algers, B. (1998) 'Hälsa och beteende i fähus och ute', Viklund, K., Fogelmark, R. and Linderholm J. (eds.) *Fähus från bronsålder till idag,* Stockholm, pp. 111-117.

BiSOS; Official statistics, H (Official Reports from the Swedish counties).

BiSOS; Official statistics, N (Arable Farming and Animal Husbandry).

Björnhag, G. and Myrdal, J. (1994) 'Nötkreaturens produktion och utfodring enligt 1500-talets kungsgårdsräkenskaper', Myrdal, J. and Sten, S. (eds.) *Svenska husdjur från medeltid till våra dagar,* Stockholm, pp. 75-96.

Cserhalmi, N. and Israelsson, C. (2004) 'Fat in summer, starved in winter?', *Bebyggelsehistorisk tidskrift,* pp. 73-84.

Ingers, E. (1956) *Bonden i svensk historia, del III,* Stockholm.

Israelsson, C. (1998) *Frithiof Johanssons dagböcker 1875-1923. Fem decennier av agrar och urban förändring,* paper, SLU, Uppsala.

Israelsson, C. (2002) 'Winterfoderordning för kor, kommentar', Cserhalmi, N. (ed.) *De oskäliga kreaturen! Något om synen på lantbrukets djur de senaste 200 åren,* Stockholm, pp. 60-68.

Johnsson, R. (1955) *Räkenskapstabeller för kontrollassistenter,* Stockholm.

Landtmannen (1898) Tidskrift för Sveriges jordbruk och dess binäringar, Linköping.

Morell, M. (2001) *Jordbruket i industrisamhället 1870-1945,* Stockholm.

Myrdal, J. (1994) 'Bete och avel från 1500-tal till 1800-tal', Myrdal, J. and Sten, S. (eds.), *Svenska husdjur från medeltid till våra dagar,* Stockholm, pp. 14-34.

Nathorst, J.Th. (1844) *Om införskrifning av Race-Boskap,* Royal Swedish Academy of Agriculture, Stockholm.

Niskanen, K. (1995) *Godsägare, småbrukare och jordbrukets modernisering, Södermanlands län 1875-1935,* Stockholm.

Peterson, G. (1989) *Jordbrukets omvandling i västra Östergötland 1810 – 1890*, Stockholm.

Spörndly, R. (ed.) (1999) *Fodertabeller för idisslare, Rapport 247*, SLU, Uppsala.

Staffansson, J-Å. (1995) *Svenskt smör. Produktion, konsumtion, utrikeshandel 1861-1913*, Lund.

Stone, D. (2003) 'The productivity and management of sheep in late medieval England', *The Agricultural History Review*, pp. 1-22.

Szabó, M. (1967) 'Svältfödningen och dess bakgrund', *Fataburen*, Stockholm, pp. 197-208.

Szabó, M. (1970) *Herdar och husdjur*, Stockholm.

Szabó, M. (1986), 'Hade djuren det bättre förr?', *Fataburen*. Stockholm, pp. 27-50.

Vennerholm, J., Dahlström, H. and Stålfors, H. (1920), *Husdjurens sjukdomar,* Stockholm.

Wiktorsson, H. (1998), 'Utfodring och skötsel av nötkreatur från bronsåldern till 1900-talet', Viklund, K., Engelmark, R. and Linderholm J. (ed.) *Fähus från bronsålder till idag,* Stockholm, pp. 104-110.

Östman, A-C. (2000), *Mjölk och jord. Om kvinnlighet, manlighet och arbete i ett österbottniskt jordbrukssamhälle ca. 1870-1940,* Finland.

4 Proud and self-conscious craftsmen. Ghent butchers and the local meat supply, 1850-1990

Maria DE WAELE, STAM-Stadsmuseum Ghent

From the Middle Ages onwards, butchers, or 'meatchoppers' as they were often called, have played a key role in the meat supply of the cities. The butchers' guilds were among the oldest in mediaeval cities and enjoyed several privileges. From times immemorial, butchers were proud and selfconscious craftsmen, who vigorously defended their interests. This proud attitude continued to live on, long after the abolition of their guilds.

I. Introduction. Records, files and figures... and the lack of them

In spite of their impressive past, it was not easy to try to portray the Ghent butchers as a social and as a professional group between 1850 and 1990. In fact, it was at times quite frustrating because the local butchers' organisations apparently didn't keep any records or documents. Their small periodical has vanished. Of course, the butchers had one excuse: time and again, I was reminded that their headquarters burned down in 1961. In some cases, records of local professional unions are kept by the national corporation, in this case by the Landsbond der Beenhouwers en Spekslagers van België. Except for a few letters and documents from the 1980s and 1990s, the Landsbond did not seem to possess many original documents either. Theoretically there was a third option, i.e. the records of the Conseil des Mines in the Belgian Public Record Office in Brussels. Professional corporations, which applied for corporate rights, were obliged to send a file to the Conseil des Mines. The files from the Ghent butchers' unions seem to have disappeared.

All of this meant that I had to make do with what I could find. Fortunately, the collections of the Ghent City Archives were quite interesting for the years between 1850 and 1914, in particular the files K (Commerce and Trade), C (Records of the Board of Alderman and the City Council) and Q (Markets, Slaughterhouse and Foodstuffs). The collections held quite a lot of documents concerning the Ghent abattoir, and the disputes between the city council and the local butchers. These records turned out to be quite complementary with the well-known collection of pamphlets and leaflets, kept in the Central Library of Ghent University.

The lack of reliable source material for the years between 1920 and 1980 obliged me to use the butchers' national professional weekly, *(La) boucherie belge/(De) Belgische beenhouwerij*, as my main source. Unfortunately, the Ghent butchers rarely made the headlines, their fellow craftsmen from Brussels, Antwerp and Charleroi were far more important. In short, the available source material was, at best, patchy and incomplete.

How many butchers were there in Ghent? Finding relatively trustworthy figures turned out to be equally difficult. For the years between 1850 and 1932, I mainly used the Manual for the City of Ghent. This Manual is a fairly reliable tool, that gives a good indication of the number of butchers and the part of the city in which they had their shops. Other inquiries at several government Departments (for the Self-Employed, for Public Health) and the National Institute for Statistics proved to be fruitless.

The (un)availability of source material severely limited the scope of my investigation. In these circumstances, how do you portray the Ghent butchers, as a professional group, as a social group and as an important link in the food chain? How do you get through to the tightly-knit world of the abattoir and the surrounding cafes and restaurants? A few interviews with retired butchers made it plain that oral history would in this case only be of limited value.

II. A rich history

The Butchers' Guild in Ghent was founded in the 12th century. Its power was based on its control of the sale of meat in the city. Meat was sold in the Butchers' Hall (*Groot Vleeshuis*), which was built in the early 15th century. Already in the 14th century, the profession of butcher had acquired a hereditary and closed character. Four families, the Van Loo's, the Minnes, the Van Melles and the Deynoodts obtained the exclusive right to practise their craft in the Butchers' Hall.

In 1795, the Southern Netherlands were annexed by and became part of France. The Guilds were abolished a year later. The two Butchers' Halls (*Groot* and *Klein Vleeshuis*) were sold by auction as 'national goods' (*nationale goederen*). After the abolition of the Guilds, in theory everyone could become a butcher, but in practice little changed. The 'traditional' butchers' dynasties remained dominant. Their members regularly intermarried and the profession remained quite closed. (Van Werveke, 1951: 363-367; Jubileumbrochure, s.d.: s.p.)

It is almost impossible to find out when and where the first butcher'shops, or 'meat selling points', appeared in the streets of Ghent. This very probably happened at the end of the eighteenth century. The professional census of 1830 doesn't mention butchers separately. The category 'bakers, millers, brewers and butchers' in Ghent runs to 435 (for a population total of 83,200). In 1846 there were 107 butchers in Ghent, and they employed 92 helpers (for a population of 107,900.[1] The Ghent Manual doesn't mention butcher' shops until 1854. In that year, the section 'General Adresses for Trade and Industry in the city of Ghent and the province of East Flanders' (*Algemeene adressen voor handel en nijverheid der stad Gent en der provincie Oost-Vlaenderen*) mentioned 29 pork-butchers (*verkensslagters* or *charcutiers*) and 46 meat-choppers (*vleeschhouwers*). In 1860 the Manual mentioned 81 pork-butchers and 142 meat-butchers. Due to the rapid rise of the population, and the slowly rising general standard of living, the number of butchers rose spectacularly during the following years. In 1890, the Manual mentioned no

---

[1] RAG, Provincial Archives, 1830-1870, Inventory 3217, file 4789/1, Staten indeling volgens beroep, o.a. Gent, 1830; inventory 4801; Staten volkstelling 1846, indeling naar beroep, o.a. Gent.

less than 620 meat-butchers. The four families Van Loo, Minne, Van Melle en Deynoodt successfully survived the Ancien Regime and remained very visible until the late 1890s. The Van Loo's in particular assumed the allure of a genuine butchers' dynasty. In the 1870s and 1880s, they owned over twenty butcher's shops in different parts of Ghent. Owing to the rapid growth of the number of butchers, the traditional families eventually lost their dominant position in the early twentieth century (Avonts, 1970: 124-125; De Vriendt, 1978: 126-144; Wegwyzer, 1854: 459, 461; Wegwyzer, 1860: 460-461, 464-465; Wegwyzer, 1870: 459, 460-461, 465-467).

III. A new abattoir for Ghent

In the early nineteenth century, the slaughtering of animals and the sale of meat was hardly, if at all, supervised by public authorities. In the following decades, the general opinion about public hygiene and public health changed quite considerably. The city council, which was responsible for the health of the city's inhabitants, gradually increased its control of the slaughter of animals, the preparation of meat and the sale of meat products. The first modern slaughterhouses were built in big cities like London, Paris and New York. Hasselt was the first city in Belgium to establish a modern abattoir in 1832.

Around 1850, it was generally admitted that the supply of fresh meat no longer met current hygienic standards and that the Ghent Butchers' Hall was no longer adequate. Fresh meat was brought into the city unchecked, a situation which was increasingly criticized by medical authorities. Many butchers slaughtered their animals at home. Every butcher decided himself whether meat was – or, as it happened, was not – fit for human consumption. In 1852 the city council decided to build a modern abattoir on the outskirts of the city. [2]

Initially, the Ghent butchers reacted very suspiciously to the plans for a new abattoir. They were not used to rules or restrictions and as a rule did not – and to this day do not – much like official snoops. The oldest professional organisation of the Ghent butchers, the *Vereenigde Vleeshouwers*, made that very clear in a jubilee booklet, published several years later. In order to find out more about the council's plans, the butchers were received by the city mayor. He assured them that the abattoir was built in order to protect public health and to make sure that animals were killed in safe and hygienic circumstances (Jubileumbrochure, s.d.: s.p.).

---

[2] SAG, F/183, Correspondance concerning the establishment of the abattoir, a.o. Cahier des charges, clauses et conditions pour l'adjucation publique, par soumissions cachetées, des travaux de construction pour fondations de l'abattoir général et dépendances, projetés sur le terrain d'exercices situé au Château des Espagnols, ainsi que pour les travaux de construction d'aqueducs dans les rues et marché avoisinants de l'Abattoir, 13 août 1852.

III.1. Rules and regulations

The abattoir was officially opened on December 13th 1857. The new slaughterhouse played a leading part in the meat supply of Ghent. The city council approved the official regulations for the new abattoir on August 8th 1857. All butchers in the city of Ghent were obliged to make use of the abattoir. The city authorities were responsible for the supervision and inspection of the slaughter of animals, the preparation and the sale of meat. In due course, the rules for the inspection of meat and meat products were considerably tightened (De Herdt, 1988: 54-56; Vandenbroeke, 1983: 234-235).

The supply of living cattle was strictly controlled and regulated. The animals were driven to the slaughterhouse along previously fixed routes. Herds were restricted to fifteen animals, and had to be accompanied by at least two adult drovers. In the early years, the abattoir was operated by a small team, consisting of a director, two meat inspectors, a porter and a few stable-hands. The director and the inspectors, vets or former butchers, carefully inspected the meat. They decided whether it was fit for human consumption and gave their approval by stamping the pieces. The director had to inform the city council of all problems concerning the public health. Many butchers didn't kill the animals themselves. This job was done by specialists, so-called 'master slaughterers' (*slagers* or *abatteurs*), who also cleaned the meat. The animals were transported to the butchers's shops in halves (*by halve beest*) on flat, open wagons.

Several stipulations were the cause of long standing disputes between the butchers and the city council. The slaughter rights remained a point of controversy for many years to come and eventually made the national Parliament. In August 1857 they were fixed as follows for live animals: 0.05 ½ frs per kilo for cows, 0.07 frs for bulls and oxen, calves, pigs, sheep and lambs. Fresh, smoked and salted meat, which was imported from the suburbs or the countryside, was taxed by a duty of 0.12 frs per kilo. The abattoir soon proved to be a lucrative business for the city treasury. On January 1st 1862 the director proudly announced that nearly 32,000 animals had been slaughtered the previous year. This number further rose in the following years, with 38,249 killed animals in 1865; 53,528 in 1875; 63,943 in 1885 and 65,237 in 1895. On the eve of the First World War around 80,000 beasts were killed each year.[3]

In the abattoir, the butchers soon made themselves known as difficult customers. The director of the abattoir and other city officials insisted that the butchers strictly observe the slaughterhouse regulations. They knew very well that butchers constantly tried to evade the rules, which they thought were far too severe. The tug-of-war between the city council and the butchers began immediately after the opening of the abattoir. The butchers clearly needed some time to get used to the new situation. They didn't intend to accept all the new regulations, and the years 1857-1860 were characterized by constant bickering and skirmishing between the two parties. During the summer of 1858, several butchers inquired whether they were allowed to breed and slaughter 'a few' pigs or cows in their

---

[3] SAG, Q 1.e, Posters 'Opening van het Slagthuis en verplaetsing van de Beestenmarkt'; Q 1.a, Etat du nombre de têtes de bétail présentés à l'abattage à l'abattoir, du 1er janvier jusqu'au 31 décembre 1904.

backyard. The director of the abattoir feared that the butchers would interpret a compliant reply as a licence for breeding and slaughtering large numbers of animals at home. This meat would be sold fraudently, i.e. without paying the necessary duties. In spite of the city authorities' strict attitude, all sorts of 'illegal' trading went on. On many occasions 'black' meat, originating from animals that had not been killed in the slaughterhouse, was discovered in the abattoir or in the Butchers' Hall.[4]

The abattoir rule book was brought up to date in 1872 and 1873. The regulations, dating from December 2nd 1872, once again clearly stipulated that only fresh meat, coming from animals which had been slaughtered in the city abattoir, could be sold in the city. On December 12th 1872, over 30 butchers, living on 'the outskirts of town' (*bewoonende de voorgeborchten der stad Gent*), once again protested vehemently against this measure. They denied that they wanted to evade the slaughter rights.[5]

III.2. Fierce competitors

Butchers were – and still are today – very individualistic businessmen. In the nineteenth century, there existed a fierce rivalry between the butchers, who lived and worked in the city center (inside the city walls or *binnen de stadsmuren*) and their fellow craftsmen from the suburbs. The butchers from the city center tried to eliminate at all costs the competition from their colleagues from the suburbs by keeping away their meat. In this, they were helped by a clause in the slaughterhouse regulations, which stipulated that fresh meat could only be imported inside the city center 'by half beast'. The import of all other pieces of fresh meat was considered illegal, and the meat, if discovered, was seized by the authorities. The butchers from the suburbs, who usually slaughtered their animals at home, felt treated unfairly, not to say discriminated against. In November 1858 they asked for permission to bring five kilos of freshly cut meat twice a week into the city. Their fellow craftsmen from the city center reacted furiously and immediately protested.[6] Some city officials showed some sympathy for the suburbians' point of view. The import of meat 'by half beast' was very much in favour of the butchers from the center – and apparently only existed in Ghent. Moreover, this measure was also to the disadvantage of many less well-to-do inhabitants of Ghent, because the suburban butchers sold cheaper meat. All of

---

[4] SAG, Q 1.a, Charles Lagrange, director of the City Taxes, to council, December 23th 1857, August 2nd 1858.

[5] SAG, Q 1.a, *Règlement pour le Service et la Police de l'Abattoir et du marché au Bétail – Reglement voor den Dienst en de Regeltucht van het Slachthuis en de Veemarkt*. Gent, 1873; Some 30 butchers to burgomaster and aldermen, December 12th 1872. In the 1980s the Ghent slaughterhouse no longer met modern hygienic standards or the demands of the European Community. By Ministerial Order of June 17th 1987, the slaughterhouse lost its EC-licence. The city council decided to end all the activities on October 1st. The slaughterhouse was demolished, only the entrance of the abattoir remains. The closure of the slaughterhouse meant that the butchers became increasingly dependent on meat traders. The cattle market and all related activities disappeared as well. The surrounding streets with their many cafés and meeting places became silent. Sadly, this very lively neighbourhood rapidly acquired a seedy, tumble-down outlook.

[6] SAG, Q 1.a, Lagrange to burgomaster and aldermen, November 25th 1857; Council to the chief commissionner, November 30th 1857; Lagrange to burgomaster and aldermen, December 3th 1857.

this meant that smuggling, cheating and the fraudulent import of meat by peasants from the surrounding countryside (*par les campagnards*) continued unabated. The city council eventually relented a bit. It gave permission for the limited import of some 'gifts' (*zenden*), i.e. sausages and other parts of a slaughtered pig (*hoofdflak, worsten, carbonades, lever, ooren, pooten*), which were offered to family and friends.[7]

Relations between the butchers and the city authorities remained tense during the following years. The butchers still thought the slaughterhouse regulations too strict and too severe. The slaughter rights remained too high. They still thundered against all competition, real or imagined, which they considered to be unfair. Butchers from the center were particularly angry that their colleagues from Ledeberg and Sint-Amandsberg were allowed to bring meat, which was not inspected, into Ghent.

III.3. A turbulent professional union

The abattoir regulations caused a lot of friction between the city authorities and the butchers. In order to defend their interests, the Ghent butchers closed ranks and on February 1st 1876 founded the United Butchers (*Vereenigde Vleeschhouwers*). The immediate cause was another dispute with the city council. In 1875 the council intended to double the slaughter rights in the abattoir. As usual, the butchers reacted furiously. One thing and another led to the formation of the United Butchers. A member of the butchers' dynasty Van Loo became the first chairman. Karel Van den Bosch, another prominent butcher and future city councillor, became secretary (Jubileumbrochure, s.d.: s.p.; Jaumain, 1995: 203-205).

The early formation of butchers' organisations was a national phenomenon, as was shown by similar developments in cities such as Antwerp, Charleroi and Brussels. In 1878 a national syndicate was formed. The early formation of butchers' syndicates was caused by some problems all butchers faced, such as the slaughter rights, conflicts of interest with the farmers and cattle breeders and the limitation of the import of cattle.

Yet, in spite of all this the new Ghent syndicate only had a small membership. It mainly represented the more well-to-do butchers from the city center. The United Butchers never acquired the importance or prestige of similar syndicates in Brussels, Antwerp and Charleroi. The national trade journal, *Belgische Beenhouwerij*, confirmed this in 1913: *Or Gand n'est pas Bruxelles, ni Anvers. Dans ces sortes d'entreprises tout est relatif et le syndicat de Gand ... ne compte pas même trois cents membres.*

From the start, the United Butchers behaved like a self-confident, even cocky professional union and a determined interest group. In the following years, this inevitably

---

[7] SAG, Q 1.a, Lagrange to burgomaster and aldermen, November 23th 1858; 92 butchers to burgomaster and aldermen, November 20th 1858; SAG, Q 1.a, Lagrange to burgomaster and aldermen, November 25th 1857; Burgomaster and aldermen to the chief commissionner and Lagrange, November 30th 1857; Lagrange to burgomaster and aldermen, December 3th 1857; A group of butchers to burgomaster and aldermen, April 28th 1858; Lagrange to burgomaster and aldermen, May 25th 1858.

led to new clashes with the city authorities. Butchers' syndicates were – and are – very pragmatic clubs. Their members were and are mainly interested in all kinds of material benefits and services. During the early years, the improvement of professional skills did not rank highly on their list of priorities. Keeping records or establishing a library was considered a waste of time and money.

On the other hand, establishing a butchers' syndicate was hardly self-evident. Butchers are stubborn and very individualistic small businessmen. The leaders of the syndicates regularly complained that butchers were very reluctant to pay their membership fee.[8] Their individualism and, in some cases, the large egos of some prominent butchers led to quarrels and several splits. In 1912 a new syndicate, 'The small butchers from Ghent and the surrounding villages' (*De Kleine Beenhouwers van Gent en omliggende*) broke away after a riotous meeting. Butchers from the city center and butchers from the suburbs and the surrounding villages obviously still found it very difficult to get along. In 1912 the small butchers' union had 125 members and published its own weekly, *Our Journal.* The 'mother syndicate' counted 225 members, but this only represented a fraction of the butchers from the city. The Ghent Manual from 1910 mentions no less than 630 butchers and 260 pork-butchers (Jubileumbrochure, s.d.: s.p.; Jaumain, 1995: 205; Wegwyzer, 1910: 571, 605-607, 613-620).[9]

Moreover, butchers and pork-butchers showed only a limited solidarity with other small craftsmen. They were not very active in the struggle of other tradesmen against the powerful local cooperatives or the pedlars. In 1897, the Ghent Aldermen decided to organise an inquiry into 'the actual economic situation of the small bourgeoisie' (*over den huidigen toestand der kleine burgerij op economisch gebied*). Oscar Pyfferoen, a barrister and professor at Ghent University, was the moving force behind this inquiry, but he wasn't fooled by the butchers' ready-made answers. He wryly remarked that some tradesmen, such as the butchers, easily got the upper hand in the competition with the cooperatives (Jaumain, 1995: 205; Kurgan-Van Hentenryk, 1983: 297-332).[10]

The formation of the Boerenbond (1890) didn't facilitate the relations with the powerful Catholic Party, its elected representatives and the catholic organisations in general. The butchers were convinced that the catholic 'pillar' primarily defended the interests of the farmers, the cattle breeders and the countryside, in a word *les intérêts agricoles.* The butchers and their helpers clashed a few times, and the helpers were supported by the local catholic workers' union, the powerful *Antisocialistische Werkliedenbonden,* led by the 'red baron' Arthur Verhaegen. On the other hand, liberal newspapers, such as *La Flandre libérale,* regularly published articles, defending the butchers. Some liberal Members of Parliament sided with the butchers and protested against the limitation of the cattle import, another sore point for the butchers. But all this by no means meant that

---

[8] *Belgische Beenhouwerij,* September 14th 1913. Le Syndicat de Gand.
[9] *Belgische Beenhouwerij,* December 16th 1936, Gent. Samensmelting der Syndikaten; November 8th 1936, 60ᵉ jarig Jubelfeest der Vereenigde Vleeschhouwers van Gent en omliggende.
[10] SAG, K, 1369, O. Pyfferoen, *Une enquête sur la petite bourgeoisie. Les cooperatives en Belgique.* Brussels and Paris, 1899, pp. 3-5. Records and documents concerning this inquiry are in SAG, fonds K, 1365-1373.

there grew a fully-fledged alliance between the butchers' syndicates and the liberals. The aforementioned city councillor Karel Van den Bosch was elected with the support of the Catholic Party, but soon preferred to adopt a politically neutral position.[11]

III.4. The butchers' problems in the National Parliament

Before 1890, the city authorities were chiefly responsible for the inspection of meat products and other foodstuffs. Public opinion and general ideas about public health, hygiene and food inspection changed considerably over the years. As a result, at the end of the nineteenth century the national government increasingly took control of this matter. In February 1891, a Royal Decree imposed general, nationwide rules for the slaughter of animals, and the selling and transport of fresh and prepared meat, poultry and game.

The health inspector, by law a vet, became the chief responsible for the abattoir, the meat trade and the meat inspection in Ghent. The butchers from the suburbs were at last allowed to bring their meat, after it was duly inspected, into the city. City officials would inspect cattle and animals, meat and all derived foodstuffs in shops, factories, supply rooms, warehouses and on markets. Meat products, imported from abroad, were to be inspected in the harbour.[12]

Between 1890 and 1914, butchers and local and national authorities quarrelled about several problems, such as the slaughter and inspection rights in abattoirs. The general dissatisfaction about the slaughter rights had led to the early foundation of a national professional organisation, the Federation of Belgian Butchers (*Fédération of Belgian Butchers*). The dispute eventually reached the Belgian Parliament. The butchers were supported by a few liberal back benchers, but the successive ministers of Finance didn't want to push the matter too abruptly. They chose a gradual approach and tried to convince the city councils that 'something' ought to be done about the slaughter rights. This approach eventually resulted in the laws of July 31st 1889 and August twentieth 1890, which limited the rights. In the mean time, enterprising butchers still tried to evade the slaughter rights. On the eve of the First World War, in March 1914, the abattoir regulations were once again brought up to date (Jaumain, 1995: 76-77).[13]

The so-called 'law Thienpont', concerning the selling of sick animals, (1894) was another hotly disputed issue. It led to a direct confrontation between the butchers on one

---

[11] BUG, Fonds Vliegende Bladen, I, B, 78, 'Lafheid'. Pamflet van de Gilde van beenhouwersgasten, 1894; BUG, Fonds Vliegende Bladen, I, B, 78, Pamflet 'Slachters en vleeschhouwers. Helpt ons! Helpt ons!' van de Gentse Slachtersknechten (1900). Karel Van den Bosch (1853-1924), butcher in the city center, city councillor for the Vrije Burgersbond (1896-1903).

[12] SAG, Q 3.7, Règlement sur le commerce des viandes, l'abattoir communal et les marches aux bestiaux, suifs et peaux, 1891.

[13] *Belgische Beenhouwerij*, September 30th 1956, Gentse Kroniek. De gedragingen der Gentse vleeshouwers na de afschaffing van het Ambachtsgild (1794), Charles Zwaenepoel; SAG, Q 1.a, Remy to the mayor; SAG, Q 1.a.1, *Verordening over den Vleeschhandel en den dienst en de politie van het Stedelijk Slachthuis. Règlement sur le commerce des viandes et le service et la police de l'abattoir communal. 23 maart 1914.* Gent, 1914.

side and the farmers and cattle breeders on the other. The farmers and cattle breeders were supported by the newly founded Farmers' Union (*Boerenbond*). In January 1893, the catholic MP Louis Thienpont protested against the old law of 1850, which very much in favour of the cattle purchaser. A year later, Thienpont introduced a new bill which was, as could be expected, in favour of the cattle breeder. In fact, the law stipulated that the breeders were not responsible when their animals were found to be suffering from bovine tuberculosis. Moreover, the new law was accepted in July 1894 in peculiar circumstances, i.e. a few minutes before the end of the parliamentary session. The struggle continued for some time, and the Ghent butchers put up a good fight against the law (Heyrman, 1991: 28).[14]

The import of foreign cattle was the third point of dispute between butchers and *les intérêts agricoles*. The law of 1887 imposed import duties on beef and lamb. (Van Molle, 1989: 128-137) The main point of contention concerned the closing of the borders to protect livestock against cattle diseases. In the 1890s Belgium concluded several bilateral agreements with its neighbours, in order to prevent the spreading of infectious diseases. The law of 1887 had tightened the inspection of imported cattle and meat. After the 1894 parliamentary elections, the Catholic Agriculture minister Léon de Bruyn further tightened the legislation on cattle import. In the 1890s, neighbouring countries such as the Netherlands were repeatedly afflicted by cattle diseases particularly foot-and-mouth disease, and the borders were closed several times. But, in all countries concerned, there was a very fine line between protecting the home market and protecting the live-stock (*Vlees*, 1988: 147).

Minister de Bruyn tried to find a compromise between the cattle breeders, who favored protectionist measures, and the consumers, who were mainly interested in low meat prices. The butchers defended the free import of cattle from Holland, a country that reputedly possessed the best cattle in the world. They were against protectionism and accused farmers and cattle breeders of wanting to keep the prices of meat high at all costs.[15]

By the mid-1890s, feelings of discontent ran very high in the butchers' world. Whenever a dispute opposed butchers and farmers and cattle breeders, the former invariably came off worst. A butcher from Malines even claimed that ' ... today the Farmers' Union is a law unto itself' (Jaumain, 1995: 84-87).

These disputes had some consequences for the (party)political positioning of the butchers' unions. Theoretically, they did not favour any particular political party. In practice,

---

[14] BUG, Fonds Vliegende Bladen, I, B, 78, Kritiek der wet van 6 juli 1894. Open brief van de Vereenigde Vleeschhouwers van Gent aan de Kamerleden en Senatoren, 19 september 1894; *La Flandre libérale,* November 28th 1904, La question de la viande de boucherie. – Louis Thienpont (1853-1932), catholic Member of Parliament for the district Oudenaarde (1887-1900).

[15] BUG, Fonds Vliegende Bladen, I, B, 44, Petition 'Zeer beminde landbouwers, vleeschhouwders en geheel den Burgerstand, zoowel arm als rijk', van 'uwen vriend, Is. Vereecken, herbergier, van In den Schaapherder, Nieuwe Beestenmarkt, Gent' (presumably 1889); Petition Landbouwersbond Oost-Vlaanderen to 'Messieurs les Membres du Sénat et de la Chambre des Représentants' (presumably 1895); Kamer van Volksvertegenwoordigers, session of August 13th 1885.

the butchers often clashed with the Catholic Party and several catholic organisations, such as the Farmers' Union, the staunch defender of *les intérêts agricoles*. In June 1903 the Ghent United Butchers reacted furiously to an article published by *De Landbouw*, the weekly of the local Farmers' Unions. The article dealt with 'the price of meat' and claimed that 'no price was ever high enough for the city butchers'.[16] But, as said earlier, this did not mean that there grew a fully-fledged alliance between the butchers' syndicates and the liberals.

III.5. Meat on the table

For a long time, historians seemed to agree that the Ghent working class ate very little meat in the nineteenth century. The local doctors Heyman en Mareska, who in 1843-1844 inquired into the living and working conditions of the textile workers, concluded that the average working class family rarely ate meat and other animal products. Other researchers confirmed these findings. More recently, historians tend to paint a more balanced and slightly more optimistic picture (Segers, 2003: 53-59; Van den Eeckhout, 2002: 373-390). Even so, during a large part of the nineteenth century, working class families seldom ate quality meat, they had to settle for bacon, fat and tripe (Anseele, 1905: 175). Whenever cheap meat was available, the shops were stormed. For instance, in 1854 some pork butchers managed to get hold of cheap salted beef (*ossenvleesch*), pork and dripping from South America. This event even made the annual report of the city administration.[17] It was hardly a coincidence that in these lean years, almost mythical forces were ascribed to meat. Serious research apparently confirmed that people, who labored hard for hours on end, needed meat. E. Ducpétiaux, the renowned pioneer in budget research, even claimed in 1855 that in Belgium 'on ne paraît pas savoir combien la viande est nécessaire au travailleur' (Ducpétiaux, 1855: 441-595; Scholliers, 1993: 46-50, 112-114).

From 1890 onwards, the average standard of living of the working class slowly but steadily rose. The quality of their food also improved. Working class families ate more beef and pork, less bacon and fat, and much less bread and potatoes. As I said earlier, the import of foreign meat, i.c. beef, was restricted repeatedly in the 1890s. As a result, the consumption of pork, already very considerable, rose further, for pork was cheap and abundantly available (*zwijnenvleesch, dat zeer overvloedig op de markt en goedkoop was*).[18]

The Central Library of the University of Ghent possesses a large collection of advertising leaflets of around 1900. Many leaflets alluded to the high prices of meat. Enterprising

---

[16] BUG, Fonds Vliegende Bladen, I, B, 78, 'Geachte Landbouwers' – Open brief van de Vereenigde Vleeschhouwers, June 19th 1903.

[17] Rapport sur l'administration et la situation des affaires de la ville de Gand, présenté au conseil communal par le collège des bourgmestre et échevins – Verslag over het bestuur en den toestand der stad Gent, door het schepencollege den gemeenteraad aangeboden, 1866, p. 115.

[18] Rapport sur l'administration et la situation des affaires de la ville de Gand, présenté au conseil communal par le collège des bourgmestre et échevins – Verslag over het bestuur en den toestand der stad Gent, door het schepencollege den gemeenteraad aangeboden, 1895, pp. 414-415.

butchers tried to lure 'thrifty housewives' to their shops by stressing their 'low prices' or by offering 'huge reductions'.[19]

How expensive was meat in Ghent around the turn of the century? Butchers sold roastbeef and beefsteak at 0.018 à 0.017 euro (1.80 à 1.70 frs) per kilo. Cheaper kinds of beef (*bolle, boef, lap, schellekens en schenkels*) were sold at 0, 011 à 0.0130 euro (1.10 à 1.30 frs) per kilo. Lean pork cost 0.015 euro (1.50 frs) per kilo, minced meat, bacon and sausages 0.014 euro (1.40 frs) per kilo, black pudding at 0.01 euro (1.00 or 0.90 frs) per kilo. Mutton, lamb and particularly veal were quite expensive. Chicken, at that time still very much a luxury product, was completely absent from the advertisements (Vandenbroeke, 1983, 232-234).[20]

IV. The Ghent butchers in the interwar period – United We Stand?

During the interwar period the number of butchers remained high in Ghent. In 1920, the Ghent *Manual* mentioned ca. 500 butchers. The 1932 *Manual* mentioned 24 horse butchers, 27 poulterers, 17 'sausage makers' and no less than 614 meat butchers (Wegwyzer, 1920: 56, 59, 74, 75, 77-80; Wegwyzer, 1932: 539, 543, 563, 567-569, 571-572).

The First World War caused a temporary reconciliation between the butchers' unions, but this *entente cordiale* didn't last long. In the early 1920s there followed another split. Only by the end of 1934 were the three existing butchers' unions ready to put aside their differences and cooperate in the 'United Butchers from Ghent and the neighbouring villages' (*De Verenigde Vleeshouwers van Gent en omliggende*).[21]

Was the reconciliation a byproduct of the economic crisis? Whatever the reasons, during the 1930s the National Union and the Ghent butchers' unions strongly emphasized the need for unity. In the early twentieth century, the butchers showed little solidarity with other small crafts- and tradesmen. This proud attitude changed in the 1930s. The National Union repeatedly insisted on cooperation and solidarity with other self-employed tradesmen (*de andere handelsklassen en middenstanders*), mainly because the butchers felt threatened by the same dangers. They lashed out against the fiscal policy of the governments, and particularly against 'the cancer of the chain stores, the trusts ...' Already on June 16th 1929, *Belgische Beenhouwerij* had published an alarming article, with the title 'The Death of the Small Butcher in the United States' (*Dood der kleine beenhouwerij in de Vereenigde Staten*). The author claimed that small butchers' shops had almost completely disappeared in the States. They had been destroyed or absorbed by chain stores and trusts. The butchers had to defend themselves against these new competitors by developing their skills and improving the quality of their meat. It was

---

[19] BUG, Fonds Vliegende Blaadjes, II, B, 35, II, B, 78.
[20] BUG, Fonds Vliegende Blaadjes, II, B, 35, II, B, 78.
[21] *Belgische beenhouwerij*, May 7th 1922; *Belgische Beenhouwerij*, November 30th 1956, Het syndicaat van Gent tachtig jaar; December 16th 1936, Gent. Samensmelting der Syndikaten; November 8th 1936, 60ᵉ jarig Jubelfeest der Vereenigde Vleeschhouwers van Gent en omliggende.

hardly a coincidence that the first trade school for meat – and pork-butchers opened its doors in May 1936 in Antwerp (Avondts, 1970: 214).[22]

More particularly, the National Union tried to respond to the widespread feelings of discontent and uneasiness among its members by emphazing the need for solidarity. From 1935 onwards, the National Union and the local unions repeatedly tried to promote 'syndicalism' and 'unity', which were to be shown during meetings and commemorations. But in spite of all this propaganda, many butchers didn't feel like marching and flag waving.[23] The Ghent butchers in particular didn't like to leave their shop. 'A true syndicalist' (*een ware syndicalist*) from Ghent was very disappointed when only a small number of butchers turned up at a commemoration in Renaix. They preferred to watch a football match, their wives didn't want them to go and over a hundred butchers had made it plain that they couldn't care less' (Avondts, 1970: 214).[24]

V.   After the Second World War – The Golden Years

On February 1st 1948 meat rationing was officially abolished. The Belgian economy recovered fairly quickly after the Second World War, and in the 1950s the general standard of living started to rise. In spite of this, the butchers repeatedly complained. Their complaints rang very familiar. Meat products were subjected to regulations and controls, which were much too strict. Government agencies controlled the purchase, the preparation, the price, and the quality of meat and meat products. The young Paul Vanden Boeynants, a very outspoken Member of Parliament since 1949 and an even more outspoken chairman of the National Union since 1952, wrote most of the editorials in *Belgische Beenhouwerij*. He thundered against the continuous 'meddling' by the 'state' and the increasingly strict regulations. According to Vanden Boeynants and the other leaders of the National Union, these 'new dangers' increased the need for a strong professional organisation.[25] At the celebration of the 75th anniversary of the Ghent United Butchers, local chairman Charles Zwaenepoel also insisted on the need for a better organisation.[26] Chain stores, food stores and supermarkets represented the main dangers for the small butchers. Food stores and supermarkets not only presented an increasingly large and varied supply of meat and meat dishes, they did so in a modern and attractive way. Butchers were obliged to follow, i.e. to modernise their shop and to adapt willy-nilly to the demands of the 'consumer'. The old-fashioned shop on the corner had to make way.

---

[22] *Belgische Beenhouwerij*, June 16th 1929; *Belgische Beenhouwerij*, November 15th 1936; *Belgische Beenhouwerij*, September 19th 1937; *Belgische Beenhouwerij*, October 17th 1937; *Belgische Beenhouwerij*, May 17th 1936.
[23] *Belgische Beenhouwerij*, September 14th 1936; *Belgische Beenhouwerij*, September 26th 1937; *Belgische Beenhouwerij*, October 3th 1937.
[24] *Belgische Beenhouwerij*, May 17th 1936.
[25] Paul Vanden Boeynants (1919-2001) was the most famous butcher in Belgian (political) history. The son of a Brussels butcher, he became the spokesman of the Belgian middle classes in general and the butchers in particular. VDB, as he was popularly known, was a Member of Parliament from 1949 until 1985, and was, among other things, minister of the Middle Classes (1958-1961), minister of Defence (1972-1977; 1979) and prime minister (1966-1968 and 1978-1979).
[26] *Belgische Beenhouwerij*, January 9th 1949 and April 17th 1949.

On the other hand, one shouldn't be fooled by the butchers' constant complaints. From the 1950s onwards, and especially during the 1960s the average standard of living rose considerably. Between January 1961 and January 1962, the National Statistical Institute carried out the first post-war inquiry into the spending habits of blue – and white collar employees and 'non actives' (pensioners, students, unemployed ...). This inquiry was repeated in March 1973-March 1974. This large scale inquiry clearly showed how much the average standard of living rose in the 'Golden Sixties'. In 1961, workers' families still spent 36 per cent of their income on food, white collar employees 27 per cent and non-actives 34 per cent . In 1974, this share had decreased to 24.7 per cent, 17 per cent and 24 per cent for workers, white collar employees and 'non actives' respectively. No less than 27 varieties of meat and meat dishes were mentioned in 1961. In 1974, this number had risen to 32.[27]

All in all, the 1960s and 1970s were very good years for the butchers and they made a lot of money. *Belgische Beenhouwerij* is very discreet about that. Over the years, the 'customer' became more demanding and turned into a critical 'consumer'. The supply of fresh meat and prepared dishes grew impressively. Many butchers' shops increasingly resembled sophisticated delicatessen. Publicity, promotion and presentation became increasingly important.

V.1. The Ghent butchers in the 1980s and 1990s

The 1980s and particularly the 1990s have not been very good years for the meat industry, the meat trade and consequently, the butchers. Meat certainly no longer enjoys the almost mythical status it possessed in the nineteenth century. From the 1980s onwards, the image of cattle breeding and the meat trade was tainted by revelations about the use of hormones. The Ghent Butchers invariably reacted furiously whenever a 'hormone scandal' was revealed in the press. They felt persecuted, not only by the press, but also by new enemies such as the local Food Inspection, consumers' organisations and the young green movement.

The hormone problem also increased the competition between the butchers. Butchers, who promoted 'hormone free' meat, incurred the wrath of their colleagues, and of the United Butchers in particular. The butchers' syndicate was as watchful and touchy as ever in these turbulent years. Even the local liberal Union for the Self-Employed was consigned to the enemy camp, when it announced a debate on the hormone problem in February 1989.[28]

---

[27] *Het gezinsbudgetonderzoek, 1961,* pp. 45-47; *Het gezinsbudgetonderzoek, 1973-1974*, pp. 45-46.
[28] Open Letter Verenigde Vleeshouwers van Gent en omliggende to Alderman Raoul Wijnakker, December 10th 1985; Butcher Valentin Poelman to Wijnakker and city counciller Gilbert Temmerman, December 28th 1986; Temmerman to Poelman, January 28th 1986; Gilbert Verschatse, chairman of the United Butchers, and M. Van Geert, general secretary, to Wim Van der Aa, director-general of the National Union, January 21th 1986; Advertisement Liberal Union for the Self-Employed, February 8th 1989; Van der Aa to Verberckmoes, national chairman of the Liberal Union, February 16th 1989; A.M. Neyts, national chairwoman of the Liberal Party, to Van der Aa, February 23th 1989.

V.2. Downhill

The number of independent small butchers in Ghent has steadily decreased since the mid-1960s. In the immediate post-war years, there were about 1,000 butchers in Ghent and the neighbouring boroughs. By the end of 1955 the city of Ghent counted 501 butchers and pork-butchers. Ten years later this figure had dropped to 407. In 1975, this figure had further decreased to 274. In the beginning of the 21st century, there are approximately 200 independent butchers in Ghent.

The membership figures of the United Butchers reflect this evolution. In 1968, the union counted 587 members, in 1975 581 and in 1983 511. In 2002 the union had fewer than 200 members. A similar evolution is noticed all over Belgium. In 1982 there were approximately 12,000 independent butchers nationwide. Twenty years later, this figure has dropped to 3,200 and keeps on dropping. These decreasing figures have various causes such as the competition form chain stores and supermarkets and a changing lifestyle. Fewer people have time – or are willing to take time – to cook regularly. The conditions and rules for running a butchers' shop have become very demanding. As a result, many butchers called it a day. Starting a new shop means investing a lot of money: the starting costs are estimated at about 300,000 euro (ca. 12 million frs).[29]

Not many young people are keen to follow a butchers' training. They are even less keen to open a butchers' shop. In recent years, the public image of meat and the meat business have further detiorated. The 1980s were dominated by hormones. The 1990s were even worse. News headlines were regularly dominated by food scares and crises, such as swine fever, the BSE-alarm, the dioxinecrisis and foot-and-mouth disease in several European countries. In response to all this, vegetarianism has become increasingly fashionable, particulary among the young and trendy.[30]

---

[29] Interview with butchers Noel Van Speybroeck and Raymond Stadius, January 18th 2001. These figures were confirmed by Marc Meerschaut, the accountant of the Verenigde Vleeshouwers. In spite of repeated questioning, it remained unclear where the borders between 'Ghent' and 'the surrounding boroughs' were drawn exactly.
[30] Verslag over het bestuur en de toestand in 1955, pp. 185; 1965, p. 192; 1970, p. 348; 1975, p. 396; *Gazet van Antwerpen*, January 25th 2002. Ieder jaar 500 slagerijen weg; M. Vanderstraeten, 'Uitbenen en tegendraads snijden. Vier slagerszonen over hun jeugd tussen het vlees', *De Standaard Magazine*, May 24th 2002, pp. 22-25.

Table 4.1   Number of (independent) butchers in Ghent, 1854-2002

Year	Butchers	Pork-butchers	Horse-butchers	Meat selling points
1846	Population: 102,000			
1854*	46	29		
1860*	142	81		
1870*	160	81		31
1880	Population: 130,000			
1890*	620			
1900*	600	140		
1910	Population: 166,000			
1910*	630	260		
1920*	500	180	26	
1932*	614	17	24	
1955° (city center)	501			
1965° (city center)	407			
1975° (city center)	274			
2002° (city center)	approx. 200			

* These figures are based on those mentioned in the Wegwyzer van Gend/Gent.
° These figures are based on those mentioned in documents / records of the Ghent butchers.

Bibliography

Manuscript sources

Verenigde Vleeshouwers van Gent en omliggende:
All original source material seems to have disappeared.

Landsbond der Beenhouwers en Spekslagers van België (Kortenberglaan 116, 1000 Brussels):
Correspondence 'De Verenigde Vleeshouwers van Gent en omliggende', 1980-1999 (1 small file); Statutes and correspondence 'De Verenigde Vleeshouwers van Gent en omliggende' (1 file); Correspondence Provinciaal Verbond Oost-Vlaanderen (1 file).

City Archives Ghent (Stadsarchief Gent):
Collection K (Commerce and industry)
Collection C (Offical records of the Mayor, the Bench of Aldermen and the City Council)
Collection L – XVI (Taxes)
Collection Q (Markets, Abattoir and Food Products)

Public Record Office Ghent (Rijksarchief en Archief van de Provinciën, Ghent), Provincial Archives, 1830-1870:
Inventory 3217, file 4789/1, Classification of professions, Ghent, 1830
Inventory 4801, population census 1846, Ghent

Central Library Ghent University – Collection Pamphlets and Leaflets:
Files I, II, III and IV, main entries 'slachthuis', 'beenhouwerijen', 'spekslagerijen', 'beenhouwersbonden'

Printed and secondary sources

Avondts, H.G. (1970) *Van gilde tot Antwerpse beenhouwersbond. Honderd jaar Antwerpse beenhouwersbond, 1868-1968*, Antwerp.

Baertsoen, M. (1929) *Gand sous l'occupation allemande. Notes d'un Gantois sur la guerre 1914-1918*, Ghent.

Baudhuin, F. (1933) 'L'importance numérique des classes moyennes', *Mélange offerts à Ernest Mahaim*, Paris.

Beer, H. de (2003) 'Krachtig en gezond, want doorvoed? Voeding, gezondheid en arbeid in de tweede helft van de 19de eeuw', *Spiegel Historiael*, 3-4, pp. 120-127.

Belgische beenhouwerij/Boucherie belge, 1919.

Boone, R. (1993) *Weer met een gerust geweten vlees eten*, Brussels.

Casier, F. (1904) 'Rapport sur l'outillage des bouchers et charcutiers', *Exposition internationale du petit outillage. Gand, juillet 1904*, Ghent, pp. 199-219.

Bulletin Communal – Gemeentebulletijn, 1868.

Collin, F. (1905) *Verslag nopens den ambachts- en handeldrijvenden middenstand*, Brussels.

Commission Nationale de la Petite Bourgeoisie. Enquête écrite. III. Monographies, Ghent.

Crossick, G. and Haupt, H.G. (eds.) (1995) *The Petite Bourgeoisie in Europe, 1780-1914. Entreprise, Family and Independence*, London & New York.

Daem, M. et al. (1978) *De provinciale Kamer van Ambachten en Neringen van Oost-Vlaanderen. 50 Jaar in dienst van haar leden*, Brussels.

De Gazette van Gent, 1857, 1876, 1884.

Devolder, K. (1994) *Gij die door 't volk gekozen zijt ... De Gentse gemeenteraad en haar leden 1830-1914*, Ghent (Maatschappij voor Geschiedenis en Oudheidkunde, Verhandelingen, XX).

Dhont, G. (1907) *Jaarboek van den beenhouwer en den slager / Annuaire du boucher et du charcutier*, Bruges.

Het gezinsbudgetonderzoek, 1961. 1963, Brussels (Nationaal Instituut voor de Statistiek, Statistische en econometrische studiën, 5).

Het gezinsbudgetonderzoek, 1973-1974. 1975, Brussels (Nationaal Instituut voor de Statistiek, Statistische en econometrische studiën, 38).

Heyrman, P. (1991) *Voor eigen winkel. Honderd jaar middenstand en middenstandsbeweging in Oost-Vlaanderen*, Ghent.

'Hoeveel slagers in stadscentrum? Verenigde Vleeshouwers van Gent en Omliggende bestaan 125 jaar', *De Gentenaar*, October 13th 2001.

'Ieder jaar 500 slagerijen weg', *Gazet van Antwerpen*, January 25th 2002.

Jaumain, S. (1995) *Les petits commerçants belges face à la modernité (1880-1914)*, Brussels.

Koolmees, P. (2003) 'Wat voor vlees hebben we in de kuip? Honderdvijftig jaar vleeshygiëne in Nederland', *Spiegel Historiael*, 3-4, pp. 148-155.

Kurgan-Van Hentenryk, G. (1983) 'A la recherche de la petite bourgeoisie. L'enquête orale de 1902-1904', *Belgisch Tijdschrift voor Nieuwste Geschiedenis*, 3-4, pp. 297-332.

Kurgan-Van Hentenryk, G. (1988) 'A Forgotten Class: the Petite Bourgeoisie in Belgium, 1860-1914', Crossick, G. and Haupt, H.G. (eds.) *Shopkeepers and Master Artisans in Nineteenth Century Europe*, London & New York, pp. 417-477.

Kurgan-Van Hentenryk, G. and Jaumain, S., (eds.) (1992) *Aux frontières des classes moyennes. La petite bourgeoisie belge avant 1914*, Brussels.

La Flandre Libérale, 1875-1876.

Lambrechts, H. and Schurmans, V. (1906) *Wat een beenhouwer of varkensslachter moet weten*, Brussels.

La boucherie belge/De Belgische beenhouwerij. Eenig orgaan van den Nationalen Bond der Belgische beenhouwers, 1894-1910.

Lefebvre, W. (2003) 'Het geografisch inplantingspatroon van voedingswinkels in Leuven tijdens de tweede helft van de negentiende eeuw (1860-1908)', *Belgisch Tijdschrift voor Nieuwste Geschiedenis*, 1-2, pp. 91-127.

Maeyer, J. De (1994) *De Rode Baron. Arthur Verhaegen, 1847-1917*, Leuven (Kadoc-Studies 18).

Molle, L. van (1989) *Katholieken en landbouw. Landbouwpolitiek in België, 1884-1914*, Leuven.

Niesten, E., Raymaekers, J. and Segers, Y. (2002) *Vrijwaar U van namaaksels! De Belgische zuivel in de voorbije twee eeuwen*, Leuven (CAG-Cahier 2).

Niesten, E., Raymaekers, J. and Segers, Y. (2003) *Lekker Dier!? Dierlijke productie en consumptie in de 19de en 20ste eeuw*, Leuven (CAG-Cahier 3).

Noppen, K. van (1990) *Slachthuis, een fijne buur*, Ghent.

Paepe, D. de (2003) 'Tussen gehaktmolen en hakblok. De Kortrijkse slager omstreeks 1900', *De Leiegouw*, 2, pp. 257-278.

Productie van gezond en veilig vlees van gevogelte, s.l., 1996.

Quelques notes pour contribuer à l'éducation des bouchers et charcutiers, Brussels, 1905.

Rapport sur l'administration et la situation des affaires de la ville de Gand, présenté au conseil communal par le collège des bourgmestre et échevins – Verslag over het bestuur en den toestand der stad Gent in 1914, door het schepencollege den gemeenteraad aangeboden, Ghent, 1836-

Salvetti, F. (1985) *De beenhouwersstraat. Het vlees en het ambacht doorheen tijden en culturen*, Tielt.

Scholliers, P. (1993) *Arm en rijk aan tafel. Tweehonderd jaar eetcultuur in België*, Berchem-Brussels.

Scholliers, P. (ed.) (2001) *Food, Drink and Identity. Cooking, Eating and Drinking in Europe Since the Middle Ages*, Oxford & New York.

Segers, Y. (2003), *De economische groei en levensstandaard. Particuliere consumptie en voedselverbruik in België, 1800-1913*, Leuven (ICAG Studies 1).

'Slachthuis wordt voedingsbeurs. Ook vis, groenten en fruit in Anderlecht', *De Morgen*, February 5th 2002.

Trommelmans, A. (1976) *De organisatie van de middenstand, 1890-1914*, Ghent (University of Ghent, Unpublished thesis).

Vandenbroeke C. (1973) 'Voedingstoestanden te Gent tijdens de eerste helft van de 19de eeuw', *Belgisch Tijdschrift voor Nieuwste Geschiedenis*, 4, pp. 109-169.

Vandenbroeke, C. (1983) 'Kwantitatieve en kwalitatieve aspecten van het vleesverbruik in Vlaanderen', *Tijdschrift voor Sociale Geschiedenis*, 9, pp. 221-257.

Vanderstraeten, M. (2002) 'Uitbenen en tegendraads snijden. Vier slagerszonen over hun jeugd tussen het vlees', *De Standaard Magazine*, May 24th, pp. 22-25.

Varenbergh, E. (1972) *La grande boucherie*, Ghent (Deel V. Souvenirs archéologiques de la ville de Gand. Uittreksel uit de Messager des Sciences Historiques de Belgique).

Vermandere, M. (1987) *Repertorium van de Belgische middenstandspers tot 1940, met casusanalyse van de Vrije Burgersbond te Gent*, Ghent (University of Ghent, Unpublished thesis).

De vereniging in de ambachten en neringen, Brussels, 1912.

'Vlees. Eet minder maar beter!', *Knack Weekend*, April 24-30th 2002.

Wegwyzer van Gent, 1770-1932, 1948 and 1972.

Willems, H. (2002) 'De lijdensweg van een rustdag: de wet op de zondagsrust (1905)', *Belgisch Tijdschrift voor Nieuwste Geschiedenis*, 1-2, pp. 73-118.

Vlees in de pan door de eeuwen heen. Jubileumuitgave van de Antwerpse verenigingen voor vee en vlees, Antwerp, 1988.

Vriendt, J. de (1978) 'De Van Loo's, van Gent, vleeshouwers van vader op zoon, van de 14de eeuw tot heden', *Vlaamse Stam, Tijdschrift voor Familiegeschiedenis*, 3-4, pp. 129-144.

Waele, M. de (2002) 'Koopkracht en levensstandaard in België en Vlaanderen, 1800-1990', *Tijdschrift voor Industriële Cultuur*, 1, pp. 3-29.

Wiele, J. van de (ed.) (1988) *De Markt. Economisch forum en kloppend hart van agrarische, stedelijke en industriële maatschappijen*, Ghent and Leuven.

Interviews

Interview with former butchers Noel Van Speybroeck and Raymond Stadius, Eeklo, January 18th 2001.

Abbreviations

RAG = Rijksarchief Gent (Public Record Office Ghent)

SAG = Stadsarchief Gent (City Archives Ghent)

BUG = Bibliotheek Universiteit Gent (Central Library Ghent University)

5   From stable to table.
The development of the meat industry
in The Netherlands, 1850-1990

Peter A. KOOLMEES, Utrecht University

I.   Introduction

Hygiene, animal welfare and consumer confidence are at the top of the modern meat industry's agenda, though this paper argues that these issues are not new. Meat, once a food exclusively for the rich, has now become a common food. Within a century, the shortage of animal protein, which characterised the consumption pattern until the second half of the nineteenth century, has changed into over-consumption of animal protein and fat by many consumers in the highly industrialised countries of today. During the last 150 years the food supply has changed dramatically. The question 'Do we have enough food?' has been largely replaced by the question, 'Do we eat healthy food?' Today, food hygiene, food safety and overweight are issues that concern government authorities, producers and consumers alike. Apparently, there has been a transition from the concern about a secure food supply to concerns about food safety (Den Hartog, 2003: 108; Henson and Caswell, 1999: 590-591; Fisher, 1998: 215-217, 221-222).

This paper presents a concise review of the development of the production, trade, supply and quality control of meat and meat products between 1850 and 1990. These developments were initiated or influenced by changes in meat consumption patterns which, in turn, were caused by social and economic shifts in society. From 1850 onwards, a gradual transition from rural society to modern welfare state occurred. This process of modernisation was attended by economic growth, industrialisation and urbanisation, which, in turn, caused a change in food patterns, in particular an increase in meat consumption. As a result of the increasing demand for meat, innovations in meat production took place while a large international meat trade and meatpacking industry developed. The latter was also stimulated by the demands of military and overseas shipping, which needed food that could easily be transported and preserved for longer periods.

During the process of transformation from agricultural and artisan societies to industrial and urban ones, fundamental changes in public health care, eating habits, foodstuff production, distribution and quality control occurred. Inevitably, these changes affected the meat supply chain. The availability and supply of safe and sound meat became one of the problems with which local and national authorities were confronted during this transformation.

In this paper the following questions will be considered: Who were the main players in the demand for and development of a professional and clean meat industry? When and under what circumstances did the emphasis on a secure meat supply shift to a concern with the supply of safe meat?

II. Mass consumption

An important change in the history of nutrition was certainly the beginning of the age of mass consumption from around 1850 onwards. The growth of world trade in meat and livestock after 1850 depended on an increase in effective demand and on developments in technology, which allowed that demand to be satisfied. The rise in personal income as a result of industrial and commercial development in the course of the nineteenth century and particularly from the 1960s onwards, meant a growing market for all animal products, including meat. Country dwellers supply a certain amount of their own foodstuffs, but modern urban consumers are entirely dependent on the food supplied or manufactured by others. The rise of the consumer society, therefore, played a vital part in the expansion of the meat market (Koolmees, Fisher and Perren, 1999: 12).

In the European countries that became industrialised, the exposure to periodic famines was finally overcome and, at least quantitatively, an adequate national food supply could be secured. Industrialisation brought about a structural change in the production, distribution, preservation and preparation of foods. The development of a large-scale food industry allowed food prices to decrease, and hence during the last 150 years the per capita consumption of most of the basic foodstuffs, including meat, showed a distinct increase. A large variety of food in sufficient quantities became available to the majority of the population. This so-called modern (western) consumption pattern is characterised by a larger variety of food products, more processed foods, more convenience food and a lessening in seasonal influences (Den Hartog, 2003: 108-109; Teuteberg, 1991: 186-190).

During the nineteenth century, knowledge of the chemical composition of food and nutrients increased. Around 1850 many physiologists, chemists and physicians stressed the importance of a high protein intake. Most authorities considered that about 90 kg of meat per year was the minimum required for optimum nutrition. They argued that a population deprived of animal-derived protein would become weak and would produce only average labourers and soldiers. Economists attributed the seemingly effortless manner in which Britain created its Indian Empire to the high meat consumption of the British and the vegetarianism of the Indians (Nitti, 1896: 31). Such notions provide an explanation as to why not only many scientists, physicians and veterinarians, but also politicians, economists and military leaders and strategists attached great importance to a high level of meat consumption. Although less extreme, these ideas still persisted during the period 1900-1940. From 1918 onwards, less emphasis was placed on meat consumption and a more varied diet was advised under the maxim of 'a little bit of everything, but not too much of anything'. The recommended meat intake dropped from 90 to 45 kg annually (Van Eekelen, 1962: 613; Koolmees, 2000: 54-55).

Akin to the more industrialised neighbouring countries, a quantitative and qualitative improvement in diet occurred in The Netherlands and the democratisation of meat consumption gradually began. Democratisation here meant that meat was no longer exclusively a food for the rich, but that sufficient meat and various kinds of meat were consumed by the majority of the population. Between 1850 and 1930, the annual per capita meat consumption in The Netherlands increased from 27 to 50 kg (Figure 5.1). Meat consumption decreased during the period 1933-1948, and it was not until the 1960s that the pre-war level was reached again. Beef consumption doubled between 1850 and 1900; from then on pork became the main constituent of the total meat consumption (Koolmees, 1997: 45-53, 90-92). Since the 1970s poultry meat (chicken and turkey), which is not included in Figure 5.1, became more popular. In 1990 the annual per capita consumption of chicken meat amounted to 11 kg and turkey meat to 2 kg. In addition, more meat was consumed in the form of meat snacks obtained from snack bars, by eating out and by the introduction of the 'barbecue cult'. In spite of criticism by health authorities, vegetarians and anti-meat lobbyists of the level of meat-eating, the consumption of meat (including poultry meat) in The Netherlands remained rather stable over the last 25 years at approximately 80 kg (Claus, 1991: 219).

Figure 5.1   Development of meat consumption (meat including bone, meat products and cans; poultry meat not included) and meat production (in million kg) in The Netherlands, 1850-1990

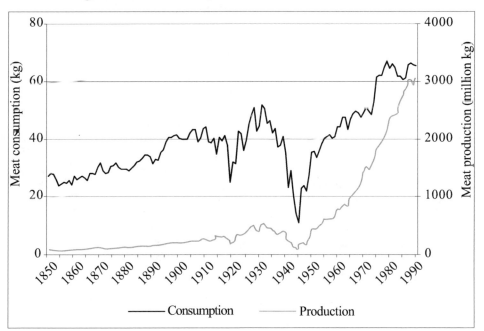

Source: Knibbe, 1993: 258-271; *Tweehonderd jaar statistiek*: 34-36.

III. Meat production

In Western Europe peasant farming predominated into the twentieth century. Extensive stock farming and cattle breeding was concentrated in the grassland regions of northwest Europe. At the beginning of the nineteenth century the use of fertilisers and the rationalisation of the rural economy made possible the establishment of agricultural surplus areas, which provided grain and animal feed for the more densely populated regions. After the economic crisis in the agricultural sector in the last decades of the nineteenth century, innovations in agriculture and hence in animal production took place. The most important issues were scaling-up, specialisation, cooperation and mechanisation in agriculture as well as in extensive and intensive animal husbandry. In the twentieth century new animal feeds, more productive cattle, pig and poultry breeds, pedigree cattle and artificial insemination were introduced. These innovations were supported by agricultural and veterinary research, education and information. Peasant farming was gradually replaced by large-scale capital-intensive farming (Koolmees, Fisher and Perren, 1999: 10, 12-14; Van Lieshout, 2001: 51-52).

Improvement of livestock for meat production was one of the main targets. The long period of domestication has resulted in a number of changes in the physical characteristics of production animals as we know them today. However, major developments in many breeds of cattle destined for meat production only occurred after fundamental changes took place in agriculture in the eighteenth century. Before that time, cattle had been used primarily for draught purposes or for milk production. From the eighteenth century onwards, several attempts were made to produce meat cattle which would fatten quickly when the skeletal growth was complete. This has resulted in a trend towards smaller and leaner animals which are slaughtered at a much earlier age. About 175 years ago the European pig with its long snout and less developed hindquarters gradually evolved when it was crossed with imported Chinese breeds. This type of cross breed was further improved for pork production, which has resulted in a longer animal such as the Landrace and Large White with a small head, light in the forequarter but heavy in the hindquarter. The same development occurred in the production of chicken and turkey meat; after the war domestic breeds were replaced by hybrids imported from the USA (Frankenhuis, 1989: 10-12; Williams, 1976: 44-48).

In the second half of the twentieth century, livestock numbers in The Netherlands increased dramatically, particularly pigs and poultry, the most important exponents of factory farming (Figure 5.2). Consequently, the production of meat increased accordingly (Figure 5.1). Already in the nineteenth century, meat production had grown faster than domestic demand, resulting in growing export figures. After World War II, this situation continued and in the 1980s about 60 per cent of the total production was exported. Animal production was strongly encouraged by the national government and from 1964 onwards also by the European Economic Community (EEC). In that year Directive 64/432 concerning the trade in live animals (cattle and pigs) became effective, facilitating cross-border trade between member states. Further specialisation, mechanisation, improvement of livestock, and a broader scientific basis characterised the development of animal production. Primary production at enterprise level shifted to a production column, which involved a broad and complex network of processing and supplying industries and services. The chain from producer to consumer, now referred to as 'from farm to fork' or 'from stable to table', became longer and more complex. Farming became 'agri-business' and animal production became 'factory farming' (Bieleman, 2001: 133-142, 227-233).

Figure 5.2   Development of livestock (in millions) in The Netherlands, 1945-1990

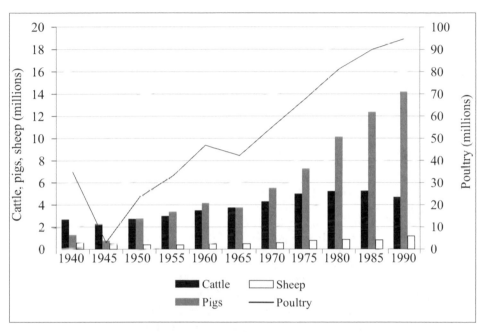

Source: *Vijfennegentig jaren statistiek*: 66.

In 1985 about 120,000 people were working in the meat production sector which had an annual production value of 13 billion guilders. Between 1980 and 1990 the total number of slaughtered bovines increased from 1.0 to 1.2 million. For pigs these figures were 13.2 and 19.9 respectively, while in the same period the number of pig farms decreased from 44,000 to 29,000. The nearly 20 million pigs produced in 1990 were slaughtered in 32 private slaughterhouses. Poultry meat production increased from 298,000 to 437,000 tons between 1980 and 1990. However, the strong growth in the meat production sector stabilised in the 1990s (Product Boards, 1994).

III.1. Contribution of veterinarians and agriculturalists

The expansion of the national and international livestock trade, however, also increased the threat of livestock diseases. The growth of livestock production then was regularly threatened and hampered by outbreaks of epizootic diseases such as cattle plague, swine fever and foot and mouth disease. This called for more effective control of animal health. Around 1900 veterinary state services and cattle disease control laws were introduced in several European countries. These measures enabled veterinarians to maintain and improve animal health and livestock production. Veterinary care slowly extended as a result of scientific developments. Experiments with inoculations in the eighteenth and nineteenth century were not very successful given the level of scientific knowledge at the time. A more effective method was the systematic killing of all infected and suspected animals, the so-called stamping out or Lancici system, which had been successfully applied since 1719. In countries where a stamping-out policy was executed rigorously, infectious livestock diseases could be controlled with minimum losses. This result was all the more impressive given that the method was applied before the advent of the germ theory. In meat-exporting countries, veterinary action was further stimulated by the threat that exclusionary policies posed to a fast-growing live animal export trade. Depending on the circumstances, the state veterinary services used a mix of exclusion and stamping out or inoculation (Fisher, 1998: 226-227; Fisher, 2003: 314-317; Koolmees, Fisher and Perren, 1999: 8-12).

By the end of the nineteenth century scientific progress enabled professional veterinarians to produce effective therapies, thereby improving the health of production animals. One of the consequences of the era of microbiological advance was the production of vaccines and penicillin and in 1904 a State Serum Institute was established in Rotterdam which provided veterinarians with vaccines. The pharmaceutical industry also started producing synthetic agents. From then on veterinarians had synthetic antipyretic, anti-infective, laxative, anaesthetic and emetic drugs at their disposal. Around 1900 effective therapies against for instance glanders, swine fever and sheep pox became available. The health status of production animals was further improved by the establishment of 11 Provincial Animal Health Services. After World War II, the amount of available animal drugs rapidly increased and the use of antibiotics enabled the widespread application of caesarean sections.

Apart from this contribution, veterinarians in cooperation with agriculturalists contributed to higher animal production through improved animal husbandry. The latter included new and improved animal feeds, selection, genetics, inspection and crossbreeding of livestock, and the introduction of pedigree cattle. Due to better breeding methods and higher quality fodder, the meat quality increased significantly. The current breeds of meat animals show the characteristics that are required in the meat industry: the ratio of meat to bone is high, the animals are well muscled, particularly in the most desirable part, they grow rapidly and produce good quality meat with a high ratio of lean to fat. Improved feed conversion efficiency, as well as improved growth rates by growth-promoting additives, hormones and antibiotics have been used to produce more meat in a shorter time. These developments, coupled with the practice of confining animals during the growing period with little or no room for movement, have resulted in considerable opposition since the 1970s. Despite this criticism, the fact remains that there have been considerable improvements in food conversion, rate of growth and animal health status over the last 150 years (Rixson, 2000: 215-224; Williams, 1976: 53-55).

IV. Meat trade

The modern international meat and livestock trade dates from the 1840s. Due to technological limitations this trade was mostly restricted to short distances, but the removal of customs barriers after 1842 provided a powerful incentive for potential suppliers. There were large numbers of surplus livestock in sparsely populated North and South America, New Zealand and Australia. Before 1870 the main product transported from these areas to Europe was salted meat; within Europe, fresh meat was carried by steamers mainly from the North Sea ports to London and the Smithfield Market in London became the centre for the world trade in meat and livestock. Indeed, the British market was crucial to meat exporters. At the beginning of the 1840s most international trade was conducted between countries in Europe. The first shipment of live cattle from the USA to Liverpool in 1868 opened the era of intercontinental trade. The transportation of meat was transformed with the development of chilling and freezing technology. Shipments of chilled or frozen meat from the USA and Argentina to Britain and France started in the 1870s and from the 1880s onwards, large quantities of frozen meat from other continents reached Europe. This technology allowed carcasses and prepared meat such as bacon to be traded internationally for the first time, and this proved much cheaper than transporting live animals. The rise in meat imports made possible by the development of refrigeration meant that meat prices followed cereal prices and started to decline after 1880. While most European countries erected tariff barriers and other import-restricting measures, Britain rejected protectionism and the British market remained open to meat imports of all kinds. With the development of refrigeration, the major flow of meat by 1914 was between the thinly populated grazing areas of the southern hemisphere and the densely populated industrial areas around the North Sea. The trade in live animals across national borders was more regional in character (Perren, 1978: 74-79, 118-119).

The Second World War caused a greater reduction in the international meat and livestock trade than had the first World War. In the period 1934-1938, world exports of beef averaged 661 million tons per annum, but were only 387 million tons in 1948-1952, and they did not recover their pre-war levels until 1960. After World War II the UK remained the largest market for meat, followed by continental Europe. In the late 1950s the USA became a major importer of beef, mainly from Australia and New Zealand; by the 1980s the USA was the second largest market for beef, accounting for about a third of beef imports, which was slightly less than the nine EU countries. The world trade in beef doubled from the early 1960s to the late 1970s. After World War II, the trade in all meats was governed to a much greater extent than before by government policies in the importing countries, which were designed to protect domestic producers. The effect of these policies in the EU has been to increase the extent of its internal self-sufficiency in meat while by no means eliminating its dependence on overseas imports, or reducing the trade between EU countries, which has in fact increased. As a result of this agricultural policy the EU had gained a position of leadership among beef exporters by 1985. In the

1980s meat demand and supply became distributed among a larger number of countries. With the growing efficiency in poultry and pork production from the 1960s onwards, this production spread to almost all parts of the world and grew at a higher rate than beef and mutton. The demand for meat and meat products in developed countries in the Middle and Far East and in developing countries has increased significantly, while meat consumption in the developed countries elsewhere has remained about the same level. Within Europe consumer preferences for meats vary considerably from one region to another. The Netherlands for instance exports bacon to Britain, fresh pork to Germany and veal to Italy. Today, the per capita pork and beef consumption averages in Europe are rather high and they appear to be constant. A significant growth has been observed in the consumption of poultry and lamb (Claus, 1991: 218-222; Koolmees, Fisher and Perren, 1999: 14; Simpson and Farris, 1982: 246).

V. Meat supply

Before the emergence of urban societies and their cattle and meat markets, consumers raised their own livestock and provided themselves with meat. In the countryside, home slaughtering remained common until well into the twentieth century when household refrigeration was introduced. In most cities, animals were slaughtered by butchers in small, privately owned butcheries. Beginning in the late Middle Ages, the slaughtering, selling and inspection of meat was regulated by butcher guilds and the local authorities. After the French Revolution, the guilds were abolished, and local meat inspection and meat trade regulations were disregarded. From around 1870 onwards, population growth and urbanisation went hand in hand with an expanded network of these small butcheries where slaughtering often took place under poor hygiene conditions. In 1861 Amsterdam for instance, had 109 butcheries for cattle, 111 for pigs, 26 for young calves, 1 for horses, 29 traders in meat products, 409 shops for meat, 38 sellers of lard, and 17 for animal offal. The city of The Hague had 50 private butcheries in 1875. The city of Utrecht, with a population of about 100,000 in 1890, had 114 registered private butcheries, 378 shops where meat was sold, and a market for cheap meat and sausages. In addition, meat from knackers' yards was marketed or processed into pies and sausages and sold to labourers and the poor. Regrettably, the meat often came from animals infected with anthrax, tuberculosis, trichinosis and tapeworms, as well as from animals that had died. A mere four inspectors were appointed to carry out meat inspection for the city of Utrecht.

It is clear that, under these circumstances, consumers were left virtually unprotected from fraud and the adulteration of meat and meat products. The calamities that followed were almost inevitable. This situation was complicated by the increasing size, reach (globalisation), and complexity of the supply chain between producers and consumers. The poor hygienic conditions which characterised meat and sausage processing were due to a lack of awareness about hygiene. The increasing numbers of butcheries represented a nuisance for the citizens. The transport and slaughtering of animals, as well as the storage and transport of offal within the city walls harmed the urban environment. Numerous outbreaks of trichinosis and meat poisonings occurred which infected hundreds and killed dozens of people. Consequently, local and national authorities were increasingly confronted with complaints about the filth that butcheries caused in the urban centres and the poor quality of the meat offered. Mass outbreaks of meat-borne diseases alarmed the authorities and clearly demonstrated the need for meat hygiene measures. Politicians and local authorities were more or less forced to develop a public health policy related to the meat supply (Koolmees, 1997: 109-112).

In the course of the nineteenth century, increased meat production and consumption necessitated a large-scale supply of sound meat. However, radical reforms in organisation and hygiene were needed to accomplish this goal. Faced with the economic concerns of meat producers and the public health and environmental concerns of consumers, the authorities found a way out by instituting mandatory centralised slaughtering in large public or private slaughterhouses located outside the towns, under mandatory professional veterinary supervision.

VI. The slaughtering scene: from craft to industry

For centuries, animal production, slaughtering and meat processing were based on tradition and craftsmanship. In the nineteenth century scientific progress provided the basis for dramatic changes in the meat supply. Home slaughtering was quite common, not only in the countryside but also in the towns where pigs were kept in small sheds and fed with household waste. Due to population growth and urbanisation home slaughtering disappeared from the towns around 1900. In the countryside, however, this situation prevailed until the 1960s when the refrigerator was introduced. In the course of the nineteenth and beginning of the twentieth century slaughtering disappeared from daily street life behind the walls of public abattoirs and private slaughterhouses which were established to supply the domestic and foreign markets with meat.

VI.1. Public slaughterhouses

From 1795 onwards, attempts were made in a few Dutch towns to establish public slaughterhouses. These attempts, however, failed. Until the 1880s, only three municipal slaughterhouses were established in The Netherlands (Maastricht, Venlo and Bois-le-Duc). Generally, there was strong opposition to abattoirs from butchers in all European countries. Having played an essential role in the municipal meat supply for centuries, the butchers feared that the municipal slaughterhouses would limit their profession considerably. For one thing, they would be forced to slaughter their cattle at the public abattoir under supervision. The competency of municipal authorities to establish a public slaughterhouse and to prohibit private slaughtering, the so-called 'slaughter warrant', became a politically contentious issue in the nineteenth century. In a number of countries, this matter was settled by the passing of a national nuisance act. In The Netherlands, for example, this warrant was implemented by the Nuisance Acts of 1875 and 1901. Most local authorities hesitated in establishing municipal slaughterhouses because of the large financial expenditure required, the possible increases in meat prices, and the question of profitability. Competition for local financing also came from other large and expensive projects in the field of sanitary reforms. Often several decades elapsed between the first proposal for establishing a municipal slaughterhouse and the actual opening of the institution (Table 5.1). In the second half of the nineteenth century, lengthy debates between supporters and opponents of municipal abattoirs took place in the municipal councils. Two factions could be distinguished: one side included hygienists with their demand for an adequate meat inspection, as well as animal protectionists and citizens with nuisance complaints about the private butcheries; the other side consisted of a coalition of butchers and meat traders whose independence was threatened by the establishment of municipal abattoirs. At first, economic interests and the objections of the butchers outweighed the arguments concerning public health and pollution put forward by the hygienists. However, by the turn of the century, the growth of socialism tipped the balance. Ultimately, though, the decision to establish a public slaughterhouse mostly depended on the financial position of the municipality and hence, on general economic fluctuations (Koolmees, 1997: 145-152).

Table 5.1   Municipal slaughterhouses established in 15 Dutch towns in the period 1883-1910

Town	Proposal*	Opening	Inhabitants	Cost (Gilders)	Animals slaughtered #
Rotterdam	1866	1883	162,000	800,000	38,000
Amsterdam	1856	1887	390,000	2,200,000	91,000
Roermond	1853	1899	13,000	110,000	12,000
Elburg	1898	1900	2,700	2,000	?
Groningen	1872	1900	67,000	430,000	36,000
Nijmegen	1875	1900	46,000	260,000	15,000
Maastricht	1876	1901	34,000	305,000	26,000
Utrecht	1866	1901	102,000	570,000	26,000
Leiden	1895	1903	55,000	328,000	16,000
Dordrecht	1894	1906	44,000	175,000	9,000
Haarlem	1878	1907	70,000	416,000	19,000
Alkmaar	1891	1908	21,000	250,000	10,000
Sittard	1892	1908	7,000	47,000	5,000
Arnhem	1902	1910	65,000	640,000	22,000
Bois-le-Duc	1898	1910	35,000	183,000	11,000

*First proposal in city council
Number of animals slaughtered in the first year after opening

Source: Koolmees, 1997: pp. 150 and 170.

The enactment of the Dutch Meat Inspection Law in 1919 indirectly encouraged the building of public slaughterhouses, since this act imposed hygiene requirements concerning the furnishing and equipment of private abattoirs, butchers' shops etc., and outlined the obligation of local authorities to establish a meat inspection service. Many existing privately-owned butcheries could not meet these new requirements and were forced to close, prompting the authorities of many cities to build public slaughterhouses in order to institute an adequate meat inspection service. Between 1883 and 1940, a network of 86 municipal slaughterhouses covered The Netherlands (Figure 5.3). Apart from the positive influence of the Meat Inspection Act and the profitability of the existing abattoirs, it was mainly the favourable financial position of the cities in the period 1922-1929 that contributed to this rapid spread of municipal slaughterhouses. In 1927, no fewer than 15 abattoirs became operative.

Figure 5.3   Establishment of municipal slaughterhouses in The Netherlands, 1883-1940

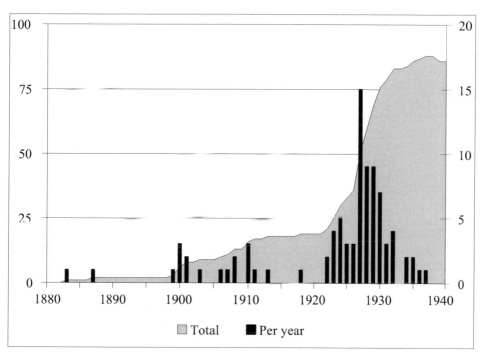

Source: Koolmees, 1997: 134-135.

The profitability of public abattoirs came under pressure during the crisis of the 1930s and the war period. After World War II, activity in the public abattoirs again increased and between 1950 and 1964 the number of slaughtered animals doubled. In 1965 public abattoirs produced almost half of the total meat production in The Netherlands, which amounted to 800,000 tons. However, the enactment of EU Directive 64/433, designed to facilitate international trade in fresh meat, proved to be the beginning of the end of the public abattoir system in The Netherlands. As of 1 July 1965, slaughterhouses have had to comply with a number of standards regarding slaughterhouse design, equipment, cooling facilities and hygienic slaughtering. Slaughterhouses unable to comply with these strict hygiene standards would lose their permit to export meat to other EU member states. Faced with increased competition from private slaughterhouses, most municipalities, reluctant to invest community money in the modernisation of public abattoirs, decided to close them. Moreover, the possibilities of balancing the profitability of public abattoirs by increasing inspection costs became more and more difficult. In the 1970s and 1980s

127

most public abattoirs closed down, or were taken over by private slaughterhouses. In 1974 the ratio between public and private slaughterhouses was 79/77; in 1980 this was 41/121 (Koolmees, 1991: 105-112; Langenhuyzen, 1989: 149-150).

VI.2. Private slaughterhouses

Besides public abattoirs, larger slaughterhouses specialising in the production of meat for the foreign market were established. The rise of these slaughterhouses for meat exports can be explained by the increase in the demand for meat from the surrounding countries, particularly Britain. They were built in production areas or close to railroad stations and seaports (Harlingen, Vlissingen). Most of these companies started as small enterprises where some 200 animals were slaughtered on a weekly base. Often it was Jewish butchers and cattle merchants who started these slaughterhouses, such as Hartog and Zwanenberg who founded Lion-Löwenberg and Wijnbergen and Levano in Oss (the meat city of The Netherlands) in 1883 and 1887 respectively. Later some of these enterprises became multinational companies such as Unilever (Langenhuyzen, 1988: 25-40; Van Puijenbroek, 1985: 205-209).

According to a report on meat inspection from 1894 there were 81 municipalities in which slaughterhouses for export and sausage factories were located. In 1901 there were 255 registered slaughterhouses for meat exports active in 132 Dutch municipalities, mainly in the provinces of Gelderland, Overijssel and Brabant. On the analogy of the dairy industry, slaughterhouses for meat exports were also established on a cooperative basis, such as the Noord-Brabantse Christelijke Boerenbond (NCB) in 1903. Meat export proved to be very profitable, particularly during the first years of World War I. By that time no fewer than 20 large, modernised slaughterhouses were active. Meat export increased from on average 2.4 to 7.7 million kg per year in the periods 1870-1879 and 1880-1889 respectively. In the period 1890-1894 this figure was 17.5 million kg, while exports further increased from on average 57 million kg in the period 1897-1906 to 73 and 100 million kg meat in 1913 and 1915 respectively. The position of the bigger companies was threatened by the activities of the many small meat-exporting companies whose meat was often of such poor quality that the importing country refused it. This was the reason why in 1902 the Dutch government started to inspect meat destined for export. Initially, this inspection was optional, but due to the fact that some small companies continued to produce poor quality meat, the inspection became mandatory in 1909. After that the smaller companies disappeared or were taken over by larger ones. Some meat exporters rented the slaughter facilities in public abattoirs to slaughter their animals destined for meat export. In the abattoirs of Rotterdam and Amsterdam these activities resulted in extra income along with that from the activities of local butchers (Brinksma, 1994: 9-14; Koolmees, 1991, 50-52; Minderhoud, 1943: 420; Reens, 1922: 81, 71-76, 101-111, 192-193, 213-216).

VI.3. Slaughtering methods

Another – still relevant – aspect of the slaughtering scene has to be dealt with here, namely the methods of slaughtering animals. In the last decades of the nineteenth century, animal protectionists drew the public's attention to the cruelty to slaughter animals inflicted by slaughterers and butchers. Such practices were found in small private butcheries, premises where home slaughtering was carried out, as well as in municipal abattoirs and slaughterhouses for meat exports where large-scale slaughtering had been introduced. Attempts to eliminate cruelty in slaughterhouses encompassed various matters. Apart from an adequate form of transport, there was the need to see that animals were sufficiently rested and calm before slaughter, as this improved the quality of the meat. The actual method of killing was not covered by specific legislation, and in the nineteenth century practices varied. Usually only cattle were stunned before being bled while smaller animals were not. The Jewish and Islamic ritual methods required both large and small animals to be bled without pre-slaughter stunning. Butchers preferred the traditional poleaxe for stunning, claiming that this produced better-quality meat. Between 1900 and 1920 the situation improved markedly with the technical development and introduction of stunning equipment and restrainers. In the newly established public slaughterhouses, shooting masks, captive-bolt guns and spring pistols replaced mallets, throat cutting, poleaxing and neck stabbing. This was due to the propaganda of the animal protection societies and the efforts of the veterinarians appointed as directors of public abattoirs. From 1900 onwards, a considerable amount of veterinary research was devoted to slaughtering methods, especially to electrical stunning. Legislation controlling the slaughtering methods was instituted in municipal by-laws and on a national level in meat inspection and animal protection acts in most Western European countries in the first half of the twentieth century (Koolmees, 1994: 19-20).

VI.4. Technological innovations

The advent of the Industrial Revolution brought about rapid changes in the way animals were slaughtered and meat was processed. As mentioned above, the introduction of the mechanised refrigeration of meat, railroads and steam shipping enabled the development of an intercontinental meat trade. Innovations in cooling and freezing techniques took place in the second half of the nineteenth century. Cooling and freezing equipment and cold-storage warehouses had been used in industry for a long time. Refrigerators and freezers were first introduced in retail stores and chain stores at the beginning of the twentieth century. Consumers however regularly had to buy their meat in smaller quantities at butcher's shops or have it delivered at home. Only after World War II did freezers and refrigerators became common household articles, enabling consumers to buy larger quantities of meat and meat products at retail shops and store them at home for longer periods.

In order to be able to satisfy the growing demand for meat, the small-scale slaughtering, dressing and curing methods, which had been used for centuries by craftsmen, had to be replaced by more efficient ways. Slaughtering on domestic premises or in small private slaughterhouses was replaced by large-scale slaughtering. The emergence of centralised slaughtering in large privately owned slaughterhouses specialising in meat exports changed the meat supply. These large slaughterhouses, which were built first in the USA and later also in Europe, proved more cost-efficient than small slaughter premises and public abattoirs and were quicker to incorporate new technology. For instance the first slaughter-line for cattle in The Netherlands was opened in 1935 in the slaughterhouse of Hartog's Company N.V. (Nijssen, 1938: 39-40).

Cincinnati became the early centre of the meat industry and by the middle of the nineteenth century was called 'Porkopolis'. Chicago took the lead in the 1860s. Stimulated by the Civil War, railways developed and the cultivation of maize replaced wheat, facilitating the fattening of pigs. With the extension of the 'corn belt' the large-scale 'corn-hog cycle' developed and, hence, Chicago became the city with the world's largest meatpacking industry. There, the meat industry started a vertical integration by taking over farming and transport and succeeded in raising the value of offal on a large scale. It was also here that the two greatest technological innovations took place which revolutionised the meat industry, namely mass production by division of labour and the conveyor belt principle. All stages in the slaughter process itself were also mechanised. In the Chicago Union Stockyards, the disassembly line including mechanised scalding, dehairing and scraping of pigs, was introduced in the 1860s. The slaughter-lines made possible the slaughtering of about 500 pigs per hour. In 1883 more than 5 million pigs, 2 million cattle and 0.75 million sheep were slaughtered in the Chicago Union Stockyards (Giedion, 1982: 113-115, 241-249; Wade, 1987: 61-66, 101-103; Williams, 1976: 51-52).

The same basic steps in a pig slaughter-line are still used today. Division of labour and slaughter-lines were also introduced in European slaughterhouses around 1900. Depending on the size of the slaughterhouse, here too slaughter rates increased from tens of pigs per hour in the non-mechanised centres to more than 400 pigs per hour in mechanised slaughter houses. In the 1960s and 1970s devices were developed to pull off the hides from bovines, to cut off claws and hoofs, and to split carcasses in two with a huge sawing-machine. The refrigeration and freezing capacities of slaughterhouses were further improved, enabling trucks for cooled transport to move the meat across Europe. Several industrial companies specialised in providing newly built slaughterhouses with equipment, machinery, refrigeration installations, etc. At first it was mainly French (Lebrun), German (Beck and Hessel) and Belgian (Ateliers Wanderwerpen) companies that provided Dutch slaughterhouses with such equipment, but from the 1910s onwards Dutch companies such as Noord-Nederlandsche Machinefabriek in Winschoten, Stork in Hengelo took over (Nijssen, 1938: passim).

One of the changes in the food pattern in the nineteenth century was the use of canned meat in addition to the consumption of fresh meat, bacon and sausages. Canned meat was one of the innovations in meat technology that developed in response to the demands of the military and of overseas shipping for food that could easily be transported and preserved for longer periods. In the early nineteenth century rather heavy 'tinplate canisters', a name later shortened to 'tin cans' or 'tins' in Britain and 'cans' in America, were used as containers for cooked meat. Some decades later the materials used became thinner and lighter. The sterilisation methods used were mainly arrived at by trial and error; losses due to spoilage were often heavy. It took more than 70 years before the theoretical background of the sterilisation method of canned food was discovered by Pasteur and others

(Goldblith, 1971: 46-50; Howard, 1949: 1-6). The first companies in The Netherlands where meat was canned - such as Tieleman and Dros in Leiden, E. Noack's Koninklijke Fijne Vleeschwaren en Conservenfabrieken N.V. in Amersfoort and Ter Marsch and Co. Fabriek van verduurzaamde vleeschwaren in Rotterdam - date from the 1870s. The consumption of canned meat increased in the twentieth century. Thomassen and Drijver N.V., the biggest supplier of cans in The Netherlands, increased its annual production from 0.2 to 277 million cans between 1920 and 1950 (De Knecht-van Eekelen and Van Otterloo, 1997: 52; Perren, 1978: 70-71).

In addition to fresh meat, slaughter animals contained other edible and valuable parts that provided marketing opportunities, such as various meat products and sausages from the remaining fat and organ meat (liver, spleen, kidneys, brains, etc) and the production of bacon, canned meat, smoked ham, sausage casings made from intestines, lard, but also soap, margarine, blood and bone meal for fertilisers and organs and offal for animal feed. For instance, sausage is one of the oldest processed foods known to man, its preparation and consumption having originated in antiquity. Over the centuries a large variety of sausages was developed and sausage processing became an important part of the meat industry. In the USA, a number of meat choppers were patented in the 1860s. In 1868 a patent was issued for the world's first steam-powered sausage meat chopper which could cut 50 kg of meat in 10 minutes. The first sausage stuffers were hand-operated machines, which were replaced by steam-pressure stuffers in the 1890s. It is estimated that today between 700 and 900 types of sausages are processed. Besides sausages, cured hams are also old traditional meat products.

Traditionally, people had looked for ways to preserve meat and, based on empirical experience, processes like cooling, smoking and drying of meat had been applied through-out the ages. Between 1830 and 1900 a number of antiseptic compounds to preserve meat were introduced by chemists, then patented and commercially sold. Examples are boric acid, sulphuric acid, hydrogen peroxide, naphtha and thymol. Around 1900 nitrate was discovered to play a positive role in preserving the red colour of meat. However, the addition of nitrate met with opposition because of its supposed effect on human health. Likewise, the irradiation of meat products was applied successfully in the US army from the late 1940s as an effective method of sterilisation. However, it has not yet been ap-plied to the civic meat supply because of consumer opposition (Koolmees, 1994: 21-22; Strahlmann, 1974: 105-113).

Today, processes like the fermentation of salami or pasteurisation or sterilisation (micro-wave, infrared, irradiation) of heated meat products are no longer based on empiricism and trial and error but on scientific microbiological and technological data and are monitored by computers. The same holds for new processes like drying, extrusion, emulsification and structuring by phase separation. Besides innovations in processing, much attention was also paid to product innovation from the 1970s onwards. This resulted in the use of new raw materials such as mechanically recovered meat, binding additives such as pork rind powder and collagen fibres, and artificial casing for filling sausages. The shelf life of many meat products could be extended by the development of new materials for (vacuum) packaging. Moreover, ready-to-serve meals, including meat or meat products, have established a strong position in the food market. These products require a new approach in quality management, based on interdisciplinary research (Van Boekel, 1998: 92-108; Goldblith, 1971: 64-68).

During the last 150 years, meat research has taken on a more systematic character and changed from an art into a science. This was prompted by the search for safe and more effective preservation methods, by consumers' concerns about the quality of meat and meat products in the wake of the increasing number of adulterations in the meat trade, and by the general advance of modern science. In several countries meat science and technology developed as a new discipline while international scientific cooperation intensified in order to find answers to issues of general concern to consumers and producers. For instance, international conferences on meat science and technology (ICoMST) have been organised annually since 1955. Meat science as such does not have a basic theory; it is an applied science that draws on many different areas and methodologies. The traditional areas are growth of meat animals, carcass and meat quality, muscle biochemistry, slaughtering, raw meat technology, cooling and freezing, meat products and their technology, by-products, food additives and non-meat ingredients, the microbiology and hygiene of meat and meat products, starter cultures for fermented meat products and analytical methods. Recently more attention has been paid to sustainable animal production and animal welfare during transport and slaughtering.

VII.   Meat hygiene

The rapid progress in the meat trade and in meat technology with its negative implications for the transmission of meat-borne diseases in man led to calls for progress in meat hygiene standards. In the first half of the nineteenth century, there was essentially no scientifically based meat inspection in Western Europe and hardly any veterinarians were involved in meat inspection. The untrained meat inspectors who were appointed in the larger towns relied almost exclusively on empirical knowledge. From 1850 onwards, veterinarians involved in health boards offered advice on all matters concerning veterinary public health based on their specific knowledge of zoonoses and their ability and training in recognising diseased animals. They focussed on a scientifically and legally based form of meat inspection, research on meat poisoning, humane slaughtering methods, and regulations for the collection and destruction of waste from slaughterhouses and butcheries. Initially, it was mainly physicians who wrote about the subject of meat hygiene. Nevertheless, in various countries, a number of veterinarians started to publish on the subject, claiming this new field of professional activity for themselves. These authors stressed the importance of veterinary medicine for society as a whole, not only in the maintenance of a prosperous livestock industry, but also because of its value in safeguarding the healthy quality of foods of animal origin. The efforts of veterinarians to promote veterinary public health as a new field were strongly supported by The Netherlands Veterinary Association. This association recognised the potential for job opportunities in meat inspection services and public slaughterhouses and the possibility of broadening the legal basis of the veterinary profession. These claims, however, required a scientific basis for meat inspection.

The standing of meat inspection as a veterinary discipline rose when the scientific backgrounds of several diseases related to meat consumption were discovered one after the other. First, there were the findings in the field of parasitology. The cycle of tapeworms was discovered in the 1860s and, with an incision meat inspectors were able to apply a preventative test. The occurrence of parasites declined, especially where trichinella tests were instituted. From 1880 onwards, meat hygiene research acquired a more scientific character and bacteriological meat research was carried out in slaughterhouse laboratories. Between 1890 and 1940, a number of meat-borne pathogens were isolated (Koolmees, 1997: 70-75, 99-105, 122-129).

Around 1900, traditional empirical-based meat inspection was transformed into an applied veterinary science. Research findings were published in handbooks and professional literature, and theoretical and practical meat inspection became part of the veterinary curriculum. National and international associations for professional veterinary food inspectors were founded. Meat hygiene became an important issue for international conferences on veterinary medicine and public health. All these factors contributed to the development of a professional meat inspection corps. Veterinarians had succeeded in providing an adequate answer to mounting concerns about the more strictly public health aspects of the meat trade. As a result of the contribution of veterinarians to the discussions in municipal and national health boards, authorities reinforced or updated their regulations on meat inspection. A large number of towns made plans to establish public abattoirs. Often, it was veterinarians who initiated these plans and who drafted reports on the feasibility of these institutions. Besides their local activities, many veterinarians were also involved in formulating national meat inspection legislation. Between 1901 and 1922 a series of laws concerning meat inspection was adopted and enforced (Koolmees, 1997: 173-202).

From the 1980s onwards governments have reviewed their approaches to meat inspection regulation. Various committees have looked for new forms of meat safety control that are more efficacious and impose a lesser burden on the meat industry. In general, end-product testing was recognised to be an inefficient form of meat safety control. The scientific rationale for food safety regulation, including meat and meat products, was incorporated into the framework of risk analysis. The latter represents a structured approach whereby risks to human health are assessed and the best means for their control identified. The principle of risk-assessment has become enshrined in the operating procedures of the international standards organisations. Requirements have been put in place for process control based on the principles of hazard analysis critical control point (HACCP). During the last two decades of the twentieth century HACCP was widely recognised in the meat industry as an effective approach for establishing good manufacturing practices in the production of safe meat and meat products (Berends and Van Knapen, 1999: 129-130; Henson and Caswell, 1999: 591-596).

VIII. Consumers' concerns

Since the 1970s, once consumers were assured of a sufficient quantity of meat, the further growth of meat production and consumption was associated with deep concern over the quality of the products and their impact on human health. The meat industry was faced with challenges from concerned consumers, animal right activists, and environmental lobbyists. Critics argued that the limits to factory farming and mass production had been reached; the perceived conflict between the needs of humans, animals and the environment became a cause of widespread concern. The meat industry acquired a poor public image due to the adverse effects of factory farming on the environment, controversial questions about the ethics in meat production, and the regular fraud and food poisoning scandals that occurred. In these circumstances, the authorities have been forced to take public action, but the knowledge of risks is often limited. As long as consensus among scientists and objective information from research are lacking, policy-making is difficult. As a result, public concern regarding meat safety flourishes; a concern that is frequently exploited by the media and anti-meat lobbyists. Modern consumers expect meat and meat products that comply with up-to-date views on health, nutrition, environment and animal welfare. In addition to economic factors, the modern meat industry also has to consider animal health and welfare, and consumer and environmental protection. However, the structure of the meat industry, which is essentially a vertical marketing system, is one of the main reasons for its rather slow reaction to changing consumer attitudes towards meat (Koolmees, 2000: 64-65; Van Otterloo, 1995: 259-262; Spitters, Tazelaar and Gerats, 1991: 275-277).

IX. Conclusions

Traditionally, livestock and meat production and exports have played an important role in the Dutch economy. From the 1840s onwards, the growing demand for meat from urban centres in the surrounding countries strongly stimulated meat production. Foreign and domestic consumers not only required sufficient meat in the diet but also safe meat. This called for a rationalisation of the production process and for sanitary and technological innovations in the meat supply chain from producer to consumer. Around 1900, The Netherlands bore witness to a gradual change from decentralised slaughtering and marginal inspection in numerous small butcheries to the centralisation of these activities in public slaughterhouses and private large meat-exporting companies. Between 1900 and 1940, the establishment of municipal and private slaughterhouses and the inspection of meat destined for foreign and domestic markets together improved the supply of sound meat considerably. The large-scale meat industry with its long and complex chain became a sector in which the differentiation of enterprises and products was to a large extent carried through. Until the 1990s, better transportation, more efficient refrigeration systems, greater mechanisation of slaughterhouses and more hygienic methods of slaughter and meat handling all improved and contributed to raising the meat industry to a higher level.

According as the meat industry underwent modernisation, state interference increased considerably. As one of the main players in the development of the meat industry, local, national and international (EU) authorities tried to harmonise the often-conflicting economic interests of meat producers with caring for the environment and the health of consumers. On the one hand production and exports were boosted, while on the other hand centralised slaughtering under mandatory professional supervision was instituted. Stressing the importance of animal and meat production for the national economy, producers claimed government support and protection. They responded to a changing market with more efficient production processes and product innovations. High-tech animal and meat production, as well as integrated quality control from stable to table are based on science. Modern science, therefore, represents another significant player in the development of the meat industry. As a result of the broadening of the meat supply and economic growth in the early 1960s, consumers could afford a sufficient amount of meat in their diet. From the 1970 onwards, critical consumers became important players, when they – supported by the media - forced government authorities and meat producers to pay more attention to the negative aspects associated with the meat industry. Due to consumers' concerns, meat safety rather than a secure meat supply became the major issue.

Bibliography

Berends, B.R. and Knapen, F. van (1999) 'An outline of a risk assessment-based system of meat safety assurance and its future prospects', *Veterinary Quarterly*, 21, 4, pp. 128-134.

Bieleman, J. (2001) 'Landbouw', Schot, J.W. et al. (eds.) *Techniek in Nederland in de twintigste eeuw*, Vol. 3, Zutphen, pp. 131-153; 227-233.

Boekel, M.A.J.S. van (1998) 'Development in technologies for food production', Jongen, W.M.F. and Meulenberg, M.T.G. (eds.) *Innovation of food production systems. Product quality and consumer acceptance*, Wageningen, pp. 87-116.

Brinksma, J.D. (1994) *Slachterijen, vleeswarenindustrie en visverwerkingsinrichtingen*, Zeist.

Claus, R. (1991) 'Meat and consumer preferences in Europe; demography, marketing issues', Smulders, F.J.M. (ed.) *The European meat industry in the 1990's. Advanced technologies, product quality and consumer acceptability*, Nijmegen, pp. 217-240.

Eekelen, M. van (1962) 'De voedingsmiddelen in Nederland thans en in de toekomst', *Tijdschrift voor Sociale Geneeskunde*, 40, pp. 609-614.

Fisher, J.R. (1998) 'Cattle plagues past and present: the mystery of mad cow disease', *Journal of Contemporary History*, 33, 2, pp. 215-228.

Fisher, J.R. (2003) 'To kill or not to kill: the eradication of contagious bovine pleuropneumonia in Western Europe', *Medical History*, 47, pp. 314-331.

Frankenhuis, M.Th. (1989) *Over het ontstaan van de bedrijfspluimveehouderij*, 2nd ed., Doorn.

Giedion, S. (1982) *Die Herrschaft der Mechanisierung. Ein Beitrag zur anonymen Geschichte*, Frankfurt am Main, pp. 111-120, 238-277.

Goldblith, S.A. (1971/72) 'The science and technology of thermal processing', *Food Technology*, 25, 12, pp. 44-50; 26, 1, pp. 64-69.

Hartog, A.P. den (2003) 'Twee eeuwen voedsel, voeding en gezondheid', *Spiegel Historiael*, 38, 3-4, pp. 108-111.

Henson, S. and Caswell, J. (1999) 'Food safety regulation: an overview of contemporary issues', *Food Policy*, 24, pp. 589-603.

Howard, A.J. (1949) *Canning technology*, London.

Knecht-van Eekelen, A. de and Otterloo, A.H. van (1997) 'De groenteconservenindustrie en de verandering van het consumptiepatroon in Nederland', *Erfgoed van Industrie en Techniek*, 6, 2, pp. 50-58.

Knibbe, M. (1993) *Agriculture in the Netherlands 1851-1950. Production and institutional change*, Amsterdam.

Koolmees, P.A. (1991) *Vleeskeuring en openbare slachthuizen in Nederland 1875-1985*, Utrecht.

Koolmees, P.A. (1994) 'Meat in the past: a bird's eye view on meat consumption, production and research in the Western world from antiquity to 1945', Sybesma, W. , Koolmees, P.A. and Heij, D.G. van der (eds.) *Meat past and present: research, production, consumption*, Zeist, pp. 5-32.

Koolmees, P.A. (1997) *Symbolen van openbare hygiëne. Gemeentelijke slachthuizen in Nederland 1795-1940*, Rotterdam.

Koolmees, P.A., Fisher, J.R. and Perren, R. (1999) 'The traditional responsibility of veterinarians in meat production and meat inspection', Smulders, F.J.M., (ed.) *Veterinary aspects of meat production, processing and inspection. An update of recent developments in Europe*, Utrecht, pp. 7-31.

Koolmees, P.A. (2000) 'Veterinary inspection and food hygiene in the twentieth century', Smith, D.F. and Phillips, J. (eds.) *Food, science, policy and regulation in the twentieth century. International and comparative perspectives*, London and New York, pp. 53-68.

Langenhuyzen, T. (1988) *Van concurrentie naar eenheid. Aspecten van de geschiedenis van Hartog's en Zwanenberg's Fabrieken en de Unilever Vleesgroep Nederland te Oss*, Oss.

Langenhuyzen, T. (1989) *...van den eerlijken handel... Een eeuw coöperatieve veehandel en vleesverwerking: ENCEBE Boxtel in historisch perspectief*, Boxtel.

Lieshout, J.A.H. van (2001) 'Kunstmatige inseminatie', *Tijdschrift voor Diergeneeskunde*, 126, 2, pp. 51-54.

Minderhoud, G. (1943) *De landbouwindustrie*, Wageningen.

Nitti, F.S (1896) 'The food and labour-power of nations', *The Economic Journal*, 6, pp. 30-63.

Nijssen, H.H. (1938) *Het slachthuis*, Eindhoven.

Otterloo, A.H. van (1995) 'The development of public distrust of modern food technology in The Netherlands. Professionals, laymen and the consumers' union', Hartog, A.P. den (ed.) *Food technology, science and marketing. European diet in the twentieth century*, East Lothian, Scotland, pp. 253-267.

Otterloo, A.H. van (ed.) (2001) 'Voeding', Schot J.W. et al. (eds.) *Techniek in Nederland in de twintigste eeuw*, Vol. 3, Zutphen, pp. 252-260.

Perren, R. (1978) *The meat trade in Britain 1840-1914*, London.

Product Boards for Livestock, Meat and Eggs (1994), *Annual report*, Rijswijk.

Puijenbroek, F.J.M. van (1985) *Beginnen in Eindhoven. Allochtoon ondernemersinitiatief in de negentiende eeuw*, Eindhoven.

Reens, A. (1922) *De vleeschexport van Nederland*, Rotterdam.

Rixson, D. (2000) *The history of meat trading*, Nottingham.

Simpson, J.R. and Farris, D.E. (1982) *The world's beef business*, Ames, Iowa.

Spitters, P.J.A., Tazelaar, R.J. and Gerats, G.E. (1991) 'The image of meat and the consumer's perception of the industry', Smulders, F.J.M., (ed.) *The European meat industry in the 1990's. Advanced technologies, product quality and consumer acceptability*, Nijmegen, pp. 271-286.

Strahlmann, B. (1974) 'Entdeckungsgeschichte antimikrobieller Konservierungsstoffen für Lebensmittel', *Mitteilungen aus dem Gebiete der Lebensmitteluntersuchung und Hygiene*, 65, pp. 96-130.

Teuteberg, H.J. (1991) 'Food patterns in the European past', *Annals of Nutrition and Metabolism*, 35, pp. 181-190.

Tweehonderd jaar statistiek in tijdreeksen, 1800-1999, Centraal Bureau voor de Statistiek, Voorburg and Heerlen, 2001.

Vijfennegentig jaren statistiek in tijdreeksen, Centraal Bureau voor de Statistiek, 's-Gravenhage, 2004.

Wade, L.C. (1987) *Chicago's pride. The stockyards, packingtown, and environs in the nineteenth century*, Urbana and Chicago.

Williams, E.F. (1976) 'The development of the meat industry', Oddy D. and Miller D. (eds.) *The making of the modern British diet*, London, pp. 44-57.

137

6 Berlin's huge stomach and the establishment of the food industry, 1850-1925

Peter LUMMEL, Open Air Museum, Dahlem Domain, Berlin

In the nineteenth century the total population of Berlin soared from 170,000 to 1.9 million and for a while its growth exceeded that of any other city in the world. The challenge of providing for the basic support of such a mass was new. Only a small part of the city's population worked as farmers. Production in the surroundings of Berlin and production from home allotments did not in the least suffice to meet the growing demand for food. The resulting problems were gigantic. Where was the food to come from that was necessary every single day, summer and winter, to satisfy the nearly insatiable 'tomach of Berlin'? How would it be possible to avoid crises in the food supply and to make food available to all groups of the population in time, in sufficient quantity and quality as well as at reasonable prices. Enormous pressure was put on all sections involved in the food supply – the agricultural economy, the retail and the food processing industries – to keep pace with the rapid changes in the developments and needs, choices and opportunities of a modern, urban society. In the rapidly maturing metropolis solutions had to be found quickly and the resulting changes can be clearly seen (Lummel, 2002: 252-258).

I. Berlin's food supply and the food industry's first steps (1820-1850)

What was the situation with the food supply of Berlin in the decades before 1850? In what way did farming, retailing as well as the mechanical processing and refining of food contribute to it? The land reforms introduced from 1806 by leading Prussian officials around Stein and Hardenberg began to have a sustained impact on farming from the 1830s onwards. Prior to these reforms, any decisive progress in agriculture was blocked primarily because of the corvée and by the absence of private ownership in rural areas. Thanks to the land reforms many farmers could, for the first time and on a large scale, purchase a privately owned plot of arable land. They began to cultivate their fields and vegetable gardens according to their own ideas and eventually according to the needs of consumers and the market. At the same time the big estates in Prussia began to cultivate grain in increasingly large quantities. There too, huge areas of land, which had not been used till then, were now also privatised and cultivated. In this way, agricultural production quadrupled in the course of the first half of the nineteenth century, that is before mechanisation and chemical fertilisers came into use. Thus, agricultural production increased even faster than the population of Berlin. Around 1850, the expanding metropolis had risen to roughly 420,000 inhabitants and covered an area of 35 square kilometres. Immigrants came especially from the Prussian provinces. In the city problems arising from the density of buildings, the construction of tenement houses and homelessness were coming to a head. Due to the growing population agricultural products had to be transported rapidly to Berlin from more and more distant areas (Aldenhoff-Hübinger, 2001: 64).

To facilitate transport, the Prussian State constructed a network of surfaced and broad national highways around Berlin during the first half of the nineteenth century. This was an important step in solving transportation problems. From 1838 onwards the railway network spread radially around Berlin, providing another means for transporting food. But until 1850 most goods came by ship to Berlin, on the better and earlier developed waterways and on the rivers Spree, Havel, Oder and Elbe. From this time on there was stiff competition in freight charges which encouraged lower costs for transportation and eventually for food.

However, the central problems in the food trade remained unsolved until 1850 (Lummel, 2001b: 82). A large part of the production was still sold in the 14 farmers' markets with about 6,000 market stalls. Due to the fast-growing urban population, more and more market stalls and market days were licensed. Nevertheless, the city's capacities didn't suffice any more. Moreover, the hygiene in the farmers' markets was criticised more and more often. This prompted the setting-up of grocery stores and specialist shops which were soon distributed all over Berlin. Within half a century the number of wine dealers rose from 37 to 149, and corn and flour dealers from 23 to 430. New specialised shops sold caviar (6) crayfish (3), butter (145), chocolate (34, including factories), coffee substitutes (11, including factories), tea (28) as well as mineral water (12). All these products were new or could be acquired now in special 'delicatessens' for the first time (On the year 1802 see Bratring, 1804: 160; on 1859 see *Allgemeiner Wohnungs-Anzeiger*, 1859).

Foodstuffs made in factories were already available before 1850. These included spirits, champagne, mineral water, rum, hot punch and grog syrup, semi-luxury items such as sweets, chocolate, marzipan, coffee substitutes from chicory and acorns, cigars and tobacco, as well as vinegar, mustard, fruit sugar, sugared syrup, starch and flour (Büsch, 1971: 109; *Berliner Adressbuch*, 1853). Most of the factories were small enterprises with a few machines and a small staff. Larger enterprises existed only for the production of tobacco (40 factories with 1,300 workers), sugar (7 sugar refineries with 425 workers) and flour.

A good example of an experimental pioneer in the early food industry was the Berlin entrepreneur Schumann (Lummel, 2001b: 79). Until 1810 the so-called mill requirement had forced farmers, brewers, distillers and bakers to grind corn in an assigned water-mill or windmill. Now freedom of trade in Prussia allowed everybody to operate a mill and Schumann availed of the liberty. He was convinced that a big steam mill for grinding corn could be profitable and in 1824 decided to construct such a mill on American models, even though it would be costly and its success uncertain. Most of the machines required had to be produced as single items in Berlin. When the flour factory was opened in 1824, the result was impressive. In the words of contemporaries, 'Everything from unloading the corn to the transport, cleaning, grinding, shaking and sorting up to bagging the flour was done mechanically' (Mieck, 1997: 566). Schumann's product was obviously more durable and of better quality than conventional flour. The entrepreneur proudly called his enterprise 'a steam-driven flour-factory'. Schumann was successful because he invested a large sum in the technical standards of his steam mill. He produced for the indigenous market as well as for export by offering his ready-made quality flour to important customers such as bakers.

A brief look at the beginning of the food industry in Berlin then illustrates that it started early from about 1825 onwards with the production of expensive alcoholic beverages and artificial mineral water as well as semi-luxury foods and quality flour for important customers. Although sometimes considerable innovations were introduced, the early food industry had no sustainable impact on the food supply of Berlin until 1850.

In the succeeding sections, we will deal with the establishment of the food industry between 1850 and 1925. In the second section a consideration of four pioneers from different lines of business will show that the breakthrough in industrial production took place at different times and followed different patterns depending on the type of business. We will look at the first public bread factory in Germany, and in the second and third parts we will examine the largest European slaughterhouse and the big dairy factory, Bolle. The international connections of the established food industry will be illustrated through the example of the Sarotti Chocolate factory. We will then discuss the strategies, successes and failures of the examples cited. In the third section we consider the significance of the Berlin examples in the national and international context. Finally, we examine the impact of the food industry in supplying the masses on the one hand and satisfying the demands of the middle classes and urban elites on the other hand. We will also return to examine other stages of the food chain.

II. Pioneers of the food industry

II.1. The first baking factory in Germany (1856)

Bread and corn were the weak points in the food supply throughout Europe. A price increase in bread traditionally meant hunger for the ordinary people (Hachtmann, 1999: 181). In the early 1850s the price of bread rose year by year although there had been good harvests. People reacted by consuming less bread per annum (1852: 268 pounds, 1854: 255 pounds), especially the more expensive wheat bread.

At this time an investor group applied for permission to set up the *Brodfabrik Aktiengesellschaft* (Lummel, 2001b: 80). Following examples in Paris and Brussels, they wanted to set up a state-of-the-art bread factory with a steam-driven mill, big ovens and powerful kneading machines. The investors believed that the factory would be able to turn up to 500 metric hundredweights of rye flour into bread. That would have been approximately one third to a quarter of the bread needed in Berlin. The concentration on rye bread was a clear indicator of the main target group of the project, the lower social classes.

Until this time small enterprises dominated the baking trade in Berlin.[1] The proposal met not only with consent, but also intense resistance particularly from the bakers' guild. The baking trade and the Prussian Home Secretary tried to stop the bread factory, arguing that competition was already too great and the bread supply already sufficient. However, the baking factory also had important supporters. The Berlin Chief of Police, the Prussian Minister of Finance and the Minister of Trade argued for the necessity of tougher

---

[1] In 1855 every bakery in Berlin employed 3.31 persons on average, see Wiedtfeldt, 1889: 132.

competition to put an end to the annoying bread price increases. At last the first public bread factory in Germany was approved on April 13th 1856 and a new era started in the baking trade. For the first time the modern form of the stock corporation was chosen, which was necessary for the large investment capital of 300,000 Thalers. The bread factory invested in three steam-driven engines with 112 hp, seven ovens, twelve stoves and it took on 65 employees. The combination of an industrial mill and a bread factory 'under the same roof' was innovative as well. Despite the doubts of the bakers, the small bakeries did not disappear. On the contrary, the first bread factory was successful only as a flour mill. We can assume that an insufficient distribution network contributed to the failure. But from this time on there was competition between the highly mechanised bread factories and the small bakeries. Since corn prices dropped at the same time, basic foodstuffs like bread and bread rolls (*Schrippen*) also became more easily accessible for the lower social classes.

II.2. The central slaughterhouse: meat processing for two million people (1870-1885)

In Berlin in 1872, ninety percent of the cattle were slaughtered in small private slaughterhouses. The cattle were guided through busy streets, a practice that facilitated the transmission of animal diseases. In most of the 780 private slaughterhouses in Berlin hygiene conditions were appalling (Lummel, 2001b: 80-82). Under these conditions it was impossible for the commune to supervise the slaughter by official veterinarians. But the Berlin butchers didn't have any alternative since there wasn't any modern slaughter-house available at that time. [2] It became increasingly clear that improvements in hygiene and continuous veterinary inspection could be attained only by constructing a central slaughterhouse. The question was whether the finances of the municipal council sufficed for such a project.[3]

The famous Berlin doctor, scientist and politician Rudolf Virchow, who was the main proponent of new hygiene standards in Berlin, fought for municipal involvement with arguments that are astonishingly still current. In 1874 he addressed the council of Berlin: '...the first task that has to be done [is] the production not of cheap but of healthy meat' (Schindler-Reinisch, 1996: 21). In 1876 the municipal authorities of Berlin decided to construct the central slaughterhouse (Blankenstein and Lindemann, 1885: 4). In 1879 the Chief of Police introduced obligatory meat inspection in response to Virchow's urging. In 1881 the central slaughterhouse with an affiliated livestock market was opened with an investment sum of 11.7 million marks. The building was placed outside the city on a 39 hectare area with a railroad station of its own. The huge livestock market consisted of gigantic halls for the sale of pigs, cattle, calves and wethers (On the buildings see Blanken-stein and Lindemann, 1885; Schindler-Reinisch, 1996). The sales hall for cattle was 217 metres long and 72 metres wide and could hold 3,700 cattle. Up to 31,000 wethers and 13,000 pigs could be held and sold in other halls. On good days 700 butchers from Berlin, 300 foreign butchers as well as 150 wholesalers assembled in the livestock market.

---

[2] For a short time until a few private livestock markets and slaughterhouses were founded.
[3] The question how to finance hygienec reforms was discussed at the same time in many European towns, cf. Koolmees in this book.

For the purposes of better control, the slaughterhouse was separated from the livestock market by a 2.50 metres high fence. It consisted essentially of three cattle slaughter-houses (148 m x 29 m) and two pig slaughterhouses (101 m x 23 m or 101 x 27 m) with accompanying stables. Cattle, pigs, calves and sheep were slaughtered and taken off to numerous chambers. A slaughter-line such as in the United States didn't exist. From platforms the meat could be loaded directly onto butchers' wagons or into railway car-riages. An unusual feature was the seasonal export of mutton to Paris (Blankenstein and Lindemann, 1885: 58).[4] Up to 1,300 cattle, 6,500 calves, 6,500 sheep and 3,200 pigs could be slaughtered daily at the central slaughterhouse. With these far-seeing measures the city of Berlin was also able to cope with the doubling of the population which would reach almost two million people.

The effects on hygiene conditions and quality of the meat were extremely positive. Now numerous veterinarians inspected the cattle and slaughter animals daily. This was essential to prevent dangerous epidemic diseases like tuberculosis, cattle plague or foot-and-mouth disease. The central slaughterhouse also changed the profession of the butchers (Wiedtfeldt, 1898: 140). Originally the master butcher had to buy the cattle from the farmer or the livestock markets, and then slaughter, process and sell the prepared cattle. Now several specialised professions developed. The wholesalers had to buy and sell the cattle. The big butchers who slaughtered most of the cattle in the central slaughterhouse became very important now and they usually sold the whole, half or quarter of a slaughtered cow to the butchers. Also so-called wage butchers did the slaughtering in the municipal slaughterhouse for restaurants and most often for butchers, who now concentrated more and more on the processing of the meat and on selling it in their shops.

Much had changed also for the consumers. Meat processing had been entirely mod-ernised by the immense investment of the commune in a central slaughterhouse. Cattle were now transported by direct rail link. Different specialists had taken over the buying, slaughtering and processing of the cattle as well as inspecting the meat for the customers. All work routines were aimed at satisfying the growing demand for better quality meat at a reasonable price. In 1884 two-thirds of the meat sold in Berlin was produced in the central slaughterhouse. The annual per capita consumption of meat was about 70-75 kilograms, approximately 20 kilograms more than 30 years beforehand (Blankenstein and Lindemann, 1885: 59; Diederici, 1854: 148).

Although the whole butchering trade had changed, there wasn't any great resistance, either from the slaughterers or the consumers. Most protagonists seemed to benefit. This was an important difference from the bakers who were sceptical of the modernising trends and who still produced predominantly in small workshops for a long time after the appearance of the first baking factory.

---

[4] In 1884 the *Compagnie d'importation de produits alimentaires Paris* imported around 18,600 slaughtered wethers by trains with freezers to Paris.

143

II.3. The company Bolle and the modern milk age (1880 - 1910)

The modern milk age started in Europe in the 1870s, evident in the breeding successes in milk cows, the beginning of milk science, the national dairy exhibitions and from 1878 the introduction of useable centrifuges (Teuteberg and Wiegelmann, 1986; Ottenjann and Ziessow, 1996; and also different articles in this book). From around 1870 onwards dairy factories producing milk and dairy products represented serious competition for farmers and dairymen. Between 1870 and 1880 the number of dairies increased from 19 to 166 in Berlin alone. The number of dairy farms also grew considerably. These so-called *Abmelkbetriebe* purchased new highly pregnant milk cows every year, giving them the best conditions and the best fodder, with the result that the cows produced almost double the quantity of milk produced by cattle on conventional farms.

At this booming time Carl Bolle decided to set up his own dairy factory in Berlin (Engel and Koop, 1995; Mitteilungen, 1908). In 1881 the trained master-builder invested in buildings, machines and three milk wagons in order to be able to sell his products on the street. The Bolle factory was successful from the very first day. In 1881 the dairy factory could extend its fleet of vehicles to around 30 cars. In 1882, 56 wagons sold 7.8 million litres of milk, which meant an average daily sale of 24,000 litres. In 1887 the expanding enterprise built a new dairy factory on a 23,000 square meter property in Berlin-Moabit, the heart of which was a three-storeyed building115 metres long and 20 metres wide.

With this Bolle established new standards in the dairy business. All the working processes 'from the stable to the table' were rationalised and well-organised. The transportation of the milk to Berlin was done by train only. Company wagons collected the milk in big cans made of tinplate. In the dairy factory all the milk was transported by lift to the third floor to be cleaned. (Führer, 1891). The ordinary milk was then sent down to the ground floor in big coolers where it was poured into locked cans for the milk wagon sale. The 'children's milk' however was heated in a modern sterilisation unit to 102°C. Bolle also produced high-quality raw milk which was put into bottles that could be kept up to eight days. The milk processing was done on the first floor, where all milk was centrifuged at first and butter, buttermilk, skimmed milk, cream and different cheese types were produced. Maturing rooms for the cheeses were located in the cellar as well as a cheese dairy that specialised in processing the day's leftover milk. A smart idea was the processing and selling of so-called waste products. For instance, lactose was produced in the factory for pharmacies by using the huge quantity of whey created in making cheese.

Distribution was efficient and well-organised as well (Engel and Kopp, 1995: 54ff). In 1886 milk wagons sold Bolle products along 70 exactly specified routes. Furthermore every milkman received a map, indicating where and when he had to stop. Dairy boys or girls who helped the milkman had to bring the products ordered in advance to the individual houses. In the meantime the milkman sold his goods on the street. The daily accounting and the ordering of products for the next day was done in the office before closing time. Such a well-organised distribution system was one of the main reasons why the Bolle business grew so much from year to year. In 1890 it sold 15.1 million litres of milk; by 1900 this figure had surpassed the 29 million mark and in 1907 it had reached 50 million. About 130,000 litres of milk were processed daily, corresponding to the milk production of about 24,000 cows and the milk consumption of a seventh of Berlin's population. At the beginning of the twentieth century the enterprise employed 2,400 workers and had over 300 milk wagons and 450 horses.

The efforts expended on milk hygiene and safety were one of the main reasons for the success of the dairy factory (Engel and Kopp, 1995: 48ff). Bolle was able to benefit from the first-class research on food being done in Berlin at that time. From 1880 onwards Robert Koch was the director of the hygiene research laboratory in the Imperial Public Health Department. In 1882 Koch discovered the tuberculosis bacillus and pointed out that it could appear also in foodstuffs like milk. From the beginning Bolle invested in exemplary laboratories. It started in 1881 with a chemical laboratory which tested the fat content, durability and condition of milk. A bacteriological laboratory in which raw milk was examined for pollution and germs was added in 1890. Pathogens like tuberculosis, cholera and typhoid fever could be diagnosed with milk tests and in 1900 Bolles' laboratories carried out 31,264 analyses.

All stages of the milk production process were subjected to the most stringent hygiene inspection. Company veterinarians checked the farms in the countryside. Every delivery of milk was examined at the railway station, during the various processing steps and even at the point of sale. The method of disinfecting the milk cans with steam was adopted by Robert Koch. In 1900 Bolle started to pasteurise milk, one of the first factories in Germany to do so. Only from 1928 onwards did this method become the obligatory heat treatment for milk traded in Germany. With its high-level quality standards the Bolle dairy factory made an important contribution to the reduction of infant mortality and the transmission of deadly diseases which were often caused by bad milk. In the European context also, the enterprise was regarded as exemplary and was classified by contemporary experts as outstanding because of its size, business organisation, hygiene standards and distribution. The transition from a family business to a big industrial diary in the form of a stock corporation was not done, however, until 1911 after Carl Bolle's death.

And what about milk consumption in Berlin during the 'Bolle age'? It was relatively high compared with other German cities.[5] Between 1893 and 1903 consumption doubled from 376,000 to 710,907 litres daily whereas the population rose only from 1.5 million to 2 million in the same period. The number of dairy factories increased significantly, from 166 (1880) to 331 (1890) up to 647 (1900). Thanks to its modern dairy factories milk acquired a good reputation in Berlin and was consumed by a large part of the population as a reasonably-priced, healthy beverage.

---

[5] For the German context see Ottenjann and Ziessow, 1996: 42.

II.4. Sarotti and the internationalisation of the chocolate industry (1890-1930)

Industrial enterprises producing basic foodstuffs such as flour, bread, meat and milk were assured guaranteed sales increases in view of the huge demand. The simultaneous development of the semi-luxury food, chocolate, proceeded differently. Until 1880 it was produced primarily by skilled craftsmen in small quantities and was very expensive, affordable only to the rich. In the following years chocolate prices experienced an explosive increase which weren't reached by any other branch of the food industry (For the following passage see: Vorstand, 1931: 14-38). In 1880, 2,200 metric tons of chocolate were processed in Germany; by 1900 already 20,000 t. were being produced, in 1905 just under 30,000 t and in 1912 just over 53,000 t. Since the beginning of the twentieth century the German chocolate industry had become the best seller in Europe.

Berlin played an important part in the industrial production of chocolate from the very beginning. In 1830 the company Hildebrand was the first German chocolate enterprise to use a steam-driven engine. Fifty years later Berlin already had 26 chocolate factories. The Sarotti company (On Sarotti see Genest, 1928; Vorstand, 1931), founded in 1868, seems to be the clearest example of the enormous possibilities for development and expansion in the German chocolate industry after 1890. In 1889 the enterprise had 90 employees. Four years later and with 162 employees, the company decided to build a new factory on the existing property. In 1903 the work force had grown to 1,000 employees so that another new building was necessary. A radical step was made in 1911 when the management decided to leave the narrow site in the centre of Berlin for a 47,500 square metre site in Tempelhof in the south of the city and in 1913 a six-storey, 4,040 square metre chocolate factory was opened for 2,000 employees. Despite the World War and the difficult conditions after 1918 the factory base was extended in 1921 by 6490 square metres. Sarotti proceeded to become the biggest German chocolate factory next to Stollwerck in Cologne.

In 1922 Sarotti experienced one of the worst fire disasters in the history of German industry when all its buildings and records went up in flames. Nevertheless, Sarotti's expansion didn't stop and the reconstructed factory was bigger than ever. It now had a business area of 65,000 square metres. Unlike any of its competitors, the enterprise owned only the most modern, efficient machines: in 1928 Sarotti had at its disposal more than 600 electric engines with 5.390 h.p. and about 830 processing machines and employed 2850 workers. The results were extraordinarily impressive: 10,000 kg of cocoa, 300,000 chocolate bars and 10,000 kg of filled chocolates could be produced each day in a choice of 1,000 different kinds, to mention only some of its achievements.

This expansion could be financed only by a corresponding flow of capital. The new building of 1903 was made possible only by converting the general partnership to a stock corporation. In this way the company founder Hugo Hoffmann and his partner Paul Tiede had a share capital of 1.5 million marks at their disposal. The 3.5 million mark cost of the new factory in Berlin-Tempelhof was also financed with the help of share capital. When both executive directors died, Max Hoffmann, the son of the company founder, took over the management to guarantee personnel continuity in the large-scale enterprise. This was interrupted after World War I in the context of the financing of the new factory building. For the first time a foreign investor group bought the majority shareholding.[6] From 1921 company policy was determined by the Swedish Kanold Group and shortly after, the Swede Anton Kanold was elected as Managing Chairman of the executive board. The 1922 fire disaster and the very up-to-date reconstruction of the chocolate factory forced further internationalisation and for the first time American dollars were invested in the enterprise as well.

All these changes and the problem of foreign infiltration were the subject of vehement discussion in German industrial circles and the media. However, the internationalisation of the enterprise strengthened its performance on foreign markets. In 1928 Sarotti was exporting to the following countries and continents: Holland, England, Denmark, Norway, Sweden, Romania, Yugoslavia, Italy, Turkey, Portugal, Egypt, Morocco, Palestine, Iran, India, Japan, Australia, North America, Central America, South America and South Africa. In 1929 there was another radical change when the Nestlé group took over the majority of Sarotti's stock (Vorstand, 1931: 43ff). In Germany Sarotti had now the exclusive right to produce and sell Nestlé's world-famous brands Peter, Cailler and Kohler. At the same time the production process was divided up. Sarotti was to concentrate on filled chocolates and special items, Nestlé on chocolate bars. The new Sarotti-Nestlé group controlled a third of the German chocolate production and with the help of Nestlé's international contacts, 'Sarotti' managed to export its products to a lot of new countries.

Perhaps no other German semi-luxury food manufacturer demonstrates more clearly how a former luxury item could now be produced at a reasonable price for the larger part of the population with the help of share capital, immense mechanical power and a global distribution network.

II.5. Strategies, successes and failures of the pioneers

The pioneering industries in flour, bread, meat, milk and chocolate all followed different patterns of development in the breakthrough to industrial large-scale production. The beginning of industrial processing varied considerably depending on the foodstuffs. Clearly not every commodity could be successful in the long run. What then did these pioneering industries have in common? Why did industrial production start much earlier with bread and flour than with meat, milk or chocolate? Why could some industrially produced foodstuffs be successful on the market and others not?

All five commodities made use of steam-driven machines and high investments, and each triggered major changes in its line of business. However, three-quarters of a century separates the beginning of full mechanisation in the production of flour (1824), bread (1856), milk (1870-80) and chocolate (1890-1900). The earliest examples were the processing of corn and bread, important basic foodstuffs for which there was a high

---

[6] Landesarchiv Berlin: Landesarchiv A Rep. 251-09 Nr. 499: Geschäftsberichte und Bilanzen der Sarotti Aktiengesellschaft, Berlin (Juli 1922 - December 1924), with different newspaper articles of that time.

daily demand. The pioneers in these businesses hadn't any difficulty in obtaining the raw materials, corn and flour, and production didn't present any great hygiene risks either. So it was one of the very early successes in the food industry.

The meat and milk businesses were different. The perishable nature of the two food-stuffs frequently led to life-threatening illnesses, particularly among small children and old people. These problems were considerably reduced in those dairy factories and slaughterhouses with equipment and testing procedures that required a knowledge of the new milk or meat sciences. The milk and meat industry started only in the 1870s and 1880s. Chocolate was already being produced with steam-driven engines by 1830, though mostly in small factories until the late nineteenth century. The breakthrough in the semi-luxury food business and the huge increase in demand started in the years before 1900. The production of chocolate then soared with the use of machines as did turnover, reaching a level of growth that wasn't reached in the German nutrition industry by any other line of business. Since chocolate - in comparison with bread, milk and meat – kept fresh for a long time and was easy to transport, it could be sold abroad. In this way it was possible to produce in large quantities, thereby keeping prices reasonable.

But what about the background of successes and failures in these instances we have examined? Why couldn't Germany's first bread factory be as successful as the industrial flour mill or the big Bolle dairy? One of the reasons may be the absence of a distribution network which Bolle had developed so well. But Schumann's flour mill also hadn't any shops or sales wagons and was nevertheless successful. Unlike the bread factory, the flour mill primarily produced for the wholesale trade and was therefore not a competitor of the bakers but a supplier of ready-ground quality flour. The bread factory, however, tried to do two things simultaneously: produce rye bread for the masses and at the same time flour for the bakers, an attempt which obviously failed.

III. The establishment of the food industry, 1870–1910

By 1850 at the latest Berlin was an outstanding centre for traffic and consumption in the centre of Europe. The metropolis was an attractive site for export-oriented enterprises as well as for a food industry with a regional sales network. Under these conditions a considerable and innovative food industry came into being in Berlin quite early, as our examples have shown. The industrial production of corn and bread in Germany started in Berlin. Schumann's flour mill from 1824 was obviously the first in Germany to use steam power for food processing. The size, excellent distribution and hygiene conditions of the Berlin milk giant, Bolle, were innovative and exemplary also by European standards. The central slaughterhouse in Berlin came into being in the context of a pan-European discussion about meat hygiene, public health and municipal responsibility. In the national and international context Berlin's slaughterhouse wasn't the first ultra-modern facility, but it was, however, regarded as the most modern and largest slaughterhouse in Europe in its time. After World War I the chocolate enterprise Sarotti belonged among the market leaders in Europe. With the most modern machinery and a distribution network extending to all five continents, the enterprise came to attract foreign investors, at first Swedish, then American and finally the Swiss global player Nestlé. On the one hand

Sarotti illustrates how advanced globalisation in the food industry was at this time. On the other hand, the Berlin site was retained despite the immense increases in production, underlining once again the function of the metropolis as an outstanding centre for production and consumption.

The years 1870-1980 as well as 1900-1910 seem to have played an important role in the establishment of the German food industry, with Berlin obviously as the most progressive site for several branches of the industry.[7] We will now examine these two decades more closely. What part did the food industry in general play in guaranteeing sufficient and reasonably priced food? What new products and luxury items did the food industry create to be able to initiate and meet fashionable trends among groups with money to spend? Since these changes are interlocked with developments in other stages of the food chain, we will come back to farming and retail again.

III.1. The years 1870-1880

With the foundation of the German Empire on January 18th 1871 Berlin became the capital of the Empire, and because of this the city continued to grow at breathtaking speed. Against the background of wild financial speculations and an enormous economic upswing, the number of food enterprises rose steeply (For the following passage see: Lummel, 2002: 252). They produced foodstuffs like baking powder, biscuits, cigarettes, coffee, new coffee substitutes, fruit juice, jams, liqueur, malt extracts, pea sausages, preserved foods, pumpernickel bread, sauerkraut, sausage, sponges and confectionary. Many food items were innovative or mechanically produced now for the first time. In particular, beer, sponges, biscuits, sweets, coffee substitutes, mineral water, tobacco, factory sausages and pea sausages (needed for military purposes) were produced in very great quantities.

Most goods produced by the food industry catered to the upper middle-class. However, rye bread, the most important basic foodstuff, became more and more reasonable thanks to falling flour prices and competitive bakeries. Meat also became not only cheaper but also considerably safer after the opening of the central slaughterhouse. For the first time the food industry had become a significant one around 1880.

What did the situation look like in the food trade at the same time? The importance of the traditional farmers' markets to the food supply was already in question from the middle of the nineteenth century. Developments in the 1870s were more critical in view of the increase in population. In 1880 there were 20 weekly markets in Berlin with 9,000 stands. The transport routes became more and more time-consuming for the sellers. But the customers also had their problems since the distances between home and market place often seemed to be insuperable. The problems with the hygiene in the weekly markets escalated particularly because of the rubbish. In this context a long debate started about the construction of municipal market halls in Berlin, but for the time being this was not financially possible because of the priority of a central slaughterhouse. However,

---

[7] For the German context of the establishment of the food industry see Ellerbrock, 1993.

the number of grocers' and speciality shops rose considerably at the same time which advertised the innovations in the food industry very effectively.

Farmers could earn good money between 1870 and 1880. Because of synthetic fertilisers and developments in machinery, they succeeded in increasing their production considerably. In the meantime most farmers decided to stop selling their products directly in the city markets. Along with the food industry and the food trade, 'agri-business' now specialised in cultivating saleable food and selling it for profit to the retailers.

III.2. The years 1900-1910

After the turn of the century the food industry provided substantially more products for the retailer and the consumer: 'Tree cakes', stock cubes, tinned fish, gelatine, yeast, kefir preparations, special children's food, brandy, liquid carbonic acid, compote, margarine, jam, 'Mondamin', sparkling wine, lard, soda, vanilla, waffles, lemon juice and rusks. It was a wide range that reached from child nutrition, special diets, preserved foods and cost-saving products like margarine or stock cubes to fancy alcoholic beverages, soft drinks and baking ingredients.

In the meantime, semi-luxury items like chocolate and cigarettes were being produced at a reasonable price and in large quantities also. Between 1885 and 1895 annual beer consumption in Berlin increased from 171 to 226 litres, and Berlin became the centre of the brewing industry with Schultheiß by far the largest German brewery. The cigarette industry also did considerable business. Many of the 200 Berlin factories had machines that produced up to 100,000 cigarettes per day whereas in the past, 50 to 100 workers would have been necessary for this. The modern flour mills reached a daily production level of up to 10,000 metric hundredweights of flour, twenty times the amount produced by Berlin's first bread factory 50 years before (Wiedtfeldt, 1898: 131) Modern bread factories and also bigger bakeries now had efficient kneading machines which produced up to 1,400 metric hundredweights of dough per week. 50 employees would have been necessary to do this work by hand. As opposed to the factories of the years 1870-1880, large food factories now predominated, with stock corporations becoming more and more frequent and multinational and global enterprises entering the scene.

In the food trade the decisive step towards modernisation had already been taken before 1900 with the opening of the first market hall (Lummel, 2001b: 84). From 1886 onwards freight wagons reached the hall by rail. They could unload up to 15,000 kg of rail freight per hour without disturbing the city traffic. Goods arriving in the centre of Berlin from all continents were taken on short routes into cold storage or directly to the market hall sales. The Berlin population loved the hall and soon called it 'the stomach of Berlin'. It was here that grocers did their bulk purchasing; they received their goods there, which they got clean, fast and reasonably priced. Around 1900 thousands of grocery shops with a high range of products existed on every corner. The grocers tried to satisfy the wishes of the customers with special offers and unusual services. Large department stores, cooperative societies, chains of retail shops and mail order firms were set up at the same time. After 1900 another stage was reached in the supply of a subtly differentiated consumer society.

The nutrition situation was not at all satisfactory for everyone, though. Even after 1900 most of the two million inhabitants belonged to the lower social classes. Their income didn't suffice to be able to purchase the quality and luxury goods of the food industry. But for the majority of the population hunger was no longer life-threatening. Moreover, affordable food had become much more varied for the lower classes also.

In 1900-1910 the food chain of Berlin consisted of modern farming, food processing which was organised manually or industrially with rail links to the ultra-modern market halls, and a food trade that was integrated into a global network. The beginnings of the consumption-oriented society of today is clearly evident in all these stages (Spiekermann, 1999: 614). In the 1870s and still more after the turn of the century the food industry played a considerable role in this development.

Bibliography

Manuscript sources

Landesarchiv Berlin:
Landesarchiv A Rep. 251-09 Nr. 499: Geschäftsberichte und Bilanzen der Sarotti Aktiengesellschaft, Berlin (Juli 1922 - December 1924), with different newspaper articles of that time.

Secondary sources

Aldenhoff-Hübinger, R. (2001) 'Agrarprodukte aus Brandenburg und ihr Weg nach Berlin', Museumsverband des Landes Brandenburg (ed.), *Ortstermine. Stationen Brandenburg-Preußens auf dem Weg in die moderne Welt*, Berlin, pp. 64-69.

Allgemeiner Wohnungs-Anzeiger nebst Adreß- und Geschäftshandbuch für Berlin, dessen Umgebungen und Charlottenburg auf das Jahr 1859. Aus amtlichen Quellen zusammengetragen. Vierter Jahrgang, Berlin 1859, reprinted 1990.

Blankenstein H. and Lindemann A. (1885) *Der Zentral-Vieh- und Schlachthof zu Berlin. Seine baulichen Anlagen und Betriebs-Einrichtungen*, Berlin.

Brandt, K. (1928) *Der heutige Stand der Berliner Milchversorgung*, Berlin.

Bratring, F.W.A. (1801) *Statistisch-topographische Beschreibung der gesamten Mark Brandenburg*, Bd.1, Berlin, zitiert nach der kritisch durchgesehenen und verbesserten Neuausgabe, ed. by Büsch, O. and Heinrich, G.

Büsch, O. (1971) 'Das Gewerbe in der Wirtschaft des Raumes Berlin/Brandenburg 1800-1850', Büsch, O (ed.), *Untersuchungen zur Geschichte der frühen Industrialisierung, vornehmlich im Wirtschaftsraum Berlin/Brandenburg*, Berlin, pp. 95-105.

Büsch, O. (1977) *Industrialisierung und Gewerbe im Raum Berlin/Brandenburg. Bd. 2: Die Zeit um 1800 / Die Zeit um 1875*, Berlin.

Dieterici, J.W.C. (1854) 'Über die Verzehrung von Brod und Fleisch im Preußischen Staat', *Mittheilungen des Statistischen Bureau's in Berlin*, 7, Berlin.

Ellerbrock, K.P. (1993) *Die Geschichte der deutschen Nahrungs- und Genußmittelin-*

dustrie 1750-1914, Stuttgart.

Engel, H. and Koop, V. (1995) *Der Spreebogen. Carl Bolle und sein Vermächtnis*, Berlin.

Führer durch die Meierei C. Bolle Berlin NW, Alt-Moabit, pp. 99-103, Berlin, 1892.

Genest, G. (1928) *Sechzig Jahre Sarotti 1868-1928*, Berlin.

Hachtmann, R. (1999) 'Ein Magnet, der die Armut anzieht. Bevölkerungsexplosion und soziale Polarisierung in Berlin 1830-1860', Pröve, R. and Kölling, B. (eds.) *Leben und Arbeiten auf märkischem Sand. Wege in die Gesellschaftsgeschichte Brandenburgs, 1700-1914*, Bielefeld.

Lummel, P. (2001) a. 'Von Kartoffeln und Austern, Destillen und Cafés. Essen und Trinken in einer wachsenden Metropole', Museumsverband des Landes Brandenburg (Hg.), *Ortstermine. Stationen Brandenburg-Preußens auf dem Weg in die moderne Welt*, Berlin, pp. 89-96.

Lummel, P. (2001) b. 'Sauber, schnell und kundennah. Die Entstehung der Lebensmittelindustrie und des modernen Kleinhandels in Berlin', Museumsverband des Landes Brandenburg (Hg.), *Ortstermine. Stationen Brandenburg-Preußens auf dem Weg in die moderne Welt*, Berlin, pp. 79-88.

Lummel, P. (2002) 'Von der Hungersnot zum Beginn modernen Massenkonsums. Berlins nimmersatter Riesenbauch. Die Lebensmittelversorgung einer werdenden Weltstadt im 19. Jahrhundert', Teuteberg, H.J. (ed.), *Nahrungskultur. Essen und Trinken im Wandel.* Der Bürger im Staat, 52, 4, pp. 252-258.

Mieck, I. (1997) 'Idee und Wirklichkeit: Die Auswirkungen der Stein-Hardenbergschen Reformen auf die Berliner Wirtschaft', IHK zu Berlin (ed.) *Berlin und seine Wirtschaft. Ein Weg aus der Geschichte in die Zukunft - Lehren und Erkenntnisse*, Berlin and New York, pp. 41-58.

Mitteilungen über die Meierei C. Bolle Berlin, N.W. Alt-Moabit 98-103, Berlin, 1908.

Ottenjann, H. and Ziessow, K.H. (eds.) (1996) *Die Milch. Geschichte und Zukunft eines Lebensmittels*, Cloppenburg.

Schindler-Reinisch, S. (1996) *Eine Stadt in der Stadt. Berlin-Central-Viehhof*, Berlin.

Schmiede, E. (1960) 'Carl Bolle', *Tradition. Zeitschrift für Firmengeschichte und Unternehmerbiographien, Heft 2*.

Spiekermann, U. (1999) *Basis der Konsumgesellschaft. Entstehung und Entwicklung des modernen Kleinhandels in Deutschland. 1850 – 1914*, München.

Teuteberg, H.J. (1986) 'Die Anfänge des modernen Milchzeitalters in Deutschland', Teuteberg, H.J. and Wiegelmann, G. (eds.) *Unsere täglich Kost*, Münster, pp. 173ff.

Teuteberg, H.J. (ed.) (1987) *Durchbruch zum modernen Massenkonsum*, Münster.

Vorstand des Verbandes der Nahrungsmittel- und Getränkearbeiter Deutschlands (ed.) (1931) *Die deutsche Kakao- und Schokoladenindustrie*, Berlin.

Wiedfeldt, O. (1898) *Statistische Studien zur Entwicklungsgeschichte der Berliner Industrie von 1729-1890*, Leipzig.

7 The North American influence on food manufacturing in Britain, 1880-1939

Ted COLLINS, University of Reading

I. Introduction: The origins and early development of modern food processing in the United States, 1860-1920

In a conversation with a famous Russian opera singer just before the First World War, Czar Nicholas II is reported to have said of American foods: 'They have a flavour of American corporations and speculation' (Levenstein, 1968: 38).

Better than any history book, this statement captures the essence of the revolution in food manufacturing and eating habits at a critical stage in the social and economic development of the world's richest nation, then just entering the age of high mass consumption. Americans were discovering that with food, as with most other goods and services, the urgency of human wants does not appreciably diminish as more of them are satisfied, and as their real incomes improve, so expenditure is determined less by basic needs as psychologically grounded desires (Galbraith, 1962: 124). The modern consumer wants above all choice and variety, and foods with built-in services – novelty, reliability, convenience. The history of the twentieth/century food industry can be written around the inter-reaction of consumers and manufacturers within a marketing system with the ability to transform inessential wants into indispensable needs.

From the late nineteenth century, 'factory foods' began to re-shape both the American diet and the American food industry. What had originated as health or, in the case of ready-to-eat breakfast cereals, as vegetarian foods for the use of minority religious groups, such as the Seventh Day Adventists, had by the 1930s become staple items of diet. That process was well-advanced by 1914. Indeed, a few American firms were then already exporting to Europe, and one or two were even manufacturing there. The globalisation of processed foods had begun.

Highly processed proprietary foods differed from basic foods not only in style and formulation but also in their economic behaviour. Standardised and mostly sold loose, the latter function in the same way as commodities, with a defined relationship between supply and demand, and food value and price. In the former, value derives from the reputation of the manufacturer or brand name. Basic industries, such as milling and sugar refining, were driven by manufacturing efficiency and scale economies, and competed on price and availability. Processed foods were driven by marketing efficiency and product innovation, and competed on perceived qualitative differences between rival brands of like products. The closer the differences between competing products, and the greater their differentiability, the greater the need and opportunity to promote them. Advertising is what is termed, a 'sunk cost', incurred with a view to persuading consumers to pay for 'added-value' (Sutton, 1991: 7-14). Relatively little is spent in promoting basic foodstuffs or on generic advertising, whereas proprietary foods require heavy and sustained expenditures to differentiate, enhance and defend the brand. By 1939, about one-half of all food items sold by grocers, and today over 90 per cent, are branded, compared with fewer than 10 per cent at the turn of the century. Notwithstanding their centrality in modern commerce, brands have little or no place in economic theory. Jones and Morgan's pioneering volume of essays on the history of brands, published in 1994, has failed so far to kindle scholarly interest among economic and business historians (Jones and Morgan, 1994).

By the late 1930s, US food manufacturing had in large measure assumed its distinctive modern form. The industry was highly concentrated and the market highly segmented, with each segment controlled by an oligopoly of brands and products. Most of the firms and brand names that were to dominate the international food industry in the post-war period were already well-established (Trager, 1996; Horst, 1974: chap. 3; Alderfel and Michel, 1950; Sutton, 1994). The Borden Company, pioneers in the manufacture of canned condensed milk, was founded in the 1850s, Heinz in the 1860s, Quaker in the 1870s, and Campbell in the 1890s. The decades around 1900 saw the formulation of most of the popular ready-to-eat breakfast cereals, and the emergence of Kellogg, Shredded Wheat, Quaker and Post Cereal Foods as the leading brand companies. Coca-Cola was first formulated in 1886, Pepsi-Cola in 1898. In canning, particularly, America was the acknowledged world leader. The great meat-packing corporations – Swift and Armour – began canning beef in Chicago in the 1870s, and soon after 1900 expanded their operations into cattle-rich Uruguay and Argentina. Canned Hawaiian pineapples, Heinz baked beans and Campbell's condensed soups date from the turn of the century. The Del Monte brand of canned Californian fruit was launched in 1915; early in the next decade Clarence Birdseye took out his quick- freezing patent.

Food manufacturing was not a high technology in the narrowly scientific sense, and research expenditures were small compared with other 'new industries'. Although the most successful products were often simple formulations capable of being produced in a small factory or even on a domestic scale, by 1900 a number were in mass-production. Speed and technical efficiency had been combined in automated high-speed production lines making extended runs of safe, affordable foods according to an exact formula (Cummings, 1941: 62-70). American technological leadership was most evident in refrigeration and canning. The first refrigerated rail cars came into service in the 1860s, and inside three decades a national cold-chain network linking the interior with the cities had been created. The huge potential market for canned foods, in the recently opened up agricultural hinterlands and the rapidly growing conurbations, was held back by the primitiveness of the early canning techniques, and ignorance of the bio-chemistry of food preservation and its implications for food safety. Technical innovations after the Civil War included the pressure retort for quicker and more complete cooking, the double-sealed can which guaranteed air-tightness, and the 'Sanitary Can', so designed that the top could be crimped by machine (Metal Box Company, undated; Statistical Abstract of the United States, 1945: 532; Sutton, 1991: 423; Thorne, 1986: 120-134; Hampe and Wittenberg, 1963: 119 and *passim*). In 1900, the most advanced machines could fill upwards of 35,000 cans per day compared with 2000 in the 1870s. The number of canneries increased over the period from 100 to 18,000. Canned foods were the fastest growing sector in the preserved foods industry until the 1960s, when they were overtaken by quick-frozen foods, whose future as a mass-consumption product had for a long time been in doubt.

II. American multinational firms in Britain

A handful of American businesses had already established sales offices and banking agencies in European capitals by the early nineteenth century. Direct investment in manufacturing became significant only from 1890, though by the outbreak of the First World War American firms were already a major force in a number of key modern industries. Food manufacturing lagged behind telecommunications, electricals, motors and petroleum. According to a US Department of Commerce report on US businesses overseas, in 1929 about 20 food companies were operational in Europe (US Department of Commerce, 1930: 10; Southward, 1931: 110-111). Apart from Canada, where US food manufacturers were already well-entrenched, direct investment before 1930 was focused on sugar, fruit, beverages and meat packing in middle America, the Caribbean, and Argentina (Lewis, 1938: chap. 13; Horst, 1974: 97). European investment grew from the early 1930s, and dramatically after 1950. However, in 1989-90, only 10 of the European Community's 50 largest food companies were American-owned, in contrast to the international lists, where US firms occupied 21 of the top 30 positions. American firms were then responsible for about 10 per cent of Community sales of branded foods, and 12-15 per cent of British (US Department of Commerce, 1990: 72). The 1970s saw growing US involvement in food retailing, carbonated soft drinks and fast foods.

In 1939, some 16 US firms have been identified as possessing a manufacturing base in Britain, namely: Borden, Chichele, Coca-Cola, CPC, General Foods, General Milk Products (Carnation), General Mills, Heinz, Horlicks, Kellogg, Kraft, Libby McNeill Libby, Mars, Quaker, Shredded Wheat (Nabisco), Standard Brands and Wrigley. The meat firms, Swift and Armour, had large wholesaling operations. An unknown number of US firms, as many possibly as 30 to 40, were exporting directly to Britain, distributing their merchandise through sales subsidiaries or local agents.

The questions arise why these firms and these products? Why Britain? Why the quickening of activity in the 1920s and 1930s, in what is widely regarded as a period of severe economic depression.

Table 7.1   American foods distributed and manufactured in Britain, c. 1939

Foods	Firms
Bakery Products and Biscuits	Weston (later Associated British Foods) cx Canada
Breakfast foods	
ready-cooked	General Mills, General Foods, Kellogg, Nabisco, Quaker
porridge oats	Quaker
Canned foods	
baby foods	Heinz
condensed and evaporated milk	Borden, Carnation, Pet
fruit and vegetables	Libby McNeill Libby
meats	Swift, Armour
pet foods	Mars
specialities including soups	Heinz
Carbonated soft drinks	Coca Cola
Chewing gum	Wrigley, Chichele
Chocolate confectionary	Mars
Corn flour, custard powders	C P C
Malted beverages	Horlicks
Processed cheese	Kraft
Quick frozen foods	General Foods/Unilever
Yeast	Standard Brands

III. British food manufacturing in the inter-war period

The traditional picture of a Great Depression in the inter-war period has been much revised. Decline and extensive unemployment in the basic industries, such as iron and steel, textiles and shipbuilding, should be set against more positive developments in light engineering, electrical goods, motor cars and consumer services. Output, labour productivity, and total factor productivity all grew much faster between 1924 and 1935 than in 1873-1914 or the mid-Victorian boom, 1853-1872. This is seen clearly in manufacturing in which total factor productivity growth was double the national average (Matthews, Feinstein, Odling-Smee, 1982: 278-279; Ward, 1990: introduction and section 1.1). Taking into account falling prices, the growth in the real economy, measured at constant prices, was as much as one-fifth larger again.

The underlying explanation for the advance is now recognised to be the growth in consumer spending power. On various measures of wages, earnings, and household income, there is general consensus that between the mid-1920s and late 1930s money incomes increased by about 10 per cent, real incomes by more than 20 per cent, and real consumers' expenditure by about 30 per cent (Ward, 1990: section 1.2; Feinstein, 1976: table 65; Mitchell and Deane, 1962: 352-353). Engel's Law assumes that as household income rises so the proportion spent on food will progressively decrease and, as a consequence, that food will under-perform the economy as a whole. Indeed, the food share of consumer expenditure held remarkably steady at around 25 per cent, and may have been larger in 1939 than in 1900 (Feinstein, 1976: table 25; Stone, 1954: 162-174; Stone and Rowe, 1966: chaps. XII, XX). The income elasticity of demand for food, about 0.6, was high for a mature industrial economy, and little changed on nineteenth-century levels. Low or negative elasticities for basic food items were offset by the more positive elasticities of higher-value items, such as milk, fruit and processed foods. In Britain, as in the United States, food manufacturing stood out as one of the best performing and dynamic industries of the first half of the twentieth century. Over the period 1907-1945 net output grew by 7 per cent per annum compared with 5.2 per cent in all manufacturing. Between the 1924 and 1935 Censuses of Production, growth at current prices averaged 3.5 per cent and real growth (at constant 1930 prices) almost 8 per cent. Given the low income elasticity of demand for basic foodstuffs, it follows that food manufacturers must have been highly generative both of new products and added-value to have captured so large a share of rising consumer expenditure. The contention that Britain ate its way through the depression is given credence by the analysis in Table 2, in which it is estimated that between 1920-1924 and 1935-1938 the consumption per head of most classes of food rose by 20 per cent, and that of 'other foods', including the new convenience foods, by more than three-fold.

Table 7.2   Consumers' expenditure per head on food at constant 1938 prices in the UK, 1920-1924 = 100

Kind of Good	1910-1914		1920-1924		1930-1934		1935-1938	
	Constant Price	Index	Constant Price	Index	Constant Price	Index	Constant Price	Index
Bread and cereals	3.32	97.6	3.40	100	3.74	110.0	3.93	115.6
Meat and bacon	7.1	96.5	7.36	100	8.21	111.5	7.96	108.2
Fish	1.07	105.9	1.01	100	1.06	105.0	1.09	107.9
Dairy products and eggs	3.05	93.8	3.25	100	3.94	121.2	4.18	128.6
Oils and fats	1.38	97.2	1.42	100	1.94	136.6	2.07	145.8
Vegetables	1.35	77.6	1.74	100	2.13	122.4	2.16	124.1
Fruit and nuts	1.11	80.4	1.38	100	1.88	136.2	1.77	128.3
Sugar. preserves and confectionary	2.11	129.4	1.63	100	1.98	121.5	2.20	135.0
Beverages	1.03	77.4	1.33	100	1.47	110.5	1.50	112.8
Other food	0.07	77.8	0.09	100	0.23	255.6	0.40	444.4
Total	21.59	95.5	22.61	100	26.58	117.6	27.26	120.6

Source: Stone, 1954: 171.

The late nineteenth and early twentieth centuries were of seminal importance in the growth and development of 'factory foods'. Most of the packaged foods occupying grocers' shelves in 1939 had not existed half a century earlier. Manufacturers were supplying new foods to meet the pressing demands of a more affluent population in a market increasingly driven by consumerism and changing social values. The new foods offered added-value in the form of ease of preparation, a long safe storage life, consistent and attractive tastes, and novelty and variety within the diet. Galbraith observed how, in a mass-consumption society, production creates the wants it seeks to satisfy and fills the void it itself has created (Galbraith, 1962: 132). Housewives, recalled Graves and Hodge in their classic memoir of Britain between the wars, came to count on certain brands of foods, foods asked for by name and sold at fixed prices, foods 'which advertisers never allowed them to forget' (Graves and Hodge, 1941: 175).

As a percentage of total manufacturing, food rose from 7.6 per cent in 1907 to 14.3 per cent in 1935, and numbers of employees by more than three-fold (*Historical Record*, 1978: table 1). A problem in measuring performance is that the food industry as defined by the Census of Production consisted of 12 different industries, initially a few basic ones, rising by 1945 to the full dozen. The statisticians had some difficulty in deciding which industries counted as manufacturing and which were essentially distribution, although the figures are robust enough to distinguish the two. The increasing number of industries and product categories covered by the Censuses reflect not just the ability of statisticians to collect and present data, but a visible change in the kinds of foodstuff being produced, and the gradual move towards higher levels of processing. It was the value added by such processes that changed an industry from one closer to the farm gate to one closer to the retailer and consumer. In 1935, convenience foods, narrowly defined, comprised about 22 per cent by value of the principal products of the food processing industry, and by a more liberal definition, about 36 per cent (Ward, 1990: section 2.1). Growth rates in breakfast cereals, canned specialities, chocolate confectionary and ice cream far outstripped those of the more basic foodstuffs such as bread, sugar, and margarine.

Table 7.3   Production of selected preserved foods in the UK; 1925, 1935, by weight (tons) and value (£000)

Product	1924		1935	
	Output (Tons)	Value (£ 000)	Output (Tons)	Value (£ 000)
Meat and fish paste in tins or glasses	-	779	2,550	497
Canned Herrings	6,300	306	6,700	281
Total pickles sauces etc	-	3,082	53,750	4,025
Marmalade, jams and fruit jellies	158,800	10,980	171,850	6,709
Fruit preserved without sugar	2,850	215	5,150	245
Total preserved fruit	173,750	12,142	207,600	8,657
Bottled jellies (except fruit jellies)	-	63	350	53
Blancmange, custard and similar powders	-	1,447	13,750	1,160
Total value of goods made	-	29,772	-	36,762

Source: Board of Trade, *Census of Production*, 1924, 1935.

Table 7.4   Estimated quantities (tons) of food purchased for final consumption or consumed by producers in the UK: 1922-1924 to 1936-1938, and index quantities

Name of Food Stuff		1922-1924	1930-1932	1936-1938
Imported canned meat	Quantity	38,600	61,850	68,700
	Index	100	160	178
Canned Fish	Quantity	42,700	63,500	79,200
	Index	100	149	185
Condensed Milk	Quantity	34	39	74
(million gallons)	Index	100	115	220
Processed cheese	Quantity	800	22,100	19,850
	Index	100	2,763	2,481
Canned Vegetables	Quantity	1,400	3,400	8,350
	Index	100	243	596
Canned and bottled fruit	Quantity	87,900	141,600	188,500
(Imported and home produced)	Index	100	161	214
Jam and marmalade	Quantity	201,550	196,850	213,750
	Index	100	98	106
Soft Drinks	Quantity	1,720	3,913	5,433
	Index	100	228	316
Ice cream	Quantity	85,550	481,950	1,492,150
	Index	100	563	1,744

Source: Stone, 1954: 50-53, 67-68, 95-96, 120-121, 160-161.

IV. The progression: initial penetration to direct manufacturing

British and American consumption patterns and industrial structures began to converge after 1920, thereby narrowing the gap that had opened up in the third quarter of the nineteenth century. US food manufacturers were quick to exploit this opportunity to test and formally enter the British market.

A guide for American firms wanting to do business in Britain issued by the London Branch of the US Chamber of Commerce in the late 1920s, described her as 'a country different from any other due to her dependence on international trade and imported foodstuffs'. North America was the largest foreign supplier of small grains, pork and pork products, and canned goods. In 1928, food and drink accounted for one-third of all US exports to Britain, while the US and Canada together supplied more than one-fifth of her food needs (Table 7.5).

Table 7.5   The US share imports into the UK: 1913, 1927 (per cent)

Grains	1913	1927
Wheat	32	32
Barley	23	34
Oatmeal	30	29
Meat		
Frozen pork	79	25
Hams	89	80
Canned beef	3	21
Animal fats	97	81
Canned foods		
All fruit (excl. pineapple)		
in syrup	92	91
in water	16	45
Canned vegetables	-	18
Canned milk	39	28
Raisins	1	25
Other dried fruits	80	80
Grapefruit	-	81
Apples	45	60
Pears	30	28

Source: US, UK Overseas Trade Statistics, 1913 and 1927.

In addition to language and tradition, Britain and America shared a diet based on cereals, meat and dairy produce, supplemented by tropical and Mediterranean fruits, beverages and sweeteners. A nation committed to the practice and principles of free trade, prior to the re-introduction of protection in 1932, importers enjoyed unrestricted access to its market place. Britain was a mature industrial country with a large urbanised, geographically concentrated population, high average purchasing power, and a highly developed business infrastructure and communications network. Literacy rates exceeded 90 per cent. A survey by Hulton Research in 1950 revealed that 74 per cent of women, who did most of the shopping, including a large proportion in the lower-income groups, read a morning newspaper, and about 50 per cent a women's magazine, which together represented a large advertising potential across the entire range of social classes (Patterns of British Life, 1950: 84-85). The distribution of goods was facilitated by the growing share of grocery sales now being transacted through multiple retailers with group pur-chasing offices and centralised warehouses. From 17 per cent in 1900, the percentage of groceries sold through cooperative societies and private multiples had risen to 47 per cent by 1939 (Jeffreys, 1954: 163).

Why some firms invested in Britain and others not is a complicated question. As was observed, firms with large advertising expenditures tended to be heavily committed to expansion abroad (Horst, 1974: 127-128). A factor here may have been the relative speed with which intensively promoted consumer goods reached the mature stage of their life cycles. The high marginal cost of increasing brand shares in mature markets was an incentive for firms to look to countries with a greater growth potential. However, investment decisions may not always have been guided by economic criteria alone. HJ Heinz purchased a factory near London in 1905 at least partly out of his great personal affection for the city stretching back to the mid-1880s when the prestigious Piccadilly firm of Fortnum and Mason was persuaded to stock his horseradish sauce and other bottled products (Alberts, 1973: 79, 189). The Kellogg onslaught in the 1930s was a classic response to depression and sluggish sales in the United States (Collins, 1994). After 1932, some firms may have begun manufacturing in Britain to circumvent the tariff, although firms exporting from Canada or Australia enjoyed the benefit of the exemptions and preferential duties accorded to Empire goods (Dunning, 1958: 44-47). Higher tariffs may have obliged some firms, though less so those with strong brands and uniquely formulated products, to withdraw from the market, or invest directly. Some of the most successful American brand companies, among them Campbell, Hershey and Mars, eschewed overseas expansion altogether, even exporting, as a matter of policy.

American manufacturers perceived their competitive advantage less in terms of the innate superiority of their products (Britain was a pioneer of convenience foods, and by 1900 boasted several hundreds of them), nor in manufacturing (albeit that American plant tended to be more automated and faster running), but in promotion, distribution and supply-chain management. The focus was narrow and extremely selective. They showed little or no interest in basic foodstuffs – flour, sugar, oils and fats, margarine or general canning. Nor did they compete where the market was already dominated by strongly branded, old-established, distinctively British products, such as, for example, biscuits, sweets, chocolate, jellies and desserts, pickles and sauces (except tomato sauce), and carbonated soft drinks (except cola). Rather, they looked to products whose appeal lay in their affordability and convenience, for which there existed a potential mass market, and which it was believed would be consumed on a more or less daily basis by all social classes. British manufacturers, in contrast, tended to specialize in niche products and dietary adjuncts targeted at better-off households. American foods were at one and the same time, distinctive, strongly branded and already successful in the United States.

The typical entrant was already a brand leader, financially secure, and with proven marketing skills. Some firms, Horlicks and Heinz, for example, began exporting to Britain inside just a few years of their founding, at a very early stage in the product life cycle. But it was normally only the established firms that could afford to create a national distribution network, mount nationwide advertising campaigns in support of new products, or build and equip a state-of-the-art factory.

Few data exist by which to measure the precise extent of US penetration in the interwar years. Data for the early post-war period indicates that it would have been extremely selective (Horst, 1974: 86, 108, 178). Unpublished Board of Trade statistics for the 1960s show foreign penetration at less than 5 per cent in flour, bread, sausages, ice cream, sugar, preserved fruits, soft drinks and fruit juices. This compares with levels of 80 per cent or above for custard powder and canned baby foods, and 'over 40 per cent' for breakfast cereals and cake mixes. In 1963, the consumption per head of processed foods in Britain was substantially higher than in any other western European country. Out of seven leading products, Britain was the largest user in four, and in second place in two. Six years earlier, she had accounted for two-thirds of all sales by European- based US subsidiaries.

V. Business strategy

US food manufacturers building a position in the British market did so cautiously, usually proceeding through a series of stages, each representing a higher level of sales activity and financial commitment. Most began by importing through independent local agents selling on commission, or in some cases exporting directly to individual stores, mainly in London, if only to meet expatriate demands for unique American foods, such as breakfast cereals. The next step was to establish a wholly owned sales subsidiary and regional branch offices. Confident of success and with a distribution network in place, the final step was to build a factory to supply the UK and markets further afield, in continental Europe and the Empire. At this stage, European branch offices were normally controlled from a head office in Britain, in most cases near London.

The following few examples serve to illustrate the above progression. HJ Heinz began importing through a local agent in the 1880s, set up a branch office in 1899, and most unusual for an American firm, in 1905 bought out a small British company in West London, and used the brand name of the previous owners until 1910 when the Heinz label was re-introduced (Alberts, 1973: 79, 189). Coca-Cola was already selling small quantities of its patent syrup directly to London soda shops in the early 1900s, and began regular sales through a commission agent about 1910. It began bottling on contract in 1924, established a wholly owned import subsidiary in 1929, and in 1935 opened its first bottling plant at Chiswick in West London (Giebelhaus, 1994: 196-197). Kellogg began importing on a small scale before the First World War, formed its own sales subsidiary in the early 1930s, and opened a factory at Trafford Park, Manchester in 1938 (Collins, 1994: 237-243). Kraft set up a London sales office in 1924, followed by a factory in the west of England in1928 (Kraftco, 2003). The Horlicks story is complicated (Ward, 1994: 263ff). The original partnership was formed in 1873 in Wisconsin by James Horlick, a pharmacist who emigrated from Britain in 1867 with a new product for infants and in-valids. The patent malted-milk powder, 'Horlicks', was first marketed in 1883; exports to Britain started almost at once, manufacturing in 1906. In the early 1920s, the business was divided into two separate companies, one based in the US with marketing rights for the Americas, the other based in Britain with marketing rights for the rest of the world. Only one example is known of a joint venture, that formed between General Foods and Unilever in 1937 to exploit the Birdseye patent (Trager, 1996: 494). Two examples are known of a takeover of a British company together with its trademarks: those by CPC of Brown and Polson, a leading manufacturer of cornflour, baking powder and custard powder in 1932 (Ward, 1994: 263-264), and by Mars of the pet foods firm, Chappel Brothers Ltd in 1935 (Sutton, 1991: 500-502). Forrest Mars broke with the family firm and established his own independent company in Britain in 1932, making Mars Bars, a product specifically designed for the British market (Ward, 2003).

UK operations accounted for only a small share of parent company revenues, and in the early years most subsidiaries ran at a loss. The Heinz London factory, for example, purchased in 1905, did not become profitable until 1919. Prior to 1930, indeed, only a handful of US foods had achieved a national distribution; many were oddities, stocked by high-class grocers and department stores (Ward, 2003). The 1930s saw a general upsurge in sales and direct investment. By 1939 US firms accounted for probably less than 10 per cent of all UK sales of preserved foods, but were brand leaders in a number of key product areas. Those quintessentially American institutions, the soda fountain and milk bar, selling flavoured milk drinks, sodas and ice cream, made rapid headway in Britain's High Streets. Around 1900, Selfridges, an American-owned department store in London's Oxford Street, installed a soda fountain, which is supposed to have made its debut at the Bath and West of England Agricultural Show in 1934 (Giebelhaus, 1994: 196-197; Royal Bath and West of England Agricultural Society, 2003). By the late 1930s many towns in the London region boasted at least one such establishment. Popular enthusiasm for vitamin foods – milk, fresh fruit, green vegetables – was inspired by American discoveries in the nutritional sciences, and glowing reports of the effects of the new American health diets in newspapers and women's magazines (Griggs, 1986: chaps. 10 and 12; Cummings, 1941: 211-212; Graves and Hodge, 1941: 189-190). Those who lived through those times testify to the powerful influence of America on the popular imagination and everyday living, epitomised at the nation's dining tables by baked beans, canned peaches and cornflakes.

VI. Breakfast cereals and canned foods

Invented in the American mid-west around 1900, most of the major brands and formulations of ready-to-eat breakfast cereals – corn flakes, wheat flakes, puffed wheat, grape nuts, shredded wheat – were obtainable in London in 1914. Rolled oats, the Quaker hot breakfast food, was already well-known (Bound, 1966: 1; Marquette, 1967: 217). First shipped in 1877, it was vigorously promoted in a famous campaign in the 1890s, when banners proclaiming the new breakfast food were draped across the face of the cliffs at Dover, visible for miles out at sea. Cold cereals were not eaten widely until the 1930s when the Kellogg Company launched an intensive countrywide advertising campaign with the declared aim of transforming what were widely regarded as food oddities into staple breakfast foods (Collins, 1994: 239-242). Annual per capita consumption of ready-cooked cereals rose over the decade from 5 to 42 grams, and by 1939 were consumed on a more or less regular basis in a majority of households, especially those with children (Collins, 1994: 224-227). The Kellogg campaign had a strong generic effect, boosting the sales of all manufacturers and encouraging the introduction of new breakfast products (J Walter Thompson Archive, Kellogg files). A 1939 survey showed 62 per cent of porridge-eating homes using Quaker Oats; and in cold cereal-eating homes, 50 per cent using Kellogg's Corn Flakes, 23 per cent Shredded Wheat, 10 per cent Force Wheat Flakes (from Canada), 4 per cent Quaker Puffed Wheat, 2 per cent Weetabix (a wheat biscuit made by a South African Company with Seventh Day Adventist (and Australian) connections based in the East Midlands), and 2 per cent Quaker Puffed Rice. Inside a decade the British breakfast had been transformed. So successful was the Kellogg campaign that, by the end of the period, 'corn flakes' had become that rare phenomenon, the eponymous product known simply as 'Kelloggs'.

Canned foods, too, were dominated by American brands, whose stranglehold, though somewhat loosened in the 1930s, was not effectively broken until after 1950. US production expanded rapidly after the Civil War, and exponentially from 1900 with the perfection of the high-speed can. In 1924, between 85 and 90 per cent of all canned and bottled food consumed in Britain – over 90 per cent of all canned fish, meat, vegetables and fruit in syrup, and over three-quarters of condensed milk – was imported, much of it from US owned firms in Canada and Argentina. The small British canning industry was largely in Swiss or American ownership (table 7.6) (Johnson, 1976: 173).

Table 7.6   US and Canadian shares of canned food imports into the UK, 1938

Product	US - Canadian Share (%)
Canadian milk	60
Pickles and canned vegetables	80
Salmon	60
Canned fruit	
apricots,grapefruit, peaches	50
pears	65

Source: US, UK Overseas Trade Statistics, 1938.

British interest in canning had flickered briefly in the First World War, but was soon snuffed out by the resumption of imports linked with falling prices from 1921. The 1924 Census of Production put the total UK output of canned foods at no more than 60,000 tons (Statistical Abstract, 1924-1928: 390-396). The home industry laboured under serious disadvantages (Johnson, 1976: 177-178; Horrocks, 1993: chap 6; Metal Box Company, undated; Thorpe, 1986; Ward, 1990: section 2.8; Plummer, 1937: 228-251). One was the small and irregular supplies of raw materials and limited range of commercially viable products. A second was the low level of canning technology: until well into the 1930s some firms still made their own cans, often by hand, rubber rings were used on can ends, and cans were still sometimes sealed with solder. Another disadvantage, seemingly insuperable, was the availability, cheapness, uniform high quality and competitive strength of the American brands. Ironically, American technology played a crucial role in the belated rise of the British canning industry after 1926, when the firm of Williamson began experimenting with American can-making machinery, and the Smedley company sent a son to America to learn the business first hand, prior to installing a canning line on the American model. The stage was set by the decision of The Metal Box Company to begin the large-scale manufacture of high-speed, open-top cans using American equipment under license. Between 1930 and 1937 can output in the UK rose from 23 million to 335 million, and the output of canned foods from 75,000 to 340,000 tons (see Table 7.7). Even so, in 1938, 70 per cent of canned goods were imported (Johnson, 1976: 184), two-thirds of home -produced canned milk was made by Swiss and American-owned firms, and over 90 per cent of canned soups bore a Heinz label. Notwithstanding protective duties and technical help in the form of specially cultivated varieties of fruit and vegetables suitable for canning, the British canning industry was overshadowed and insecure. Unflattering comparisons were drawn between imported and home-produced canned foods, the former offloaded in bulk at the major ports packed to meet wholesalers' needs and graded to a uniform standard; the latter sold in hundreds of markets in small lots of varying quality under unfamiliar brand names. How, it was asked, could American canned loganberries sell at 7 shillings per 12 cans where the British sold for 9 shillings. And why was the proportion of rejects in British can-making plants so much higher than in American? British canners lobbied for a special marketing board to promote British produce under a universal brand name, with a quality guarantee. The gap was closing, to be sure, but American firms continued to set the standard.

Table 7.7   Canned food production in Britain: 1924, 1937 (1924=100)

	Output	Volume	Growth: average per annum %	
1924	100	100	1924-1930	9,8
1937	554	171	1930-1937	35,7

	Output				
	Meat	Vegetables	Fish	Fruit	Milk
1924	100	100	100	100	100
1935	212	3.982	105	559	476

Source: Board of Trade, *Census of Production* 1924, 1937.

VII. American business practice

American firms brought with them a business culture that pervaded every aspect of company activity and sharpened their competitive edge. American entrepreneurial capitalism in the first and second generations threw up some very successful businessmen who, through a combination of innovativeness, shrewdness, boundless energy, self-belief and willingness to take risks, founded what were to become large and successful food corporations and whole new industries. The corporate firms that succeeded them, whilst less spontaneous, were more single-minded, more ruthlessly competitive, and more driven by profit than the typical British firm. They demanded of their employees unquestioning loyalty to the company and its ideals (Turner, 1969: chap. 7). Far less is known about their manufacturing and financial aspects than about sales and marketing. Few company records appear to have survived for the period, or where they have done, are lodged with the parent company, as in the case of Heinz, or are closed to researchers. Fortunately, there exists the clients' archive of the London branch of J Walter Thompson, the New York advertising and market research agency which from the 1920s worked for a number of leading US and British food companies. This provided the basis for A. V. Ward's research on the history of convenience foods, and the author's on ready-to-eat breakfast cereals (Ward, 1990; Collins, 1994).

American firms were in the main more hierarchical and inflexible than British firms. They normally functioned not as independent units but as replicates of the parent (Southward, 1931: 133, 143). The whole approach to decision-making, observed J. H. Dunning, was divided along US lines, with frequent referrals to head office for approval, and clearly laid-down procedures for all main aspects of the business from the purchase of materials through to after-sales service (Dunning, 1958: 107-108, 170-171, 199-200). A training in American business principles and company philosophy was obligatory for all British managers joining American firms. British subsidiaries had little discretion as to what they sold or manufactured. Apart from a few concessions to British taste, these were essentially the same in style, formulation, even packaging, to the home product. Research and development was carried out mostly in the United States. Up to the Second World War, Heinz produced only canned beans and tomato ketchup in Britain, importing all other lines from its American plants. American firms used mostly American machinery and even American raw materials, in order, so Kellogg claimed, to be independent of domestic suppliers who did not always deliver to schedule or to the agreed level of specification.

Company histories tend to highlight the most successful products and to pass quickly over the failures. That a large proportion of promotional budgets was wasted is a truism, many products disappeared from wholesalers' lists within a short time of their introduction. What is not entirely clear is why so relatively few of the enormous number of products made for the American market were distributed in Britain, or built up a strong position there. Among the breakfast cereals, popular American brands such as Sugar Puffs and Post Toasties, met a lukewarm response; a modest selection only of the Heinz portfolio of over 100 canned and bottled specialities reached British stores. Many leading US food manufacturers were reluctant to expand overseas, preferring instead to consolidate their position at home, or delayed doing so until after 1950, when the lure of the now rapidly growing and increasingly affluent European and Japanese markets proved irresistible.

VIII. Advertising and promotion

US firms set new standards in advertising practice. The Quaker campaign of the 1890s, one of the most aggressive and unorthodox ever waged in Britain, the Kellogg campaign of the 1930s, which re-shaped the British breakfast, and the Horlicks 'sleep campaign' of 1933, were models of their kind. The hugely successful campaigns commissioned by growers' associations to increase sales of Californian walnuts and raisins to Britain in the glut years after the First World War inspired a succession of British-run generic campaigns promoting bananas, milk, herrings, bread and Empire fruit.

'The real secret of these firms' success', concluded Horst, 'was their willingness to lavish large sums on advertising campaigns, promotional schemes, and other methods of brand promotion' (Horst, 1974: 123). Britain in the 1920s was said to be 20 years behind America in sales management techniques, sales having only recently become recognised as a separate function, whereas in America it was a primary activity. American firms' sales were driven by their advertising budgets, whereas most British firms regarded advertising as an option, and judged it as much by artistic as commercial criteria.

In a letter to *The Times* (24 July 1942), the economist, Sir Stanley Jevons, made this observation. 'Some 70 years ago, manufacturers discovered that advertising was an easier road to profits than investing in ways of lowering costs, competition took the form of more and more elaborate advertising and attractive packaging devices'. 'Persuasive', as against 'informative' advertising, allowed big producers with large budgets to dominate the market at the expense of smaller rivals (Bishop, 1944: 110, 155). Food manufacturers were operating in an increasingly imperfect market, in which competition was based not on price-cutting but on advertising. Many of the most successful products – breakfast cereals, meat extracts, cola drinks, malted beverages – were conceived originally as health foods, and were promoted in much the same way as patent medicines (Corley, 1994: 218). Up to the 1920s, the manufacturers' first priority was to reassure a suspicious public as to the purity of the ingredients and intrinsic goodness of their products. As safety fears receded, the promotional thrust shifted towards nutrition and convenience. Advertising expenditure as a percentage of GDP rose from an estimated 1.3 per cent in 1907 to more than two per cent in 1937 (Ward, 1990: 205). In 1938, food accounted for about 13 per cent of all advertising expenditure, more than patent medicines and similar to household supplies. The staple foods – bread, meat, milk, potatoes, green vegetables, sugar and rice – together accounted for over 80 per cent of food expenditure, but less than 20 per cent of all food advertising (Bishop, 1944: 82-103).

Table 7.8 Manufactured foods as a percentage of press advertising expenditure and total food expenditure in the UK, 1938

	% of total press advertising expenditure (A)	% of total expenditure food (B)	ratio A/B %
Breakfast cereals	4.0	0.2	20.0
Meat extracts	4.0	0.3	13.3
Sauces and pickles	2.2	0.2	11.0
Health foods and drinks	15.5	2.0	7.8
Cheese	1.8	1.4	1.3
preserved cheese	1.6	0.3	5.3
Jams and preserves	5.6	1.5	3.7
Sugar and chocolate confectionary	15.3	4.5	3.4
Canned products	5.8	3.0	1.9
Tea, coffee, cocoa	6.0	4.5	1.3
Biscuits	2.6	2.2	1.2
Butter, margarine	4.1	6.3	0.7

Source: Kaldor and Silverman, 1952, passim.

Kaldor and Silverman's analysis of advertising expenditure in 1935 confirms the dominant position of proprietary foods in the national lists (Kaldor and Silverman, 1948: table 7, 144-145). Malted health foods, breakfast cereals, canned foods and chocolate confectionary, followed by custard powders, blancmanges, sweets, sauces and pickles, baby foods and beef extracts and essences, all had large outlays relative to their share of food expenditure. Spending by breakfast cereal manufacturers alone exceeded that on bread, flour and dairy products (excluding processed cheese) combined. The rapid growth in sales over the 1930s bore a direct correlation with the sizes of advertising budgets. Kellogg for example, spent nearly $4 million, about 20 per cent of its revenue, on Corn Flakes, sales of which rose from 284,000 cases in 1931 to 1,834,000 cases in 1939 (Collins, 1994: 240). Between 1928 and 1938, sales of Horlicks' malted beverage rose threefold, from £88,000 to £336,000 while advertising expenditure rose fourfold (Ward 1994: 264-270). Although primarily an 'up-market' product directed at AB class homes, by 1937 the proportion of sales of Horlicks to class CD purchasers exceeded 85 per cent. At Brown and Polson (acquired in 1932 by CPC), sales of custard powder rose from £5,600 in 1924-1925 to £74,700 in 1937-1938 which, though far smaller than those of the market leaders, Birds and Foster Clarke, reflected the more aggressive marketing policy of the new owners (Ward, 1990: 312-25). Advertising-sales ratios exceeded 30 per cent in gravy mixes, custard powder and malted health drinks, compared with less than 3 per cent in food and drink overall (Kaldor and Silverman, 1948: 144-145).

Table 7.9   Advertising expenditure on food products and health drinks in the UK, 1935

	Manufacturers net sales (£000)	Manufacturers total advertising in the UK (£000)	Advertising as % of net sales
Malted health drinks	2,300	950	41
Custard powder and blancmanges	1,200	450	38
Gravy mixtures	480	160	33
Beef extracts and essences	2,600	570	22
Canned and packet soup	730	160	22
Cocoa and chocolate powder	1,400	290	21
Proprietary breakfast cereals	3,900	700	18
Dog biscuits and food	1,400	190	14
Baking powder	380	40	£11
Mustard	1,000	100	10
Table jellies	1,450	140	10
Sauces, pickles and salad cream	2,500	230	9
Coffee extracts and essences	1,100	100	9
Meat and fish paste	1,600	60	4
Chocolate confectionary	16,800	630	4
Soft drinks	9,400	340	4

Source: Corley, 1994: 223.

Individual Americans had been a dynamic force in British advertising since the 1860s, and had pioneered a number of innovations, such as multi-sheet posters, and full-page advertisements in national newspapers. In the belief that British advertising would benefit from an injection of American know-how, American 'ad-men' settled in London in search of commissions (Corley, 1994: 69; Heindel, 1968: 194-195; Southard, 1931: 156-158).

Though of a high artistic standard, British advertising was criticised as eclectic, poorly focused and lacking punch. Display advertisements commissioned by American firms tended initially to be similar or identical to those at home, which diminished their appeal to an audience used to a more graphic approach. From the 1920s, brand companies began using the services of British branches of US agencies whom, they believed, understood better the mood of consumers, and had closer working relationships with their clients than British agencies. Without question, the most important development of the inter-war years was the marriage of advertising and market research, signifying the beginnings of a more scientific approach to marketing and product development (Corley, 1987). It became normal for large brand companies to commission market research to inform promotional campaigns. Advertising was adapted to the product and vice versa, and supported the brand rather than, as before, being a mere marketing tool. Through research into the socio-economic profiles of buyers and their reactions to various forms of advertising, firms were better placed to decide which products to promote and which to discard. This was a first stage in the development of 'socially engineered' foods designed to meet the needs of scientifically defined and identified classes of consumer.

J Walter Thompson, a successful New York advertising broker, pioneered the science of market research (West, 1987: 198-217; Nevett, 1984). He formed a research department, staffed by academics including a behavioural psychologist, to develop scientific models, and a consumer panel consisting of families whose buying habits were surveyed and the results supplied to clients. A London branch office was established in 1899, and subsequently a network of offices across Europe. From the 1920s the client base grew to embrace a number of leading US manufacturing companies, such as General Motors, Kodak and Gillette, and food companies, such as Horlicks, Coca-Cola, Kellogg, Quaker Oats and Libby McNeill Libby. Some firms, among them Horlicks, the malted beverage manufacturer, were provided with a complete service, from advising on new products and marketing strategy to designing and managing entire advertising campaigns. The agency's test-marketing facilities were said to be the best in Britain (Corley, 1994: 225). Three organisations – The London Press Exchange, the British Market Research Bureau, and J Walter Thompson – carried out probably two-thirds of all market research done in Britain between the wars. Better known to social and economic historians is the British firm of W S Crawford, responsible for the Milk Marketing Board's 'Drink more Milk' campaign, which commissioned a celebrated study of British eating habits based on market research, *The Nation's Food,* published in 1938 (Crawford and Broadley, 1938). The following year saw the arrival at Oxford of the Chicago firm of A C Nielsen, a pioneer in the development of accurate techniques for tracking sales and calculating market shares of all leading brands in proprietary foods and other consumer products (A.C. Nielsen, website).

IX. Conclusion

American principles of 'scientific management' took hold only slowly in Britain, but to a greater degree probably in the food industries than any other. Sir Thomas Lipton, the most successful of the new breed of multiple grocery retailer, had as a young man worked for four years in a New York food store (Matthias, 1967: 111; Heindel, 1968: 192). 'During my travels in America' he wrote in his diary, 'I have observed that the firms that were making good were all regular advertisers'. He had been greatly impressed by the spirit of genial combativeness, 'the go get it' approach which stamped the American spirit all the way through. The Swiss firm of Nestlé (which along with the Anglo-Dutch firm of Unilever was later to challenge American international leadership in manufactured foods) had American origins (Horst, 1974: 14-15). Charles H Page, the US consul in Zurich, co-founded the Anglo-Swiss Condensed Milk Company, which merged with Nestlé in 1866. An employee, John B. Meyenberg, emigrated to the United States where he set up the Helvetia Condensing Company, the forerunner of the Pet Milk Company, and later played a key role in the early development of the Carnation Company. In the 1930s, the British ice-cream makers, Lyons and Walls, had looked at advertising and distribution in the US before setting up their manufacturing plants (Food Manufacturing, 1939: 190-199). In meat distribution, too, American companies set the pace, running their ships to schedule and demanding 100 per cent payment and weekly settlements, a procedure that British companies were to imitate (Horrocks, 1993: 64).

We will consider briefly the effects of the processed foods industry at other points of the food chain, in agriculture and food retailing. Being unimportant in Britain, meat canning had little direct impact on the livestock industry. The increase in canned milk production after 1928 reduced the dependence on imports, and after 1933 provided the newly established Milk Marketing Board with alternative outlets for its growing milk surpluses. Breakfast cereals conferred few benefits on British grain growers; much of the raw material – maize, high- gluten wheat, rolled oats – was imported from North America. The growth of the canning industry after 1926 did not have quite the stimulating effect on British horticulture as was first hoped (Astor and Rowntree, 1938: chap. X). A major constraint, difficult to counter, was that British consumers preferred canned peaches, pineapples and sweet juicy pears grown in warmer climates, to canned indigenous fruits. Soft fruits were either sold to the consumer direct, or used for jam, and while jam production increased, a growing proportion was made from imported fruit pulp. The principal fruit used in canning was the plum, but canners were constantly complaining of shortages of the preferred variety, the Victoria, and of having to use the inferior Pershore instead. Progress was held back by the unreliability, variable and indifferent quality, and high cost of the popular soft- and stone - fruits.

Canning benefited vegetable-growers more than fruit-growers (Astor and Rowntree, 1938: chap. IX; Ward: 1990: section 2.8). The cultivated area under the one grew by nearly 80 per cent between 1922 and 1936; the other stood still. The range of vegetables suitable for canning was limited, however, and the growing season short. The output of canned peas exceeded that of all other canned vegetables combined. Asparagus and tomatoes were luxury items, sold direct. Large quantities of vegetables were used in sauces, pickles and soups, although those employed in the manufacture of the great Heinz staples, Baked Beans and Tomato Ketchup, were mostly imported. With a very few exceptions, such as the Cadbury's and Horlick's dairy farms, the Horlick's maltings at Sudbury, and Colman's, the mustard- makers, who grew their own mustard seed, there was little integration from manufacturing back into primary production. A few large retailers had their own wholesaling facilities, and some, such as the Cooperative Wholesale Society, had their own jam and canning factories. Branded and packaged foods played an important but unacknowledged role in the re-structuring of food retailing. Preserved foods were sold by grocers and provision merchants rather than by butchers, greengrocers and other specialist trades dealing in fresh produce. The grocer was no longer an expert intermediary between producer and consumer, selecting and blending produce for sale, but increasingly the distributor of pre-packaged foods asked for by name and sold at a pre-determined price. The growing importance of processed foods was to the particular advantage of the multiple grocers, skilled merchandisers with their own dedicated supply chain, who purchased in bulk, often directly from the manufacturers. Tin-plate producers, can makers and packaging firms also would have benefited from the growth of the processed food industry, as much perhaps as the farmer.

By the late 1930s, most of the elements that were to underpin the food revolution of the post-war period were already in place. Food manufacturing was a modern consumer-goods industry, convenience foods an integral part of the nation's diet, and packaged branded foods the grocers' chief stock-in-rade. The contribution of manufacturing, the secondary sector, to the added-value created by the UK food chain had steadily risen, while that of agriculture had shrunk. At the present day, primary production generates about 13 per cent of added-value compared with 31 per cent by manufacturing, and 21 per cent each by retailing and food services (Table 7.10).

Table 7.10   The cost of agricultural raw materials as a ratio of the output value of selected foodstuffs, 1935

	Cost of agricultural raw materials (A) £ million	Output Values (B) £ million	B/A %
Biscuits	4.09	16.87	312.00
Bread and cakes	28.80	63.99	112.20
Butter. cheese. condensed milk and margarine	28.74	14.57	97.20
Chocolate and Sugar confectionary	8.68	36.80	324.30
Grain	65.13	46.50	41.30
Preserved foods	8.43	36.76	336.30

Source: Board of Trade, *Census of production*, 1935.

By the outbreak of World War 2, a core of leading American food firms had established themselves in Britain and to a lesser extent in western Europe. The second stage in the globalisation of the food industry, first agriculture then manufacturing, had been reached. North America – its companies, food concepts, product design, technology, merchandising skills – played an all-important enabling role in the process.

Bibliography

I owe a special debt of gratitude to Dr Vernon Ward formerly of the Roehampton Institute, University of Surrey, for his advice and assistance with the sections on the inter-war UK food economy, especially convenience foods, based on his PhD thesis and related research.

A.C. Nielsen website: http://www.acnielsen.co.uk. September 2003.

Alberts, R.C. (1973) *The good provider*, London.

Alderfer, E.B. and Michel, H.E. (1950) *The economics of American industry*, 2nd edn., New York.

Astor, Viscount and Seebohm Rowntree, B. (1938) *British agriculture*, London.

Ball, E.N. (1970) 'Multi-national firms', *Business History*, 18, pp. 76-93.

Bishop, F.P. (1944) *The economics of advertising*, London.

Board of Trade (1931) *Third Census of Production of the UK, Final Report*, London.

Board of Trade (1936) *Final Report on the 1935 Census of Production and the Import Duties Inquiry*, London.

Board of Trade (1939) *Statistical Abstract for the UK, 1924-38*, Command 6237, London.

Board of Trade (1940) *Final Report of the Census of Production, 1936*, London.

Round, J.A. (1996) *A brief marketing history of Quaker Oats Limited*, 2nd edn., mimeograph, Southall, London.

Braznell, W. (1982) *California's finest: history of the Del Monte Corporation and the Del Monte brand*, privately published by the company.

Collins, E.J.T. (1976) 'The 'consumer revolution' and the growth of factory foods: changing patterns of bread and cereal-eating in Britain in the twentieth century', Oddy, D. and Miller, D. (eds.) *The making of the modern British diet,* London.

Collins, E.J.T. (1994) 'Brands and breakfast cereals in Britain', Jones, G. and Morgan, N.J. (eds.) *Adding value: brands and marketing in food and drink*, London.

Collins, J. (1924) *The story of canned foods*, New York.

Connor, J.M. and Mather, L.L. (1978) *Directory of the 2000 largest U.S. food and tobacco processing firms, 1975*, U.S. Department of Agriculture, Washington, D.C.

Corley, T.A.B. (1987) 'Consumer marketing in Britain 1914-1960', *Business History*, 29, pp. 65-83.

Corley, T.A.B. (1994) 'Best practice marketing of food and health drinks, in Britain 1930-1970', Jones, G. and Morgan, N.J. (eds.) *Adding value*, London.

Crawford, W.S. and Broadley, H. (1938) *The people's food*, London.

Cummings, R.O. (1941) *The American and his food*, Chicago.

Dunning, J.H. (1958) *American investment in British manufacturing industry*, London.

Feinstein, C.H. (1976) *Statistical account of national income, expenditure and output of the U.K. 1855-1965*, Cambridge.

Foley, J. (1972) *A history of General Foods*, Banbury, England.

Giebelhaus, A.W. (1994) 'The pause that refreshed the world: the evolution of Coca-Cola's global marketing strategy', Jones, G. and Morgan, N.J. (eds.) *'Adding value'*, London.

Graves, R. and Hodge, A. (1941) *The long weekend: a social history of Great Britain, 1918-1939*, London.

Griggs, B. (1986) *The food factor*, Penguin, London.

Hampe, E.C. and Wittenberg, M. (1964) *The lifeline of America*, New York.

Heindel, R.H. (1940) *The American impact on Great Britain, 1898-1914*, New York.

Herr, J. (1966) *World events 1866-1966: the first hundred years of Nestlé*, London.

Historical Record of the Census of Production (1978), Government Statistical Service, London.

Hoffmann, A.C. (1940) *Large scale organization in the food industry*, Temporary National Committee, Washington D.C.

Horrocks, S.N. (1993) *Consuming Science: Science, Technology, and Food in Britain*, unpublished PhD thesis, University of Manchester, 1993.

Horst, Thomas (1974) *At home abroad*, Cambridge, Mass.

Jeffreys, J.B. (1954) *Retail trading in Britain 1850-1950*, Cambridge.

Johnson, J.P. (1976) 'The development of the food canning industry in Britain during the inter-war period', Oddy, D. and Miller, D. (eds.) *The making of the modern British diet*, London.

Jones, N.B. (1998) 'Food in south Wales', Williams, S. and Taggert, F. (eds.) *British foods*, Cardiff.

Kaldor, N. and Silverman, R. (1948) *A statistical analysis of advertising expenditure and of the revenue of the press*, Cambridge.

Kraftco (2003): Information supplied by the firm.

Kellogg' and Horlicks files, c. 1920-1950, J. Walter Thompson Archive, J. Walter Thompson, Berkeley Square, London.

Lewis, C. (1938) *America's stake in international investments*, Washington, D.C.

Marquette, A.F. (1967) *Brands, trademarks and goodwill: the story of the Quaker Oats Company*, New York.

Matthias, P. (1967) *Retailing revolution*, London.

Metal Box Company (n.d.) *A history of canned foods*, Reading.

Mitchell, B.R. and Deane, P. (1962) *Abstract of British historical statistics*, Cambridge.

Monopolies Commission (1973) *Report on the supply of ready-cooked breakfast foods*, London.

National Commission on Food Marketing (1966) *Studies of organization and competition in grocery manufacturing*, part 2, Washington D.C., pp.55-240.

Nevett, T. (1984) *Advertising in Britain: a social history*, London.

Patterns of British Life, London, 1950.

Plummer, A. (1937) *New British industry in the twentieth century*, London.

Royal Bath and West of England Agricultural Society (2003), *Ex inform.*

Smith, B.A. (1994) *The food industry*, London.

Southard, F.A. (1931) *American industry in Europe*, Boston, Mass.

Stone, R. assisted by Rowe, D.A. et al. (1954) *The measurement of consumers' expenditure and behaviour in the UK, 1920 – 1938*, vol. 1, Cambridge.

Stone, R. and Rowe, D.A. (1966) *The measurement of consumers' expenditure and behaviour in the UK, 1920 – 1938*, vol. 2, Cambridge.

Sutton, J. (1991) *Sunk costs and market structure*, Cambridge, Mass.

The Economist, 10 June 1939, 'Consumption and trade in canned food'.

Trager, J. (1996) *The food chronology*, London.

Thorne, S. (1986) *The history of food preservation*, Kirby Lonsdale, England.

Turner, E.S. (1965) *The shocking history of advertising*, London.

Turner, G. (1989) *Business in Britain*, London.

U.S. Department of Agriculture (1990) North Central Regional Project, NC-194, '*Economic Studies*', no.1, Washington D.C.

U.S. Department of Commerce (1930) *American direct investments in foreign countries*, Trade Information Division, Washington D.C.

U.S. Department of Commerce (1957) *American direct investments in foreign countries*, Trade Information Division, Washington D.C.

Ward, A.V. (1990) *Economic Change in the UK Food Manufacturing Industry 1919-1939; with special reference tot convenience foods*, unpublished PhD thesis, University of Reading.

Ward, A.V. (2003) *Ex inform.*

West, D.C. (1987) 'From T-square to T-plan: the London office of the J. Walter Thompson advertising agency 1919-70', *Business History*, 29, pp. 198-217.

Woodcock, F.H. and Lewis, W.R. (1938) *Canned food and the canning industry*, London.

8 The history of the sardine-canning industry in France in the nineteenth and twentieth centuries

Alain Drouard, CNRS, Paris

I. Introduction

Sardines were the first kind of fish to undergo the factory-canning process. Until the middle of the nineteenth century the catch from fishing, an age-old activity on the coasts of Brittany, was eaten fresh or salted by the coastal populations, but was mostly transported to the towns by cart. What was left over was sent to be pressed. Sardines were packed into barrels and pressed so as to extract oil for lighting. The barrels of pressed sardines could keep for up to two years and were exported to Nantes, Bordeaux and Spain. All of this changed with the process called 'appertisation'. From then on the fishing industry supplied the canning industry which mushroomed on the Vendean and Breton coasts.

In France, the sardine-canning industry was above all the result of invention and innovation. The invention belonged to a confectioner named Nicolas Appert who perfected a safe method of preserving food at the beginning of the nineteenth century. At the same time, the technical innovation of using tin cans instead of glass jars was a deciding factor in enabling the rapid development of the sardine-canning process. This new industry, combining modern and traditional methods, originated in Nantes in the 1820s and spread to the Atlantic coast, mainly to the Vendée and Brittany, in the second half of the nineteenth century.

A highly labour-intensive industry, the canning factories employed thousands of men and women during the summer fishing season, giving rise to major shifts in population as migrant workers moved from the hinterland to the coasts and between ports. For over a century and a half, the canning industry experienced varying rates of growth, depending on the risks and uncertainties of the fishing. Growth and prosperity alternated with recession and poverty, before an irreversible decline following the Second World War (Boulard,1996). Only ten of the two hundred factories operating at the end of the nineteenth century still exist today. For many years, the canning industry had a marked effect on the way of life in Brittany, to such a degree that sardine fishing had its own culture, with its own ceremonies, festivals, traditions and historical conflicts. This has totally disappeared today.

II.   Two centuries of sardine-canning industry in France

II.1.   A revolutionary invention

Until the end of the eighteenth century, the various ways of preserving food – by drying or smoking, or by the use of salt, sugar or vinegar – were not considered satisfactory as they altered the nature of the product concerned. Moreover, the period of preservation was often extremely limited. Supplying the sea-going Navy and armies abroad with foodstuffs was not only a difficult undertaking but also the source of serious problems, as Duhamel du Monceau (1759) noted: 'The salted meat eaten by the crews seems to be one of the principal causes of scurvy'. He added: 'The crews' health improves when they can remain a while at ports of call where the fishing is good, and they are able to enjoy fresh fish'.

The invention of a man who was not a scientist but a mere confectioner would become a major landmark in the history of food preservation. Nicolas Appert was born on November 17th 1749 in Châlons-sur-Marne, where his father was an innkeeper. In 1772 he became catering steward to the Palatine Duke Christian IV, and to the Princess of Forbach in 1775.[1] In 1784, he opened a confectionery business in the Rue des Lombards in Paris. At that time, a confectioner was a caterer who prepared foods using sugar, fat or salt to preserve them. He was, therefore, in an ideal position to see the limitations of such methods of preservation. One should remember that at the time no one knew why food decomposed.

While he was testing a new method that involved heating food in a hermetically sealed container, Nicolas Appert accidentally discovered the principle of sterilisation through boiling.Thus, a century before Pasteur, he gave the first definition of pasteurisation: 'Heat destroys, or at any rate neutralizes, the fermenting agents that in the normal course of nature produce those modifications which, by changing the constituent elements of animal substances, alter their qualities'.

At first he used champagne bottles from his native region, and later had bottles made with a wider neck that were easier to fill. In 1795 he set up a workshop in Ivry-sur-Seine where he spent about ten years improving his method and perfecting his invention. In 1803 he moved to Massy, where he acquired a plot of land to grow fruit and vegetables for making preserves. The business employed up to fifty workers.

In addition to bottles, Appert used tin cans from England that were hermetically sealed by welding the lid. This discovery was acclaimed by a leading gastronomer, Grimod de la Reynière (1805) who wrote in the third edition of his culinary guide *L'Almanach des gourmands*: 'We shall speak only of the result, namely to find in each jar an excellent and inexpensive dish, that recalls May in mid-winter, and that one might even mistake for fresh when dressed by a skilled cook. It is no exaggeration to say that peas, in par-

---

[1] The catering steward was not a cook. His role was to supervise the supply of foodstuffs for his employer's table. Nicolas Appert, like Vatel, was a catering steward.

ticular, prepared in this fashion, are as green, tender and tasty as those one eats in high season (...). In the same manner, certain fruits have been successfully preserved, such as redcurrants, cherries, raspberries, greengages, plums, apricots, damsons, nectarines and even peaches. Naturally, the latter especially cannot be preserved whole in view of the narrowness of the bottle-neck upon which the success of the operation depends and so they are cut neatly into sections. They are served in a fruit dish with the addition of a little sugar; they may even be used to make sorbets. In fact, these vegetables and fruits possess the invaluable advantage of staying in the same condition for a very long time; some have been preserved for several years and were as delicate as after six months...'.

In the sixth yearly edition of his *Almanach,* Grimod reported on the first test of Nicolas Appert's preserves by the French Navy. On December 2nd 1806, eighteen of his canned products were taken on board ship. Three months later a tasting session was organised on board and a report dated May 22nd concluded that the contents were perfectly preserved.

A few years later, an official report by Guyton-Morveau, Parmentier and Bouriat, all members of the French Royal Society (1809), stressed the merits of Nicolas Appert's procedure: 'M. Appert's methods are as safe as they are useful; they offer a means of enjoying throughout the year and throughout the whole Empire foodstuffs which are produced in one of its parts, without the fear of their having been spoiled in transit or by the lapse of time since their season of freshness ...'. The report went on to conclude: 'The Society (for the Promotion of National Industry) praises highly the inventor (Nicolas Appert) for his contribution to the advancement of the art of preserving vegetable and animal substances'.[2]

It was not until 1810 that Nicolas Appert wrote and printed at his own expense a precise description of his process. This was done at the request of the Home Office Minister, the Count of Montalivet, who on January 30th awarded him prize-money of 12,000 Francs for inventing an efficient solution to the problem of provisioning the armies. However, he attached a condition to this award: M. Appert had to agree to make the secret of his invention public. In the work entitled *The book of housekeeping or the art of preserving animal or plant substances for several years*[3], Nicolas Appert described the preserving process known today as *appertisation*[4], which consists basically of the following steps:

---

[2] A report by M. Bourriat, on behalf of a special Commission on the animal and plant substances preserved according to the principles of M. Appert, in Massy, near Paris . This report is reproduced in Appert's book (1810).

[3] Three other editions would appear in 1811, 1813 and 1831.

[4] The decree dated February 10th 1955 defines the expression *appertised preserve.* It concerns perishable foodstuffs of plant or animal origin whose preservation is ensured by the combined use of two techniques:
  - processing 'by heat or by any other authorised method, with the aim of destroying or inactivating all enzymes, micro-organisms and their toxins, whose presence or proliferation might spoil the foodstuff concerned or render it unfit for human consumption';
  - a form of 'conditioning in a waterproof, airtight, germ-proof container'.

1) 'Place the substance to be preserved in bottles or jars;
2) Seal the containers with great care, as success depends mainly on correct sealing;
3) Place the sealed containers in boiling water for the time necessary, according to the nature of the substance and in the manner that I shall indicate for each type of foodstuff;
4) Remove the containers from the bain-marie at the required time (1810)'.

In the absence of a patent to protect the invention, Appert's discovery was immediately copied in England and America. Then in 1814 Napoleon was defeated and during the subsequent Prussian invasion Appert's workshops at Massy were looted. In 1817, he received in compensation the use of the Musketeers' enclosure, a walled garden at the war hospice, where he carried out his research until 1828 with the support of the Society for the Promotion of National Industry, which awarded him a gold medal for his culinary preserves in 1820 and a grant of 2,000 Francs in 1824. Nevertheless Nicolas Appert ended his days in poverty and died on June 1st 1841.

II.2. Tin plate

From the beginning of the nineteenth century, the English used tin plate to make the first canned preserves. Sheets of iron were cleansed in acid, then coated with tin.[5] This light, flexible material was widely used in France well before national production began in the middle of the nineteenth century (Cornu et de Bonnault Cornu,1989).The lids of tin cans are easy to seal after filling. They have a narrow aperture that allows the air and steam to escape during the heating process in a boiler or pressure cooker. When the container has been adequately heated, a drop of welding metal – lead, or later, tin – is dripped on to the aperture, sealing it safely and simply. From the nineteenth century on, tin plate was printed or varnished.[6] The technique for manufacturing tin plate was to undergo considerable developments in the twentieth century with cold lamination, which permitted the tin plate to be shaped in long bands, wound onto spools.

II.3. The emergence and expansion of canned sardines (from 1824 to the mid-nineteenth century)

On June 8th 1822, the *Journal de Nantes et de la Loire-Inférieure*, a regional newspaper, published an article that may be considered as the first ever advertisement for canned sardines. On his return from a long voyage on which he had taken Appert's preserves, Captain Louis de Freycinet declared: 'These (preserves) withstood perfectly and beyond my expectations all the trials to which they were subjected during thirty months at sea'. The article goes on to add that the foodstuffs mentioned in Captain Freycinet's letter were provided by Colin, a confectioner in the rue du Moulin, who manufactured them

---

[5] The manufacture of tin plate was probably perfected in Bohemia towards the end of the 13th century.
[6] Before printing, a varnish is applied to the inner surface of the sheet which has been cut into plates. This is to protect the tin plate from corrosion by the contents of the can. A gloss finish is applied to the outer side. After 1870 tin plate was printed using the process of chromolithography or colour lithography. Today, the offset process is used.

according to the Appert method, to the satisfaction of several Nantes sea-captains who had tested them on board. It added, 'Colin stocks a full assortment of all these preserves and in addition certain items one finds in Paris, such as sardines in butter, or in oil'.[7] A new and great industry had been born.

Colin extended Appert's sardine preserving technique to produce sardines 'preserved in butter and oil', according to the traditional fishermen's wives' recipe. With the opening of a factory at 9, rue des Salorges, Colin's business changed from being a cottage industry to a manufacturing industry. He brought in sardines from La Turballe using the express coach service normally reserved for travellers. Colin's range of canned foods was varied, and included meat, vegetables and fruit as well as sardines. The newspaper *Le Breton* described his establishment in the following terms in an article that appeared in 1836: 'The equipment in the workshop is impressive. There are pressure cookers that can boil down the bones of four oxen at a time. There are steamers that can cook an entire calf. There is a powerful tallow press. A curious kind of oven works day and night without interruption throughout the year. The workforce consists of 50 to 60 people to prepare the meat, and 30 tinplate workers to weld the cans. When peas are in season, over a period of 10 weeks, 300 women are taken on to work a 14-hour day. (...) The firm has a stock of 300,000 cans, including 100,000 cans of sardines'.

After Joseph Colin's, two other canned food businesses were set up in Nantes. In 1830, a restaurant owner named Millet set himself up in the sardine-canning business with a 'fish-fry' in the rue de Santeuil. Frying fish in the town centre was considered a nuisance, so the local authorities made Millet move his business first to the outskirts of town, and then to Le Croisic in 1836. Colin and Appert were both confectioners, Millet a restaurateur. A third type of profession – that of pork butcher – was the founder of another sardine cannery. *Leydic & Bertrand* made salt meat for the Navy and in 1833 they branched out into food preservation, particularly sardines. The company became *Bertrand, Philippe & Canaud* in 1839. Philippe and Canaud entered into a partnership in 1841 and set up a sardine-canning factory in Port-Louis in the same year. In 1847, there were 5 canneries in Nantes, 2 in La Turballe, 1 in Piriac and 4 in Le Croisic. Establishing factories on the coast ensured top-quality ingredients and eliminated transportation costs.

II.4. Expansion (1850 – 1880)

From 1850 and into the 1880s the canning factories established themselves firmly in Nantes and on the Breton coast. The best-known companies made their appearance at this time: *Cassegrain* (1856), *Amieux frères* (1866) and *Saupiquet* (1877). In 1861 a publication for the National Exhibition in Nantes records that the Nantes canners produced between 100 and 150 thousand cans of sardines. The fishing industry employed 15,000 to

---

[7] Pierre Joseph Colin (1785-1848) was the pioneer in the industry. His father, Joseph Colin (1754-1815) was a confectioner at 4, rue du Moulin in Nantes. Apparently father and son tried Nicolas Appert's technique together. The earliest writer on the subject of sardine preserves, Jeune Caillo, wrote in 1855: 'M. Joseph Colin, of Nantes, by subjecting this recipe to the Appert process of food conservation, was the true inventor of canned sardines in oil'.

18,000 men during the season, and 4,000 to 5,000 women worked in the Nantes canning factories. Similar factories multiplied on the Breton coasts, in Douarnenez[8], Audierne, Concarneau, Belle-Ile-en-Mer, Quiberon and Port Louis.

II.5. Foreign competition and the first sardine shortage (1880-1914)

Although the variable fortunes of the fishing industry led many producers to set up factories in Portugal and Spain where catches were more plentiful and regular, French production continued to expand thanks to exports and its reputation for quality. About 160 sardine-canning factories were in operation in the 1880s: 14 in the lower Loire region, 59 in Finistère, 59 in Morbihan, 25 in Vendée, 1 in Charente-Maritime and 1 in Gironde.

In 1879 an order from the prefect stipulated that for health reasons pure tin had to replace lead as the welding material for cans. The canning factories were forced to take on skilled welders. These would soon be replaced by crimping machines at the beginning of the twentieth century.In 1900, French production of canned sardines was at its peak. 5,214 fishing boats were registered from the Breton coasts down to Saint-Jean-de-Luz in 1898. In that same year 23,000 fishermen had landed over 50,000 tonnes of fish. The canners were extremely well represented at the Universal Exhibition in Paris. At Amieux Frères, 4,000 workers produced 12 million cans per year in 11 factories; Saupiquet had 9 sites – in Nantes, Les Sables d'Olonne, on the Ile d'Yeu, in La Turballe, on Belle-Ile-en-Mer, in Port-Louis, Concarneau, Kérity-Penmarch and Audierne – and produced around 10 million cans a year.

After 1902, the sardine deserted the French waters. The poverty that ensued led to population migrations to inland Brittany. The fishermen of Guilvinec moved south in pursuit of a livelihood and their move to Quiberon with their families aroused the hostility of the local population.[9] The shortage and production crisis was coupled with a major social crisis: in 1905, in Douarnenez, the female workers – called *penn sardines* in Breton[10] – went out on a long and bitter strike. Another aspect of the shortage was that the industry in France lost its position as a leading exporter of preserves, and of canned sardines in particular. At the end of the crisis, in 1909, the French factories came up against foreign competition, which in the meantime had gained a share of the export market.

---

[8] The *Chancerelle* canning factory was founded in Douarnenez in 1853; it has recently celebrated its 150th anniversary.

[9] The local press covered the events surrounding the arrival of the people from Guilvinec in Quiberon. The paper *Le Progrès du Finistère* reported: 'The regrettable scenes of violence that took place unexpectedly in Quiberon during the night of September 5th - 6th 1909 on board Finistère fishing-boats were unforgettable. After surrounding our fellow countrymen and taking them by storm, the Morbihan fishermen smashed everything and caused serious injuries'.

[10] In French, this means 'sardine heads'. The workers usually wore a Breton head-dress, with smaller side-pieces so as not to be hampered while they worked. This head-dress was called *penn sardin* in Breton. Another possible interpretation is that the workers were paid by the number of sardine heads they cut off , hence their nick-name.

II.6. 1914-1955: a turning-point in the canning industry

Foreign competition from Portugal and Spain continued to have an effect before the worldwide economic crisis made it even more difficult to find market outlets due to the protectionist policies introduced by many governments. Just before World War II, several canneries set up operations in North Africa, mainly in Morocco, where reserves were plentiful and labour was cheap. A report by the French Union of canners of sardines and other fish (1932) stated that there were 180 canning factories on the Breton and Vendée coasts, 7 in the Basque region, 12 in the area between La Rochelle and Arcachon, and 3 in the Mediterranean. These factories operated from June 1st to November 1st, i.e. about 155 days a year, with an average workforce of 75 women per factory. The report continued:

'The French producers come up against 3 main competitors on the French market:
1) Portugal and Spain, which export approximately 400,000 to 600,000 crates of sardines to France every year;
2) The United States of America. Until 1914, production of canned American sardines or 'pilchards' was almost non-existent. Since the war, this industry has expanded considerably, and the US is beginning to penetrate the French market;
3) Norway, which exports sprats or 'brislings' to France'.

Assuming that production would recover as fishing intensified, the factory owners encountered various difficulties in their attempt to improve productivity. Removing fish-heads by mechanical means was complicated, as the fish were easily broken. The automated hot-air drying process in general use was a time-saver. In some factories the sardines were cooked in ovens, but frying in oil resulted in a better-quality product, though this process required a longer draining time. The productivity of workers filling cans by hand could be improved by installing conveyor-belts: 'Nearly all French factories are now equipped with automated crimping machines, some of which achieve an output of 2,600 sealed cans per hour. For a factory to be able to maximise the use of its equipment, it must receive a reliable and regular supply of fish. The producers have shown interest in the idea that keeping surplus fish in cold storage might help to achieve positive results, and have supported the various experiments undertaken over the last few years'. The experiments were inconclusive for sardines, which, unlike tuna, do not withstand refrigeration beyond 24 hours.

II.7. The current situation

The 1950s and 1960s were prosperous years: fishing catches and the tonnages destined for canning equalled and then exceeded pre-war levels. At this time, yields from fishing came to approximately 25,000 tons per year. In 1950, the amount used by the canning industry was 17,500 tons out of 23,000 tons caught, a significant increase on 1947. Two technical innovations took hold at this time: the hoop net or 'bolinche' which caught larger quantities of fish at a time than the straight net, and the sounder, which eliminated the need for bait – salted fish eggs – to attract shoals of sardines to the surface.[11]

---

[11] 'Nets' weighing several hundred kilos became commonplace.

The situation was turned around in the 1970s. After a few bad years in the mid-sixties, the sardine supply disappeared completely, bringing about the gradual decline of the factories. Today, only about ten sardine-canning factories remain on the Breton coasts which in 2002 processed 15,774 tons of sardines. Even though 9,771 tons from French producers represent 62 per cent of the supply to the canneries in 2001, the so-called Atlantic sardines are not fished in Brittany but off the coasts of the Basque region by the Saint-Jean-de-Luz fishermen, with the remainder coming from Italy, Morocco and Portugal. In 2002 the canning factories produced 86.5 million cans with a net weight of 9,956 tons. The production of whole sardines in olive oil and of fillets is rising, whereas all other products (whole sardines in vegetable oil, tomatoes and other sauces) are declining.The export market is negligible – barely five hundred tons a year – but France has to import nearly 13,000 tons of sardines from Morocco, Portugal and Spain to meet consumer requirements, estimated at 22,500 tons.

III. A great 'small-scale' industry

Despite being open to technical innovation, the sardine-canning industry began as a labour-intensive industry, distinguished by the gender-allocation of tasks and the role of craft-based techniques and knowledge. Traditionally, women worked in the sardine-canning factories while the men fished, transported the goods, welded the cans or operated the boiling and steaming equipment. The earliest form of sardine-fishing used vertical drift nets. The double-masted sailing boats – called '*pinasses*' – usually remained close to the shore to fish the sardines. The fishermen used *rogue* as bait – fish eggs mixed with groundnut flour. As the sardines came up to the surface, they were caught in nets with a mesh corresponding to the size of the fish required by the canners. With the arrival of motors in the 1930s, the boats became larger. The fishermen could venture further from the coasts and took dinghies to throw the bait to the sardine shoals.The use of the hoop net or *bolinche* has only been allowed in France since the 1940s, although countries such as Spain and Portugal have been using such nets for many years. The fine-mesh net surrounds the shoal of sardines that has been drawn towards the bait and the bottom end is closed by a moving cable. The crew heaves the net up and the fish are dumped onto the boat. Regardless of the fishing technique used, the fishermen are always keen to return to harbour, as the first to arrive can usually sell their catch at a better price to the canners.

IV. Production

Sardines are processed as quickly as possible on delivery to the factory. The various stages of the process are repeated as each new delivery arrives.

Removing the head

With one movement of the knife, the workers cut off the head and innards of the sardine. Attempts to mechanise this procedure have been unsuccessful, so it has remained a manual task to this day. The sardines are first washed, then marinated in brine; the duration of the marinating is variable and depends on the size and quality of the fish.

Placing on grids and drying

After another rinse, the sardines are placed by hand on special grids. For a long time, drying took place in the open air, but today ventilated air tunnels heated to 37°C are used to evaporate moisture from the fish.

Cooking

On their grids, the sardines are fried in groundnut or olive oil between 130 and 180°C. Then they are cooled and drained before canning. They are canned 'white-side up', i.e. belly-up, or 'blue-side up', i.e with the back facing upwards. The can is filled with oil before it moves to the crimping machine.

Sterilising

The cans are placed in pressure-steamers and heated to 110°C for sixty to ninety minutes. New methods of cooking have been introduced including hot-air ovens, or cooking raw sardines in their cans by sterilisation. The canners with a reputation for quality remain faithful to the traditional method, i.e. frying in oil.

V.   Luxury goods or standard products?

Until World War I, canned foods and particularly sardines were considered to be luxury goods for export or for a small, elite market. The 1913 *Amieux frères* catalogue extols the practical virtues of canned goods, for motorists in particular: 'We would like to draw our customers' attention to the sale of these small cans, which have become an everyday feature, due to the growth of the travelling, touring and motoring public. They are highly favoured by the military for use during manoeuvres'.

However, canned food was beyond the means of the workers producing the cans at the end of the nineteenth century. In 1890, an unskilled female worker earned 0.10 francs per hour whereas a tin of sardines in oil cost 0.60 francs. She would have had to work six hours to be able to buy a can of sardines.

Price was not the only reason for the low consumption of canned sardines and other foods. As Martin Bruegel illustrates (2002), the fear of new ideas and cases of food-poisoning caused by faulty preservation techniques explain people's cautious reaction to this mysterious new food. Two institutions – schools and the Army – would reconcile the French to eating canned food, and the 1914-1918 War accelerated the process for both domestic and industrial preserves. In the twentieth century, the can of sardines became an item so commonplace that it began to suffer from an unfavourable brand-image. In 1990, a can of sardines in oil cost around 6 francs whereas a worker on minimum wage earned around 32 francs per hour before tax; thus about 20 minutes' work would buy him a can of sardines. Today, the status of the can of sardines is undergoing a new change with the appearance of 'vintage' cans, and a red label (a symbol of quality) given to sardines from Saint-Gilles-Croix-de-Vie. Efforts are being to make sardines in olive oil or canned sardine fillets be seen as gourmet foods.

VI. Family firms

Set up on the Atlantic coasts usually by men from other regions, the sardine-canning factories have remained family-owned until recently. An instance is the Chancerelle family, who came originally from Nantes. Two brothers, Toussaint-Laurent (1806-1889) and Robert (1808-1868) set up a partnership in 1828 to run a fish business under the name *Chancerelle Frères*. The two brothers opened factories in the ports of South Finistère and the Lower Loire region around 1830. In the middle of the nineteenth century, Robert settled in Douarnenez and his sons went on to found canning factories in the ports of Tréboul, Audierne, Concarneau, Gujan-Mestras and L'Herbaudière. The sardine shortage of the 1880s led the Chancerelle family to set up canning factories in Portugal. Today, the fifth generation of the family manages the Wenceslas Chancerelle company.

Another example of continuity in family ownership is that of Jules Joseph Carnaud. A tin-can manufacturer born in 1840, Carnaud established himself in Nantes in 1894 and became the main customer of the *Forges de Basse-Indre,* the local metalworks. The two companies merged in 1902 under the name *J.J. Carnaud et Forges de Basse-Indre* and rapidly became the leading tin-plate manufacturer in France. The firm was managed by Jules Joseph Carnaud's son-in-law from 1911, and by the latter's son in 1945 (Rochecongar, 2003).

VII. Working conditions

As a rule, the labour force in the canning factories was female, Breton, from a rural or coastal background. Wages were related to the seasonal character of the job. In 1890, for example, the can welders were paid piece-work: 1.50 francs for every 100 cans of sardines (side, bottom and lid). The quickest and most skilled workers managed to seal up to 325 cans in ten hours, enabling them to earn up to 4.90 francs a day. At the same time, a metal-worker for Carnaud earned 0.40 francs per hour, or 4 francs for a 10-hour day. The female workers on the other hand earned only 0.10 francs an hour, or 1 franc for a 10-hour day. Until 1905, most workers were not paid by the hour, but by the number of sardine heads they removed – from 1.50 to 1.75 francs per thousand.

In June 1905, a major strike – the first of its kind – broke out at the Chancerelle works in Douarnenez in protest against this kind of remuneration, and also against the low pay and working conditions. Faced with the determination of the workers, the cannery owners gave in and agreed to hourly wages. Further strikes and stoppages disrupted the factories: in June 1905 the welders at the Chanterelle factory took action against the crimping machines, equipment was destroyed and production halted. A few years later in 1909 there was another strike against low wages, and in 1924 a further strike in Douarnenez was marked by the presence of Charles Tillon, one of the Black Sea mutineers, who came to address the strikers on behalf of the CGTU (the General Confederation of Workers' Union).[12]

The working conditions were extremely hard. The short fishing season and the large quantities of fish to be processed meant that the female workers had to work long hours, often late into the night. At the beginning of the twentieth century, the canning industry obtained special dispensations authorising women and children to work on condition that they did not exceed 72 hours a week. The 72 hours could be worked day or night regardless, and up to 48 hours at a time. In 1936 at the height of the social crisis that rocked the entire country, the canning factories not only did not strike, but even worked on July 14th to process a particularly plentiful catch! After 1936 and with the introduction of the 40-hour working week, the canning factories maintained certain special dispensations such as having recourse to overtime after 44 hours of work, or night working for women.

Over and above the strikes and wage-claims, the greatest tension within the sardine-canning industry was that between sea and land, between the fishermen on the one hand and canners, fish-merchants and tin-plate manufacturers on the other. This struggle gave rise in the 1920s to so-called sardine communism that united fishing boat owners and sailors against the people on land who were considered as exploiters; several fishing ports such as Douarnenez, Lesconil, Kérity and Saint-Guénolé elected communist town councils.

VIII.   A lost civilisation

The sardine is not merely a fish like any other, as for many years it was celebrated through various festivals, ceremonies, songs and traditions. A memorial was built in its honour in Brittany, a church named *Notre Dame de la Sardine* at the tip of Cap Sizun. Since the beginning of the twentieth century, life in the sardine ports has been punctuated by numerous traditions and festivals. In Saint-Gilles-Croix-de-Vie, in the Spring, the boat-owners kiss the first sardine to come out of the sea. On June 24th in Notre-Dame-de-Larmor, a sardine mass is celebrated. On September 29th the town celebrates the sardine festival, which coincides with the return of the fishing fleet. Every year since 1905, the 'Blue Net' Festival, the oldest folklore event in Brittany, has been celebrated in Concarneau. The festival takes its name from the colour of the nets used by the fishermen since the end of the nineteenth century. It was originally a charity event organised to help the families of fishermen impoverished by the disappearance of the sardine from the Breton coasts. Groups of Breton musicians from all over the region parade through the town each year in traditional costume, playing the traditional Breton bagpipes. A 'blue net Queen', chosen from among the sardine factory workers, is elected each year, and a song entitled *Les filets bleus* by the poet and songwriter Albert Larrieu, a friend of Jean Richepin, is sung. Sardines and sardine factory workers are the theme of many other popular songs,

---

[12] The CGTU appeared in 1922 after a split with the CGT, which had itself been the result of the split of the Congress of Tours and the birth of the Communist Party.

such as *Les petits sabots, La sardinière, La friteuse, Marche bretonne, Vivent, vivent nos filets bleus,* and *Pour les filets bleus.*

IX. Conclusion

Although invention and innovation were without doubt the foundation stones of what used to be a great industry, much remains to be written about the history of sardine-canning in France and many questions remain to be answered by historians. How was capital raised by companies that often began on a very small scale and remained under family ownership? How did the sardine factory owners learn their trade and what were its characteristics? How was the workforce recruited and trained? What were living conditions like in the sardine ports? Most of the time, historians must make do with information and data derived from press articles or from eyewitness accounts. Direct accounts by the companies, however invaluable they may be, are difficult to obtain, as many factories have not kept records. The history of fishing may be better known, but its historical vicissitudes are still unclear, as are the reasons for the recent disappearance of the sardine supply from Breton waters.

Bibliography

Printed sources

Bulletins de la Société d'Encouragement pour l'Industrie nationale, volumes VIII, XIII, XV, XVIII, XIX, XXI, XXIII, XXIX.

Exposition universelle de 1855 (1856) *Rapports du jury mixte international publiés sous la direction de SAI le Prince Napoléon, Tome 1, Classe XI. Préparation et conservation des substances alimentaires*, Paris.

Chevalier M. (ed.) (1862) *Exposition universelle de Londres de 1862. Rapports des membres de la section française du jury international sur l'ensemble de l'exposition, Tome I*, Paris.

Chevallier, M. (ed.) (1868) *Exposition universelle de 1867 à Paris. Rapports du jury international. Tome XI, Groupe VII ; Classes 67 à 73*, Paris.

Exposition universelle de 1889 (1891) *Rapports du jury international publiés sous la direction de M. Alfred Picard, Groupe VII. Produits alimentaires. Classes 67 à 73*, Paris.

Exposition universelle internationale de 1900 (1902) *Rapports du jury international. Groupe X, Aliments Première partie Classes 55 à 59*, Paris.

Largillardaie, M. de (1878) *Fabrication des conserves de sardines à l'huile*, Paris.

Mercier, M. and Heuze, G. (1880) *Exposition universelle internationale de 1878 à Paris, Groupe VII, Classes 72 et 73. Rapports sur les viandes et poissons, les fruits et légumes*, Paris.

Merlant, A. (1932) 'Rapport sur l'écoulement des conserves des sardines françaises sur les marchés étrangers', *Note présentée par l'Union des syndicats français des fabricants de conserves de sardines et autres poissons à la Commission pour l'étude des questions sardinières lors de la réunion du 14 juin 1932*, éditeur inconnu.

Secondary sources

Anginot, P. (2002) *La sardine*, Paris.

Appert, N. (1810) *L'art de conserver pendant plusieurs années toutes les substances végétales et animales*, Paris.

Boulard, J.-C. (1996) *L'épopée de la sardine*, Rennes.

Bruegel, M. (1996) 'Un sacrifice de plus à demander aux soldats: l'armée et l'introduction de la boîte de conserve dans l'alimentation française, 1872-1920', *Revue Historique*, 596, pp. 259-284.

Bruegel, M. (1997) 'Du temps annuel au temps quotidien: la conserve appertisée à la conquête du marché, 1810-1920', *Revue d'Histoire Moderne et Contemporaine*, 44-1, pp. 40-67.

Bruegel, M. (2002) 'How the French learned to eat Canned Food, 1809-1930s', Belasco, W. and Scranton, P. (eds.), *Food Nations-Selling Taste in Consumer Societies*, London & New York, pp. 113-130.

Cornu, R. and Bonnault-Cornu, P. de (1989) *Pratiques industrielles et vie quotidienne: conserveries et ferblanteries.* Rapport de recherche au conseil du Patrimoine ethnologique, ministère de la Culture et de la Communication, Paris.

Duhamel de Monceau, H.L. (1759) *Moyens de conserver la santé aux équipages des vaisseaux avec la manière de purifier l'air des salles d'hôpitaux et une courte description de l'hôpital Saint-Louis*, Paris.

La Casinière, N. de (2002) *Sardines à la clef*, Rennes.

Lachèvre, Y. (1994) *La sardine, toute une histoire*, Quimper.

Libaudière, F. (1910) *Des origines de l'industrie des conserves de sardines, 1824-1861*, Nantes.

Rochcongar, Y. (2003) *Capitaines d'industrie à Nantes au XIXe siècle*, Nantes.

9 The industrialisation of seafood. German deep-sea fishing and the sale and preservation of fish, 1885-1930

Hans Jürgen TEUTEBERG, UNIVERSITY OF MÜNSTER

Fishing, that is the catching and appropriation of all edible animals and plants from the water, belongs like hunting and gathering among the oldest forms of food provision (Sarghage, 2002; Gunda, 1984; Radcliffe, 1921). In Europe it is a pursuit that reaches back to Ancient Rome. For instance, in early Christian art the fish was a symbol for Jesus Christ as we know from amulets, glass vessels and gravestones (Achelis, 1888; Dölger, 1910-1922). During the Middle Ages the monasteries developed many fish ponds for the provision of typical Lenten fare. Since the beginning of the early modern era with the spreading garden economy and growing research on animals and plants, there has been an increase in studies on the 'perfect fish-pond economy' as well as in government regulations on lake and river fishing. The sovereigns of the emerging modern states recognised, like the churches and landlords, that they could increase their income with artificial fish-breeding, ceding fishing privileges or selling fresh fish at the market. The new 'fish books' as well as accounts from the courts of princes, cloisters and manor houses indicate that between the mid-sixteenth and the nineteenth centuries there was an upswing in inland fishing in Central Europe.[1] Besides this, of course, there was also considerable illegal wild fishing similar to poaching. Most common was the farming of carp and trout, later on also of pike, perch, bluefish, whiting and other species of fresh fish. Special devices were created to keep track of eel and salmon as they travelled upstream during the spawning season so as to catch them more easily later.

For many centuries sea fishing was done only from small rowing and sailing boats or on foot from sandbanks along the coast using baskets, creels and small hand-held nets or later larger dragnets. This kind of fishing was possible only during the summer. The small catches were primarily used by the coastal population for their own consumption. Because of poor road transport facilities, fresh sea fish as well as oysters, crayfish, lobsters and shell fish (mussels) could be transported to the interior only at extremely high cost. Since inland freshwater fish remained a most expensive 'diet of the lord' (*Herrenspeise*), fresh fish was not part of the common daily diet in the pre-industrial period throughout the whole period of the 'Holy Roman Empire of the German Nation' (Hitzbleck, 1971; Kuske, 1905). Seafood that had been dried, cured, salted or pickled reached regions distant from the coast only in small amounts.

The simplest way was to dry the fish for a long time in the open air by threading them on sticks and then dispatching them pressed as bale goods. A more complicated way was

---

[1] See, for instance, Mangolt, 1557; Gesner, 1598; Heger, 1727; Bloch, 1782-1784; Mehler, vol. 3, 1787; R., 1845; Landau, 1865; Biermann, 1865.

to smoke the fresh sea fish (for instance mackerel, sprat, plaice and herring) under intense heat in a special oven and pickle them with salt or vinegar, which was also possible with roasted fish. All these preserving methods were intended solely to coagulate the fluid protein and to extract the water as otherwise the fish would quickly become rotten and cause dangerous fish poisoning. Drastic changes occurred in the late nineteenth century when the railway network was extended and new methods of catching, preserving, processing and selling created new inland markets for sea fish.

The aim of this contribution is to outline the causes of this structural change that influenced the food economy of the former German Empire and large parts of Central Europe. The sources used include contemporary books on German deep-sea fishing since 1885, as well as books about new methods of preserving, processing and selling fish, which have never been evaluated by historians. This article wishes to encourage more source-oriented investigations in this direction in order to scrutinise hitherto hypothetical results, to formulate new questions and to initiate international comparative studies.

I.   The origin and development of deep-sea fishing with steam trawlers

In the late Middle Ages, the first commercial fishery had already been established on the lower Elbe between Hamburg and the North Sea which primarily provided provisions for the growing Hanseatic town (Lübbbert, 1925; Schümann, 1931). Around 1710, fishermen from the little river village of Blankenese, near the smaller neighbouring town of Altona and Hamburg's harbour area of St. Pauli, began to fish continuously between May and September every year in the region around the mouth of the Elbe and on the North Sea coast. Sailing under the Danish flag they landed their hauls in Dutch harbours where fish auctions had been introduced in 1780. These fishing expeditions were successful seemingly because they used the older Dutch trawl nets that better preserved the quality of the fish. The number of small sailing boats for coastal fishing (*Ewer*) increased between 1740 and 1805 from 60 to 172. With Napoleon's blockade of the continent the fishermen could live only from smuggling to the English-owned island of Helgoland. Many of them then sold their ships or after the second Vienna Peace Treaty (1815) turned to the transport of goods.

Then the fishermen of Finkenwärder, with their more favorable location at the tip of the mouth of the Elbe, took over the leading position in German North Sea fishing for the next decades. But in 1885 the Finkenwärder fishermen were in turn faced with new competition. At the mouth of the river Weser the first deep-sea fishery was set up with steam-powered ships that could operate the whole year round. By contrast, the small *Ewers* were not capable of withstanding a rough sea. After a few years the older seashore fishing with its sailing boats experienced a crisis and dwindled steadily (Stahmer, 1925: 99). Only a few Finkenwärder fishermen survived by changing over to motorboats and new dragnets. They confined themselves henceforth to daily landings of high-quality fish from the coastal waters and so they continued to get good prices (Senst, 1918: 111; Abraham, 1931: 79).

The revolutionary introduction of steam power into sea fishing started first in England in 1880. Following some initial difficulties, this proved so highly profitable that it was imitated very soon after in The Netherlands, Belgium, France and Germany. The great advantage was that steam-driven fishing was independent of the quickly changing weather conditions. While a sailing boat could not go to work during a dead calm or in a stormy sea especially in autumn and winter, the steamer could make hauls the whole year long. The traditional three-month season for sea-fishing came to an end for the first time in history.

The fishmonger *Friedrich Busse* from Geestemünde at the mouth of the Weser, a learned ship's carpenter who had become acquainted with the large fish markets in Baltimore and New York, together with a partner from Bremen founded a fish trading centre in 1870. He then persuaded some Finkenwärder fishermen to deliver their complete hauls exclusively to his firm and promised to buy all their catches on credit in advance. So these fishermen could not sell their fish to others and were tied permanently to him. But they had the great advantage of being able to sail again shortly after unloading, while all the other fishing boats had to stay in the harbour for a longer time trying to sell their fish. Busse's new business relations were so successful that he was able to build the first German steamer *Sagitta*, based on an Anglo-Saxon model, which was put to sea on February 7th 1885 at Geestemünde (Stahmer, 1925: 84; Duge, 1898: 43; Agatz, 1919: 155). Although it suffered at the beginning from some mismanagement, his firm went on to make so much profit that he was able to build two more ships in 1887 and 1889. His success became well-known very quickly. By 1888 the Bremen companies Schilling & Pust and Barthling put two other steamers into operation in Bremerhaven. The first four modern trawlers landed 836,6t fish after 117 voyages. Geestemünde's deep-sea fishing fleet had already been increased to 17 trawlers when in nearby Bremerhaven the first hall for fish packaging and fish auctions was opened in 1892. How efficient this new technical development was can be demonstrated by a comparison: the many old sailing boats of the *Norddeutsche Fischerei Gesellschaft* in Hamburg could catch only 438t fish on 117 trips. No wonder that in 1886 the first steamer was taken into service in Hamburg also (Schümann, 1931: 49-50). Within a few years the number of these new ships increased rapidly and then continuously till 1914. Sea fishing was partly suspended during the First World War because of the English blockade, and most of the German trawlers, which were in the service of Imperial War Navy, were lost. By 1920, however, the shipping traders had re-established their deep-sea fishing fleet. This was also connected with the building of several new ships because the new steam-powered ships had in the meantime become much more efficient. The production of sea fish therefore continued to increase although there were fewer ships than in 1914 (Höver, 1936: 227-228; Rogowsky, 1922: 13; Tern, 1924: 39).

The industrialisation of German deep-sea fishing was marked by two tendencies: a change in the entrepreneurial structure and a rapid improvement in the production quality. The old method of fishing from sailing boats was organised on a small scale, like handicrafts. However, the investment costs for a modern steamer were so extremely high that no single fisherman could gather such capital. In 1892 the costs of building a steamer were between 116,000 and 125,000 marks, which then with the growing gross register tons (G.R.T) rose to 140,000-170,000 marks in 1910. Such huge investments could only be made by commercial shipping traders or a few big fish traders. This means in other words a quite new separation between skippers, captains, navigators and sailors on the one side and investors, financiers and shipping-traders (*Reeder*) on the other. The normal fisherman changed from being an independent small-scale craftsman to a dependent salaried employee of a shipping trade company which usually became a public limited company (plc, in German *Aktiengesellschaft*). In 1930 there were upwards of 347 deep-sea trawlers, 243 of which were already involved in such a new type of enterprise (Tilman, 1933: 24; Duge, 1898: 13; Goldschmidt, 1911: 21; Rohdenburg, 1975: 151).

II.   New types of ships in deep-sea fishing

The technology of the larger steamers required a reasonable division of labour. On the small sailing boats each skipper could do everything because the work was manual, but the operation of an engine and deep-sea navigation required technical knowledge. Contrary to the former situation then, simple fishermen were now involved only with on-board fish preservation (Rohdenburg, 1975: 156; Schümann, 1931: 51-52; Höver, 1936: 551).

International treaties before World War I had laid down that anybody could fish in the North Sea, except within a state's territorial waters defined as an area three miles from the coast (Gutmann, 1912). At first the German boats' best catches came from the area around the East Frisian Isles in the eastern North Sea. From 1896 the new steamers began fishing in the middle and northern parts of the North Sea also, but not until 1908 was this done intensively. The reason for this extension was that the haddock (*Schellfisch*) hauls had dwindled in the traditional fishing regions and trips from 1903 to the Skagerak and Kattegatt Straits on the Danish coast did not solve the problem. The German fish traders therefore shifted more and more to catching cod and moved north towards the Arctic (Gutmann, 1912; Geiser, 1918: 21; Duge, 1911: 24-25). There were also some attempts, for instance by the *Nordsee GmbH,* to fish in the Mediterranean Sea along the coast of Egypt and in the Adriatic as well as along the seashores of Portugal, Spain and Morocco. But after a while such voyages had to stop because of high tolls, prohibitions on landing fish and the size of the fishing vessels. In order to be free of such fluctuations in the catch, the German Sea Fishing Association (*Deutscher Seefischer-Verein*) then explored the Lofoten Islands and the Arctic Sea and started fishing also near Bear Island and in the southern Barent Sea. This new extension of German deep-sea fishing began then after the end of the First World War (Rohdenburg, 1975: 144; Nordsee GmbH, 1996).

Fishing in such different regions also required different types of ships and improved technology. The seaworthiness of the traditional open boats (*Ewers*) of the Blankenese fishermen, which were 14-16 m long, 4-5 m wide and had but one small cabin, was very limited. Because of their smooth bottoms they had drop-keels (fins) on each side to allow cruising against the wind and one mast with a square sail, which could be supplemented with a small foresail (*Focksegel*). For the traditional fishing on streams and at the seashore these *Ewers* were entirely sufficient.

The Finkenwärder fishery used the larger *Kutter-Ewer* with two masts and a bigger sail span for greater speed. From the 1880s they began for the first time to build cutters on the English model. The most important feature of these was that the old smooth bottom was now detached, thereby raising the sailing speed and hence shortening the length of a voyage. The old cambered stem of the ship (*Vordersteven*) was now a straight shape and the stern (*Heck*) was raised. The new cutter was equipped with three main rooms on deck: a cabin, an ice-room and the *Bünn*, i.e. a room with water to hold all the fish that would be landed alive, such as plaice which fetched particularly high prices in the markets. The *Bünn*, which took up one third of the ship's area, was not without its dangers: it got its seawater through holes in the side and this strengthened the dip of the ship when the waves were high, so that there was a great chance of keeling over. With the increasing competition in deep-sea fishing, more and more voyages were made in wintertime and consequently the loss of ships and sailors was high. The inshore fishery began in 1877 with experiments to use small petroleum motors, but the German maritime authorities prohibited such installations in timber ships because of the fire hazard. Diesel-propelled cutters were used later in coastal shipping but not in deep-sea fishing (Grotewold, 1908: 52-58 and 64; Timmermann, 1957; Goldschmidt, 1911: 23-24; Timmermann, 1962; Dittmer and Buhl, 1902).

The tendency to search for more distant fishing grounds influenced ship-building. Following English models, a ship was judged primarily for its seaworthiness, its efficiency in dragging nets and its loading capacity, together with its capacity to process the catch as quickly as possible on board. The first German deep-sea trawler, the *Sagitta*, had a length of 33 m, a breadth of 6.5 m and a draught of 3.5 m. To assist the engine with wind power an additional tackle (*Takelage*) was attached (Grotewold, 1908: 75; Timmermann, 1902: 71; Rohdenburg, 1975: 146; Höver, 1936: 210). The relation between the size of the ship's hull, its engine power and the fish rooms was determined by the amounts of coal and ice necessary for a long voyage. The problem was that the haul could not be kept fresh with natural ice longer than 10-12 days. In order to provide the fish with adequate water, the fish processing room had to be as large as possible, but on the other hand the ship's tonnage could not be enlarged at will. A compromise was reached by building three types of trawlers for different distances, for short, average and long voyages. The main room of each ship was reserved for the steam engine together with the boiler and the coal-bunker, which for a normal voyage (of up to 12 days) could take 70 t coal, and for an Iceland tour (up to 20 days) up to 120 t coal. The most space had to be allotted to the fish room which needed 50 t of natural ice. On deck was the room for the sea-charts (*Kartenhaus*) and the helmsman (*Rudergänger*) and on larger ships the captain's cabin also. The other rooms for the crew (10-12 persons), the kitchen and the provisions store were below deck. Each ship now had a steam-propelled winch (*Dampfwinde*) for lading, mechanical devices for launching and hauling down the nets and finally a steam pump for washing the gutted fish. From the beginning of the 1930s all ships were equipped with wireless telegraphy and a sounding device (*Peilanlage*), enabling the captain to report the results of the haul to other steamers or to call them in cases of emergency. The fish trader at home could also get information in advance about the catch and make arrangements at the market (Andresen, 1934: 24-26).

III. The trawl net as the most important catching device

For mechanised deep-sea fishing the traditional trawl net was the only hauling device that could increase fish production quickly. This net was a funnel-shaped sack which was dragged with thin ropes by a fishing boat above the bottom of the sea. The sack was held open with a beam 11 m long, which the fishermen called a *Baum* and the net was therefore called a *Baumnetz* or in English a trawl net. Fishing with this trawl net was, contrary to the former *Treibnetz*, a form of active fishing, because the fish were truly hunted into the net. They went through the opening in the net into the inner funnel-shaped sack (*Steert'*) which blocked every possibility for escape. When the nets were hauled on board with the help of the steam winch, the sack was opened and the whole catch immediately fell out. It was no longer necessary to pick every fish out of the net by hand as before (Höver, 1936: 272; Andresen, 1934: 25; Grotewold, 1908: 102; Dittmer, 1902: 33).

The trawl net was already known in England since the fourteenth century and there all fishing boats were called 'trawlers'. On the European continent this method of hauling came into use in The Netherlands in the eighteenth century and at the Finkenwärder fishery in the early nineteenth century. The steamers of some decades later had much larger trawl nets, which speeded up the dragging and so reduced the time for one haul from 7 to 4-5 hours. How much the catch could be increased in a short time can be demonstrated by an example: on her first voyage, the earliest German steamer, the *Sagitta* from Geestemünde, used old fishing gear like that used by Dutch cod fishers. This was a fishing line 100 m long which had shorter lines of 60 cm long with fish-hooks attached. Friedrich Busse's ship made 388 marks profit from this voyage, but after the introduction of the improved trawl net profits rose to more than 1.000 marks (Rohdenburg, 1975: 153).

Because there were technical and financial limits to improving the trawl net, fishermen became interested in a new form of this net that had been developed in the meantime again in England (*Scheerbrettnetz*). The Elbe fishermen took it over in 1895 when the patent expired. This new hauling device was in form and function similar to the older net. But instead of one beam on the dragging lines it used two slanting planks that, when towed, would try to move apart under pressure of the water thereby opening out the mouth of the net fully. The inventor had called this the 'dragon effect'. Because this device was heavier the net came closer to the sea bottom. It was now possible to fish not only in larger ocean areas but also in much deeper water, even 300m under the surface and that meant the opening up of new fishing grounds outside of the continental shelf. From 1903 on further efforts were made to improve the trawl net even more which had in the meantime become a fixture of German deep-sea fishing.

IV. On-board processing of fish

When the net had been hauled on board, the processing of the catch began (Grotewold, 1908: 93; Goldschmidt, 1911: 31; Timmermann, 1962: 74). The fish were sorted according to species, killed by a cut in the neck, the innards removed and the fish washed. With the exception of the liver which was cooked for the production of cod-liver oil, all the other fish waste was thrown back into the sea. The helmsman ('mate') examined the quality and species of fish before they were taken by basket to the fish-room below deck and packed. The fresh fish were put into a partitioned container under a layer of ice and a layer of matting was put on top to keep the ice from melting too quickly. The careful cleaning of the fish-room with a disinfectant became the rule after hygienists discovered the role of germs in fish spoilage and official food inspection had been introduced in Germany.

The use of natural ice for cooling meat had been well-known for many centuries, especially in the large estates in Germany, but it was not fully applied until the second half of the nineteenth century. The fast-increasing demand of the big breweries, 'beer houses' and meat factories as well as large diaries in the surroundings of larger towns led to the development of a new business. Cargo freighters now began to regularly transport large blocks of ice from Norwegian to German harbours where they were stored in special 'ice-cellars' (*Eiskellern*). Later special 'ice-factories' were founded in some seaports, designed to deliver natural ice to the fishing industry for the whole year. The United States served as the pattern here. As far we can see, the Finkenwärder fishermen were the first in 1868 to fit out their ships with ice-boxes This did not meet with approval everywhere, but in the 1870s cooling with natural ice was already a common practice and changed the old troublesome system of lading and transportation. Decisive progress was now made in the preservation of fresh fish, which had always been a major problem in the past (Teuteberg, 1995: 51-61; Tilmann, 1933: 21; Stahmer, 1925: 152; Tülsner, 1994: 94; Rohdenburg, 1975). Because natural cooling methods did not give uniform durability to the different species of fish, there was a call for more efficient technology in this area from early on. But the first 'freezing machines' (*Kältemaschinen*) that were developed after 1877 were initially too expensive and it was not possible to install them on the steamers. Their owners therefore rejected this new invention at first.

V. Deep-sea fishing hauls between 1888 and 1929

We now have to ask how all these factors influenced sea fish production during the late nineteenth and early twentieth century. The answer is clear: the initial increase in and subsequent modernisation of the fish trawlers, the important improvement in the nets and the development of on-board fish processing resulted in a large increase in production. As the hauls of fresh fish in German sea fishing from the North Sea between 1888 and 1929 demonstrate, up until the outbreak of the First World War the catch increased by 46 per cent, almost parallel to the increase in the number of fishing vessels. This growth was even stronger from 1924 up to the beginning of the world economic crisis in 1929.

This increase in the catch is remarkable because it happened, by contrast to the increase in the late nineteenth century, despite a drastically reduced number of fishing ships.[2]

The most significant hauls consisted of cod, herring and haddock, with the greatest increase in herring. As the hauls of *Heringslogger* between 1872 and 1924 show, annual average production up until the introduction of steam- powered ships in 1898 – with the exception of one record year in 1894 – was relatively small in comparison with the period up to 1924, though the war years and the subsequent period of inflation and currency reform cannot be included in this. The *Dampflogger* (modern steam trawlers) could travel independently of the wind much faster than the shoals of fish, could fish also in rough seas and did not need to cut the nets if the waves were too high, which in pre-industrial times had often meant a loss of the whole haul. The number of voyages had increased from four to six per year and this alone contributed to a production growth of 3 per cent. The improved expert knowledge of the captain and the crew's sharing in the profits contributed also to this economic success. But it is true, the total haul on average remained very changeable, even with the bigger hauls. Because the market demand for fresh herring during the 1920s was so high, part of the herring trawler was devoted to the on-board processing of the haul to 'salted herrings' which could on landing be put into cold storage and thus sold over a longer period.[3]

VI. The traditional sea fish trade

As already mentioned the trade in fresh sea fish before the industrial age was irregular and seasonal. Because fish spoiled quickly and because of poor transport, consumption was limited to coastal areas. The fishermen began to sell their haul from board ship immediately on landing or, like hawkers, attended the next urban weekly market. They adjusted their haul to local demand but often had to fix very elastic prices. This type of direct personal sale to the consumer was very time-consuming and their ships often remained for days at the harbour pier before the whole catch was sold. If the fishermen could not sell all their fresh wares on time they are forced to get rid of the rest very cheaply at a loss. Sometimes the profits barely covered the production costs. When there was a crowd of people at the harbour, the fresh fish, which was seen as a luxury, was a favourite object of thieves. Some customers who purchased a larger lot of sea-fish tried to sell it again in other places, and others who bought fish for their own use could sample the wares thoroughly and compare prices. But neither they nor the fishermen had any general view of the real market situation. Moreover, the fishermen were dependent on the weather and were not able to the varying demand in the dispersed local markets. Only Hamburg's large fish market offered better possibilities, though even here price discounts were likely when the hauls were too large (Schildt, 1897: 1; Goldbohm, 1935: 3; Mohlen and Köser, 1962: 15; Stahmer, 1925: 94; Arens, 1927: 16). The fish-women at Hamburg's fish market were known for their loose tongues and were specifically trained

---

[2] See the statistics in Höver, 1936: 324-328 and also Reichsministerium, 1924-1930.
[3] See especially for the herring fishery Reid, 1929 and Jaspers, 1921. The sources give different figures for the yearly hauls which began in 1894: Lindemann, 1888: 166; Grotewold, 1908: 142; Reichsministerium, 1924-1930.

to sell there. After the expansion of deep-sea fishing with steamers and the enormous increase in the size of the hauls, selling prospects became more difficult for those fishermen who had no steady purchasers and had to sell their catch fish by fish. However most of the inshore fishermen who did not take part in the mechanisation of sea fishing stuck to their old system of selling up until the early twentieth century.

After 1885 fish middlemen, or in German *Reisekäufer*, emerged as a solution to these problems. Their name had to do with the fact that these traders purchased the whole haul in bulk for a flat sum in order to sell the fish to other small fishmongers or directly to individual customers. Because the simple fishermen traditionally hadn't any firm business partners, they always considered the bids of several fish middlemen before they made a bargain. But the 'Reisekäufer' reacted here with their own strategy: they hurried to accost the returning fishermen on the Elbe before they landed to prevent them from comparing offers before selling. Some of the middlemen even went as far as to sail to the mouth of the Elbe to Cuxhaven or even into the North Sea to fetch a haul for as cheap a price as possible.

If a deal was made, the fishing boat had to remain in the harbour until the middleman had sold the whole haul. This was also necessary because until 1880 there was no possibility of putting the fresh fish into cold storage. Later on the harbour administration allowed fishing boats to remain at the pier of Hamburg's fish market for only two days. This intermediate trade had advantages and disadvantages. The middleman knew the market situation much better than the fishermen and on visiting the ship could estimate roughly the real amount of a haul, the quality and the species of fish. But because the true quantity of the catch could not be determined absolute precisely before the actual sale, a certain price reduction and a lump sum for the risk were usual thus reducing the fisherman's profit. However, this sort of dealing was seen on both sides as positive until the emergence of deep-sea fishing with steamers.

A later form of this middleman was the so-called fish commissar. In this case the fisherman who landed in Hamburg, Altona, Bremerhaven or Geestemünde, handed over his complete haul to a *Commissär* who sold the wares for him, for instance to big hotels on the North Sea islands. But this was based more on a purchase in good faith than on commercial calculations. Many other wholesale fishmongers had begun to supply high-priced sea fish such as turbot to the new seaside resorts as well as to large cities with good railway connections. But a mail-order business with cheaper sea-fish did not develop for technical and financial reasons. A huckster with herrings could do better in the traditional way at coastal harbours (Stahmer, 1925: 98; Arens, 1925: 15).

VII.   The centralisation of the modern fish trade through auctions

Many of the North Sea fishing voyages had led very early to Dutch ports, where in 1780 the first fish auctions had been introduced and the individual selling of fresh fish prohibited. There 170-pound lots of fish were now sold from time to time. The procedure followed was that the auctioneer following his experience named an opening bid which subsequently went down; the customer to first express interest in this bid was the winner. In this way, not only was the best price obtained, but also a reduction in time (Lübbert, 1925: 26-27; Schümann, 1931: 32-33).

When deep-sea fishing with steam powered boats began and the catches consequently increased, the dispersed local market system turned out to be sufficient. The merchant Gustav Platzmann from Hamburg, who probably knew of Friedrich Busse's activities at Geestemünde, in 1886 proposed to the Finkenwärder fishermen that they deliver their whole hauls to his planned new fish auction in Hamburg. They agreed and the city's Senate gave its consent on 30 March 1887. Platzmann and a second wholesale fishmonger, who had already sold two hauls from the steamer *Sagitta* very successfully on a private commission basis, were appointed as auctioneers. The Senate also made available a large fish hall in the harbour quarter of St. Pauli, and on 1 May of that year the fish auction in Germany could start. Both auctioneers, who became official employees and had to administer an oath to the 'Free and Hanseatic Town of Hamburg', acted now as neutral intermediaries between the fishermen and the buyers who consisted of wholesale fish traders, fish manufacturers and small fishmongers. Hamburg's new institution was copied a few weeks later in the neighbouring town of Altona, and then at Geestemünde (1888), Bremerhaven (1892) and Cuxhaven (1908) (Schildt, 1897: 3 and 5-6; Mohlsen and Köser, 1962: 23; Möhring, 1904; Vereinigte Fischmärkte, 1937). The great fish traders could now concentrate their operations on providing the fish auctions with hauls and unlike auctions in The Netherlands, the sale went to the highest bidder. These merchants transported their wares now by railway or freight ships to the auctions. Furthermore local wholesale fishmongers also used the auctions when they wanted to dispose of fish that was not purchased immediately.

By bringing suppliers and customers together at a central place, large quantities of fish could be sold in a minimum of time, clearly a demonstration of the efficiency of the free market. Yet, although supply and demand continued to change and auction prices to fluctuate, it is clear overall that prices for fresh sea fish were declining in the long run as a result of growing mass production. The large fish traders lost the influence they had had on the market, because it was impossible to hold back much fresh fish for a long time from the auctions. On the other hand, they could not immediately order new hauls when the price increased as a result of too low offers. The time when minimal prices could be demanded so as to avoid the risks of a sale and other expenses came to an end now. Bringing together so many buyers in one spot had also the effect of creating a high demand so that the wholesale fishmongers tried to outbid one another. The seller then got much higher prices for his fish by comparison with a direct sale. The fish traders as owners of the fishing vessels always tried to reduce the number of unproductive layover days at the harbour and sometimes auctioned parts of their next fish catch even before the actual voyage had begun, a highly risky manoeuvre of course. Through these auctions, sea fishing acquired the character of a continuous production. This had to do with the fact that only a few powerful fishmongers and fish manufacturers could participate at an auction. Even in earlier times the purchase of a complete haul also needed large financial resources, but now the purchase had to be strictly adjusted to the market. Because the fish trading companies oriented their strategy more to an increase in the quantity rather than to the quality and variety of their products, they became more and more dependent on the wholesalers who had an advantage by being solely focused on quantity. The owner of the fishing companies tried to minimise the decrease in fish prices by laying off some of their ships during the summer months or by buying the surplus for their own fish manufacturing (Göben, 1966: 44-46; Kostka, 1953: 40-41; Stahmer, 1925: 101-103).

The auction turnover in the German fish markets between 1888 and 1930 indicates that the shipment of fish was increasing continuously, except for the World War I period. Geestemünde's auction had the biggest turnover, which even in the third year of its operation already exceeded that of Hamburg and of Bremerhaven, and it kept its leading position. Between 1920 and 1925 this seaport (meanwhile renamed Wesermünde) could step up its turnover of fresh sea fish by 2.5 per cent. In 1930 it sold 126,650t which was 43,569t above the sum of all the other named fish markets (Göben, 1966: 199 and 201; Höver, 1936: 324-328; Vereinigte Fischmärkte, 1937: 364). Geestemünde's top performance on the lower Weser had to do with the fact that the Prussian House of Parliament (*Preussischer Landtag*) in 1893 granted the gigantic sum of 5.57 million marks for the extension of this fishing harbour, because the old facilities were no longer adequate. In 1896 the biggest Prussian fishing port was opened. The fishermen and the fish traders created a common 'Business Co-operative Ltd' (*Betriebsgenossenschaft GmbH*) which was the basis of the modern and still existing enterprise and of Europe's largest fishing port. Hamburg achieved its highest turnover between 1911 and 1912 with 17,000t fresh sea fish, but this figure was never reached again, and in 1930 only 11,575t were registered. Altona's development was rather similar, but this harbour was able to increase its turnover after the collapse of the German Empire in 1918 and in 1929, before the outbreak of the world economic crises, reached 48,675t, its top result since the introduction of the fish auction. It was then superseded by the Elbe port Cuxhaven which moved to second place in the German fish market. Bremerhaven's turnover remained far behind, so that it was forced to join up with the nearby harbour of Wesermünde (Geestemünde). Wesermünde and Cuxhaven were now the real centres for the cheaper species of fish and more than a hundred fish dispatching firms settled there. The main place for fish processing, however, remained Altona, because its harbour had the best traffic connections, especially direct main railway lines to the major towns in the hinterland. Moreover there were short routes from the ships in the harbour to the packaging and processing halls. Hamburg's sea port could not expand in this direction and its auctions shifted to dealing primarily in high-priced fish which had much lower haul quantities but in the end fetched greater profits.

VIII. Methods of fresh fish preservation and the origins of the fish industry

The quick spoiling of fresh sea fish, due to the very short period of *rigor mortis*, had always been mentioned as an obstacle to long storage and transport to distant interior markets. All the traditional arts of preserving had certain time limits and also changed the taste and appearance of the fish for the worst. These old preservation methods became more and more inadequate as fish production increased. Although these preserving traditions were retained with some technical improvements, a totally new preservation technology gradually developed. During the Napoleonic Wars the French master chef Nicolas François Appert already had had the idea of enclosing foodstuffs in a container with an airtight plug and then heating the contents in water with the result that all the air was expelled and the germs killed. Through this process of heat sterilisation, foodstuffs could be preserved much longer than hitherto without totally changing the taste and appearance. Some decades later this invention became the basis of the modern preserves industry. From the 1840s in Germany experiments were done on preserving gourmet vegetables and fruits by this heat sterilization method, but in the beginning it did not have any influence on the preserving of fish, because many types of fish lost their consistency and original taste in the process. The first canned fish could not be stored very long and had to be consumed rather quickly. The products were therefore called 'half-preserves' (*Halbkonserven*) and were mainly in use up to 1930. The more durable sterilised fish preserves in Germany were primarily imported from foreign countries and prevailed much later as a mass product (Tern, 1918: 4; Stahmer, 1925: 224-255; Hilmer, 1942: 43).

The first durable canned fish were the oil sardines which had been sterilised by the Spanish method. After being sorted, killed and their innards removed, the fresh fish were put in a strongly salted brine for 1-2 hours, steamed in a special 'steam locker' (*Dampfschrank*) at 89°C, then put in oil for three hours and finally packed into tins in olive oil seasoned with thyme, cloves and bay leaves. The tin was sealed by machine and then sterilised in a tightly closed cauldron (*Autoklav*) under high pressure and at a heat of 117°C. After a cold-water bath and a final check of the tin's seal, the process of curing (*Reifung*) began which gave the canned sardines their typical taste. This preserve from the beginning could be kept a minimum of ten years (Stahmer, 1925: 280-282; Hilmer, 1942: 18-20; Biegler, 1960: 534; Winter, 1909: 31; Seumenicht, s.d.: 79-80). With this 'full preservation' the fish industry broke through the nature's barriers and became independent of the results of the changing and seasonal hauls. Fish as foodstuff was now in principle freely available all year and everywhere. This revolutionary technical innovation had, however, to adapt to economic circumstances.

The emerging German fish industry saw its main task as meeting the increasing demands of the domestic market. Since transport times had been significantly shortened by the extension of the railway system, initially the shorter lasting 'half-preserves' met the demand and so, with few exceptions, the old methods of curing, pickling and marinading could still continue. Only small amounts were processed as 'full preserves' and exported overseas. Although the 'German Agricultural Association' (*Bund Deutscher Landwirte*) and the government tried to insist on greater production of 'full preserves', only 5-6 enterprises were engaged in this up to the mid 1920s.

The German fish industry differs here from that of other European nations. But this can be easily explained. The Scandinavian coasts, for instance, continuously yielded large shoals so that very rich hauls were possible, but since the population was too small to consume all this fish, these countries and especially Norway were pressed to produce 'full preserves' for export. Similar pressures existed more or less in Portugal, Spain and France. Because sardines were not to be had along the German sea coast and because importing them from distant areas seemed unreasonable, the German fish industry began to process sprats from the Baltic Sea in the bays of Kiel and Eckernförde. This small species of herring with a length of 10-15 cm regularly surfaced between May and June along the coast of Holstein where it had been cured for centuries. This little fish seemed appropriate for preserving in oil like the sardine. But the hauls were irregular and not large enough for mass production, partly because the local curers always bought the small catches of the fishermen at high prices. Sprats from the Elbe had therefore to be sold for processing. On the world fish markets the small catch of *Kieler Sprotten* (Kiel sprats) could not compete with other foreign 'full preserves'. The German fish industry, which had its first centre in the fishing village Schlutup on the right bank of the Trave near to the famous Hanseatic city of Lübeck, therefore held fast to the traditional method of curing fish; only later did the commercial frying of fish frying take over. At the heart of this art of processing, which originated along the coast of Pomerania, was the *Bückling* (red herring or kipper). This fish was cured so completely that in the late nineteenth century it could be transported for 2-3 weeks without a problem. The sale of kippers with their unique taste and attractive yellow-brown colour was so successful, especially in the growing large cities, that Schlutup's fishmongers imported large amounts of Scandinavian herrings in specially hired freight ships and so founded a new branch of the German fish industry. The number of curing firms increased from 8 in 1870 to 20 in 1889. In that year they dispatched 59t fish, most of it cured, to interior markets. During the 1880s also herrings were fried and then packed in a marinade in tins. This new preserve began to boom when in 1895 this fishing was connected by railway with Lübeck and the *Schlutuper Brathering* (Schlutup fried herring) now became another brand name of the German fish industry (Ludorff and Meyer, 1973: 130; Hilmer, 1943: 10; Kreutzfeldt, 1968: 121-123).

While in Lübeck new fish firms were set up on the Schlutup model, other entrepreneurs in Flensburg began to cure eels for sending by mail. After 1850 the processing of Elbe fish began in Hamburg and Altona, especially eel, salmon and sturgeon, initially only to serve local demand. With the improvement in transport Hamburg participated then in the emerging herring trade to Lübeck-Schlutup and to Kiel. With the introduction of deep-sea steamers the fishing industry tried to adjust its production more and more to certain fish species. Because herring remained the major raw product, the factory owner in Altona at times of high demand sometimes even imported British herrings. The success of deep-sea steamers finally led to the establishment of another centre for the fish industry at Geestemünde. Up to 1880 at Bremerhaven as well as at Geestemünde and in many other seaports there was only limited and seasonal fish curing for the local market, but then large industrial companies set up here to profit from the rapid increase in the catches.

Thanks to the processing of fresh sea fish into a much longer lasting foodstuff, changes took place in the household products offered. The fish industry realised that it would be very profitable to free the housewife and chef from the troublesome work of preparing a meal and to prepare ready-to-eat food that was at the same time long-lasting. Tasks such as cutting off the heads, removing the scales, skins, innards and sometimes also the fish bones, as well as filleting the fish needed new methods of preservation. So from the 1880s the number of marinated fish products increased.[4] Here too primarily cheap products were produced for the mass consumer market in the large cities. A very small number of expensive gourmet marinated products for delicatessens appeared around 1900, the most popular being Norwegian anchovies, appetisers and other titbits (*Gabelbissen*), but also fish in aspic and fish sausages. The raw material used here were *Brisling*, or the sprats caught in the fjords of Norway, or very fat and tasty herrings which were pickled with salt and garnished with various spices and sugar. The production and packaging in containers was mechanised, so that up to 200t of fish per day could be processed. After a cure of several weeks in order to enrich the flavour, the fish was put into small tins or glass containers weighing between 100g and 220g. The production of such fish delicacies required specially trained personnel, and therefore the processing was done in Norway and the packaging in Germany. If the raw marinated products were sterilised at a low temperature and boric acid added to prevent alterations in the taste and appearance, they could be kept for four weeks. The North German delicacies soon attained the quality of Scandinavian products and took over the leading market position, though unlike Scandanavia their popularity remained limited, except for canned shrimps and prawns (*Krabben / Garnelen*). The marinating establishments, often annexed to larger curing firms, were situated mainly on the coast in order to make use of the landed fish. However, later they also were situated in inland towns, though here they restricted themselves to the production of cold marinades in vinegar. This sort of fish processing profited from the fact that fresh sea fish were no longer transported in caskets, but in cooled sheet metal packages.

The reports of the German Sea Fishing Association in 1899-1900 show there were between 447 and 457 fish processing enterprises in the country. This number had increased to 650 by 1913-1914, but the total number of all the registered small curing and marinating centres was not certain.[5] Since the turn of the century the region Hamburg-Altona has been the most important location of the German fish processing industry with approximately 100 enterprises.[6]

---

[4] Stahmer, 1925: 265-266; König and Splitgerber, 1909: 59-74 (here detailed descriptions of curing, roasting, pickling, marinading, preserving with jelly, fish sausages and other titbits).
[5] Deutscher See-Fischerei-Verein, 1899: 627-634; 1900: 305-312; 1913: 575-583; 1914: 632-640; Goldschmidt, 1911: 110. Winter estimates this as being too low; see Winter, 1909.
[6] Fischereidirektion der Stadt Altona, 1923: 19-22. Wesermünde, Cuxhaven, Bremerhaven and Nordenham could not achieve such a position. On the Baltic coast, Lübeck-Schlutup, Eckernförde and Kiel remained the most important locations for fish processing. In the inland fish industry Berlin was at the top for the whole period and also had the highest fish consumption per head and year in Germany. See Kretzschmer, 1902 and Gossner, 1901.

IX. The transition to artificial deep-freezing of fish

From research on the history of cooling technology we know that in earlier times winter frost was used to freeze fish catches and to store them in cool places and later in special 'ice cellars' (Teuteberg, 1995: 51-61). From the middle of the nineteenth century, the erection of cool storage warehouses closely connected to the central market halls in the major cities was a major step forward, because the cheaper sea fish cooled by natural ice could be transported faster to the big inland markets and stored there for a while at a low temperature. Russia and North America then began to export huge amounts to European countries using railways and steamers that preserved the fresh fish with a mixture of salt and ice. The USA was the most successful by using for the first time specially constructed refrigerator cars and wagons (Heiß, 1924: 197-198; Heiß, 1937: 62; Dunker, 1908: 171-218). But the general public in the beginning disliked imported frozen fish, because like frozen meat from America its appearance was very unusual and raised fears.

The breakthrough in new cooling techniques was made by the Danish fish importer Ottensen. Between 1911 and 1913 he invented a totally new rapid freezing system which was much better than the gradual cooling with natural ice and cold air. He had realized that sea fish cool best in salt water, their habitual medium. His first tests in this direction failed, because dipped in a salt solution the fish absorbed too much salt and were inedible. After further experiments he discovered that every sodium chloride solution has, depending on its concentration, a specific freezing point at which ice is suddenly emitted instead of salt. Ottensen now chose a solution with a salt degree of 28,9 per cent where freezing takes place at -21.2°C. At this degree the fish now did not absorb ice. A carp weighing 8 pounds needed nearly 36 hours to be frozen with cold air, whereas with this new method the freezing time could be reduced to 2 hours with the same temperature. This procedure also had another advantage over earlier methods. Products frozen with cold air when thawed lost some of their freshness and flavour. After certain negative effects (alteration of consistency, discolouring and tendency to rancidity) had been removed, the new technology became as successful in the production of long-lasting products as the invention of airtight heat sterilisation.

Ottensen's procedure was as follows: after the heads were cut off and the innards removed, the fish were cleaned, packed into wire containers and dropped into a shaft at -21°C. After two hours the fish were dipped in fresh water so that they got an ice glaze. In this way the cold water could not evaporate and the fish could be stored for several months in cool storage. This was a major development, because fresh fish could be stored with natural ice for only 6-8 weeks at -9°C and their quality was impaired. This was the reason why the longer storage of fresh fish was limited up till then.

In Germany in 1924, the Nordsee GmbH, the Deutsche Fischerei AG Wesermünde and the famous machine factory Humboldt in Cologne-Deutz together founded the affiliated company *Deutsche Ottensen Gefrieranlagen-Lizenz-Gesellschaft* (German Ottensen Freezing Company) into which the Kühltransit AG Hamburg/Leipzig was later incorporated. One year afterwards at Cuxhaven and Wesermünde similar enterprises began to operate. With all these establishments the age of deep-freezing fish had truly arrived (Classen, 1924: 287-299).

Because the steamers made increasingly long voyages to Iceland and the Arctic Sea, on-board freezing became increasingly important and the idea developed of equipping ships with the new freezing units. The hope was not only to maintain the quality of the fresh fish from net to plate, but also to ensure a longer storage time and thus minimise expenses. Whenever the fish supply decreased and prices rose, it was now possible to fall back on these frozen reserves. Market fluctuations could never be avoided, but through this new technology their impact was diminished, and the fish supply became independent of the season and more stable. The deep-freezing technology contributed in this way to a structural variation in the whole fish economy.

Another technical stage was then achieved by the American biologist Clarence Birdseye who developed the Ottensen procedure further and constructed his hydraulic plate froster (*Plattenfroster*) which could mechanically freeze fish fillets. This invention made possible for the first time the production of a package suitable for household use (Teuteberg, 1995; Nordsee GmbH, 1996: 48). From the end of the 1950s with the spread of household and shop freezers, the deep-freezing technique together with the transition to self-service in the supermarkets and the invention of plastic packaging brought great changes in the Federal Republic of Germany as in other European countries. But unlike British trawlers, German steamers were not equipped with deep-freezing units for a long time yet. Evidently this had to do with the different fish consuming habits in both countries. A frozen whole fish was harder to sell in Germany, because German consumers had become accustomed early on to ready-to-eat frozen fish fillets. Until the 1950s German fishing ships produced fish fillets packed in natural ice and salt while still at sea; the real deep-freezing process began then after landing in specific firms in the harbour. As statistics from Bremerhaven indicate, the landings of normal fresh sea fish decreased very significantly between 1957 and 1963, though the landings of frozen fish did not exceed them until 1985.[7] Similar developments probably happened at other German locations for fish processing. The rather late introduction of deep freezing while at sea and the on-board production of fish fillets cooled only by natural ice and salt had a big economic advantage. The quantity of raw fish that could be hauled on one voyage was tripled because the fillet was made from only a third of the whole fish and two thirds of the fish was then thrown back as waste into the sea.

The success of the new deep-freezing system had to also to do with the fact that the cooling chain from catch to consumer could be nearly closed. Special refrigerated railway wagons and later trucks as well as cold-storage-warehouses and large freezers in supermarkets were absolute necessary to maintain a constant storage temperature, because a slow thaw meant a loss of quality.[8] The last gap between the shop and the private household today can be bridged through the use of the portable cooling bag or by special firms dispatching the frozen wares directly to the home freezer.

---

[7]   These figures were very kindly made available by the Fischerei- Betriebs-und Entwicklungsgesell-schaft mbH Bremerhaven. On the introduction of the Danish plate-froster in the German deep-sea fishery see Ernährungswissenschaftlicher Beirat, 1969: 458; Ludorff and Kreuzer, 1956.
[8]   See the detailed description of methods for keeping fish deliveries fresh around 1900 by König and Splitgerber, 1909: 79-85.

X. Final considerations

The catching, preserving and trading of fish, like the subject of seafood in general, is a broad field and difficult for the historian to review. This survey therefore is limited to certain periods, regions and details, and focuses on analysing only some selected key factors. So the study concentrates on deep-sea fishing in the North Sea during Germany's first main phase of industrialisation between approximately 1885 and 1930. This makes it feasible to draw connections and comparisons with similar structural changes during this period in other West European countries, especially in the Netherlands and England. Following the general theme of this book fishing is here seen to be, like agriculture generally, a branch of the larger food economy. The revolutionary change in the methods of preserving, processing and selling fish are the focus of consideration and for the first time the plentiful contemporary writings on these aspects have been evaluated and compared with the statistics available. The abundant historical sources would allow another chapter specifically about the development of the herring catch; which was for centuries always the favoured fish of the Germans. Several comparisons between Northern and Western Europe with regard to herring consumption are possible.

Some important problems regarding transport and trade have been described here as necessary background information to explain the causes and results of various technical and economic innovations. However, many other questions in this area remain neglected, for instance, the transport of fish by railway from the seashore to the interior markets or the government policy on freight, revenues and tolls as well as the steep increase in fishmongers in large cities and the fluctuations in the price of fish. A more detailed investigation would also deal with the variations of fish consumption in private households and restaurants. In this context, inland fresh water fishing should also be examined. Finally all the physiological, medical and ecological aspects have been omitted. There is, of course, much to say primarily about the growing knowledge on fish with its high amounts of protein, minerals and vitamins D and A, its ability to prevent human diseases like heart coronaries and blindness or the never-ending discussions about the dangers of fish poisoning and the treatment of fish waste. All these problems, which are well-documented, await a more scholarly study. A few central questions and issues which might be of general European interest may be briefly noted.[9]

(1) In the frame of this book we should first ask how far the fish industry, as it emerged in the late nineteenth century, can be compared with other branches of agriculture-based food industries and food processing. This is not an easy question to answer. As this case-study on Germany indicates, deep-sea fishing acquired the latest in advanced technology probably more quickly than agriculture. The transition to steam energy, modern devices, new methods of preserving, and selling by new types of entrepreneurs was done in a very short period, while the modernisation of farming on the whole needed many more decades.

(2) On the other hand we also have to note some major disadvantages for sea fishing.

[9] For inspiring ideas I have to acknowledge here Adel P. den Hartog (University of Wageningen) and the discussions at the CORN conference at Leuven.

The on-board processing of fresh fish began to develop very early on, but fish industries could be established only at a few harbours with railway connections. Contrary to other food enterprises, fish manufacturers did not spread to other interior regions and always had to send their packaged wares to large urban markets, Berlin excepted. Geographical and transportation factors then determined the locations of fish processing plants.

(3) Further it is of interest that at first the traditional coastal fishermen with their small open sailing boats could not participate in this revolutionary modernisation process, because the huge capital investment in a steam-powered trawler was too much for the ordinary fisherman. Industrial and pre-industrial production methods ran parallel for some time, as in the countryside. But after a while the coastal fishermen too found a niche in the new economic system in the fast, daily supply of high-priced fish to tourist hotels and gourmet seaside restaurants. That implied quality instead of quantity. It would be perhaps worthwhile to compare this development with similar activities of farmers who in the suburbs of big industrial towns founded new firms for food processing, for instance, dairies, vegetable gardening and orchards, etc.

(4) There is no doubt that the use of natural ice (produced by a quite new specific trade!) was a decisive breakthrough not only for preserving fish, but also fresh meat. The introduction of this innovation opened the gate for further transitions to artificial cooling and then deep-freezing, thereby establishing a cooling chain from the beginning of production to the last stage of consumption. There is already some research about the technological apsects of this new sub-zero innovation but much more research is needed on how, when and where food habits have varied because of it.

(5) Yet, despite the increase in fish production, it is remarkable that fish consumption per head and year in Germany remained rather low compared with that in other neighbouring countries. The regional differences in meat consumption were probably even greater. The variations in fish consumption evidently were due not so much to religious causes or differences in price, but more to the traditional deep-rooted 'food landscapes' which are based on natural surroundings, climate and the agrarian economy.

Despite its over 80 million inhabitants, Germany's fishing fleet today ranks in ninth place in the European Union, behind Spain, Italy and Great Britain and just above the small nations of Ireland, Sweden and Belgium. Moreover, in 2001 its consumption of 15.1 kg of fish per head per year remains low despite an increase elsewhere in Western Europe. The reasons for this lag can only be found in the other historical background of food culture.

Bibliography

Abraham, E.A. (1931) *Die Organisation der deutschen Hochseefischerei in der Nordsee*, Leipzig.

Achelis, H. (1888) *Das Symbol des Fisches*, Leipzig.

Agatz, A. (1919) *Die technische und wirtschaftliche Entwicklung der deutschen Hochseefischereihäfen*, Tech. Diss, Hannover.

Andresen, H. (1934) *Der Hochseefischereibetrieb im Elbe- und Wesergebiet*, Diss. rer. pol., Mannheim.

Arens, W. (1927) *Die Absatzorganisation der deutschen Seefischwirtschaft der Nordsee (unter besonderer Berücksichtigung des Frischfischs)*, Diss. iur., Würzburg.

Biegler, P. (1960) *Fischwaren-Technologie. Fabrikationsmethoden zur Konservierung von Fischen*, Lübeck.

Biermann, A. (1865) *Neuestes illustrirtes Fischereibuch*, Hamm, Westfalen.

Bloch, M.E. (1782-1784) *Naturgeschichte der Fische Deutschlands*, 4 vols., Berlin.

Classen, T. (1924) 'Eine Kühlhaus AG für die Fisch-Industrie Deutschlands', *Der Fisch. Mitteilungen über Fisch, Fischindustrie, Fischhandel und allgemeine Fischverwertung*, vol. 2, Lübeck, pp. 287-299.

Deutscher See-Fischerei-Verein (1899) *Deutscher Seefischerei-Almanach*, Leipzig.

Dittmer, H. and Buhl, H.V. (1902) *Seefischereifahrzeuge und –Boote ohne und mit Hülfsmaschinen*, Hannover and Leipzig.

Dölger, J. (1910-1922) *ΙΧΘΥΣ*, 3 vols., Leipzig.

Duge, F. (1898) *Die Dampfhochseefischerei in Geestemünde*, Geestemünde.

Dunker, E. von (1908) 'Die kommunalen Einrichtungen Deutschlands für die Fischversorgung', Fuchs, C. J. (ed.) *Gemeindebetriebe. Schriften des Vereins für Socialpolitik*, vol. 128, Leipzig, pp. 171-218.

Fischereidirektion der Stadt Altona (1923) *Die Altonaer Fischwirtschaft, Zur Einweihung der neuen Altonaer Fischmarktanlagen*, Altona.

Geiser, W. (1918) *Die Islandfischerei und ihre wirtschaftsgeographische Bedeutung*, Diss. phil., Münster.

Gesner, C. (1598) *Vollkommenes Fisch-Buch/ das ist / Ausfuehrliche beschreibung vnd lebendige Conterfactur aller vnd jeden Fischen [...]*, Frankfurt am Main (reprint Hannover, 1981).

Göben, H. (1966) *Marktstruktur und Preisbildung bei Fischen und Fischwaren in der Bundesrepublik Deutschland*, Bonn.

Goldschmidt, H. (1911) *Die deutsche Seeschiffahrt in der Gegenwart und Mittel zu ihrer Hebung*, Berlin.

Gosner, E. (1901) *Über die heutige Entwicklung und Organisation des Berliner Fischhandels*, Diss. phil., Berlin.

Grotewold, C. (1908) *Die deutsche Hochseefischerei in der Nordsee*, Stuttgart.

Gunda, B. (ed.) (1984) *The Fishing Culture of the World*, 2 vols., Budapest.

Gutmann, F. (1912) *Die internationalen Verträge über die Nordseefischerei*, Diss. iur., Würzburg.

Heger (1727) *Landwirthschäfftliche Teich- und Weyher-Lust. Oder gründliche Information zur Edlen Fischerei[...]*, Frankfurt am Main and Leipzig.

Heiß, R. (1924) 'Kühlhäuser', *Handwörterbuch der Kommunalwissenschaften*, vol. 3, Jena, pp. 197-198.

Heiß, R. (1937) *Die Aufgaben der Kältetechnik in der Bewirtschafung Deutschlands mit Lebensmitteln, vol. B: Frischhaltung von Fleisch, Frischhaltung von Fischen*, Berlin.

Hilmer, C.O. (1942) *Geschichte, Entwicklung und Produktion unter besonderer Berücksichtigung ausländischer Fischvollkonserven*, Hamburg.

Hilmer, C.O. (1943) *Standort und Eisenbahngütertarifpolitik. Abgeleitet am Beispiel der deutschen Seefischindustrie*, Diss. rer. pol., Hamburg.

Hitzbleck (1971) *Die Bedeutung des Fisches für die Ernährungswirtschaft Mitteleuropas in vorindustrieller Zeit*, Diss. rer. pol., Göttingen.

Höver, O. (1936) *Die deutsche Hochseefischerei*, Oldenburg.

Jaspers, H. (1921) *Die große Heringsfischerei Deutschlands*, Diss. iur., Würzburg.

Kietzmann, U. (1969) 'Amtliche Lebensmittelüberwachung der Fischereihäfen und fischverarbeitenden Betriebe', Ernährungswissenschaftlicher Beirat der deutschen Fischwirtschaft, *Fisch - das zeitgemäße Lebensmittel*, Berlin.

König, J. and Splitgerber, A. (1909) *Die Bedeutung der Fischerei für die Fleischversorgung im Deutschen Reiche*, Berlin.

Kostka, E. (1953) *Der Markt für Frischfische (Struktur und Preisbildung)*, Diss. rer. pol., Mannheim.

Kretzschmer, (1902) *Der Fischhandel von Berlin,* Berlin.

Kreutzfeld, B. (1968) *Der Lübecker Industrie-Verein. Eine Selbsthilfeeinrichtung der lübeckschen Bürger 1889-1914*, Hamburg.

Kuske, B. (1905) 'Der Kölner Fischhandel vom 14.-17. Jahrhundert', *Westdeutsche Zeitschrift für Geschichte und Kunst*, 24, pp.1 ff.

Landau, G. (1865) *Beiträge zur Geschichte der Fischerei in Deutschland. Die Geschichte der Fischerei in beiden Hessen*, Kassel.

Lindemann, M. (1888) *Beiträge zur Statistik der Deutschen Seefischerei*, Berlin.

Lübbert, H. (1925) *Vom Walfänger zum Fischdampfer. Hamburgs Fischerei in zehn Jahrhunderten*, Hamburg.

Ludorff, W. and Kreuzer, R. (1956) *Der Fisch vom Fang bis zum Verbrauch. Vorschläge zur Qualitätsförderung aus dem, Institut für Fischverarbeitung der Bundesforschungsanstalt für Fischverarbeitung Hamburg*, Bremerhaven.

Ludorff, W. and Meyer, V. (1973) *Fische und Fischerzeugnisse*, 2nd rev. edn, Berlin and Hamburg.

Mangoldt, G. (1557) *Fischbuch*, Zürich.

Mehler, J. (1787) *Die Landwirthschaft des Königreichs Böhmen [...]*, vol. 3, 6: Vom Fischfange überhaupt, Prague and Dresden.

Möhring (1904) *Cuxhaven als Fischereihafen und Fischmarkt*, Hamburg.

Mohlsen, K. and Köser, H. (1962) *Fischexport - Fischimport - Fischversand 1862-1962*, Hamburg.

Nordsee GmbH (1996) *Hundert Jahre Frische. Jubiläumsschrift zum 100jährigen Bestehen der Nordsee GmbH am 23. April 1996*, Bremerhaven.

Plank, R. (1954) 'Geschichte der Kälteerzeugung und Kälteanwendung', Plank, R. (ed.) *Handbuch der Kältetechnik*, vol. 1, Berlin, Göttingen and Heidelberg.

R., H. B. (1845) *Handbüchlein der wildern Fischerei und Beschreibung der vorzüglicheren in Deutschland vorkommenden Fische*, Quedlinburg and Leipzig.

Radcliffe, W. (1921) *Fishing from the Earliest Times*, London.

Reichsministerium für Ernährung und Landwirtschaft (ed.) *Jahresbericht für die deutsche Fischerei*, 6 vols., Berlin.

Reid, H. (1929) *Die deutsche Trawlingsfischerei*, Diss. rer. pol., Köln.

Rogowsky, B. (1922) *Die Organisation der deutschen Fischwirtschaft im Krieg*, Berlin and Leipzig.

Rohdenburg, G. (1975) *Hochseefischerei an der Unterweser. Wirtschaftliche Voraussetzungen, struktureller Wandel und technische Evolution im 19. Jahrhundert bis zum Ersten Weltkrieg*, Bremen.

Sarhage, D. (2002) *Die Schätze Neptuns. Die Kulturgeschichte der Fischerei im Römischen Reich*, Frankfurt am Main.

Schildt, H. (1897) *Der Fischhandel in Hamburg und Altona*, Altona.

Schümann, F. (1931) *Der Fischhandel Deutschlands unter besonderer Berücksichtigung der Hochseefischerei speziell des hamburgischen*, Diss. rer. pol., Hamburg.

Senst, O. (1931) 'Die Entwicklung der deutschen Dampf-Hochseefischerei', Flügge W. von (ed.) *Die Fische in der Kriegswirtschaft*, Berlin, pp. 79-111.

Seumenicht, K. (s.a. / 1958) *Fischwarenkunde*, Hamburg.

Stahmer, M. (1925) *Fischhandel und Fischindustrie*, 2nd rev. edn, Stuttgart.

Tern R. (1918) *Die deutsche Fischindustrie, ihre bisherige Entwicklung und weitere Förderung in gewerblicher und wirtschaftlicher Hinsicht*, Berlin.

Tern, R. (1924) *Die deutsche Seefischerei in volkswirtschaftlicher Bedeutung unter besonderer Berücksichtigung der Fischabfallverwertung*, Berlin.

Teuteberg, H. J. (1995) 'History of cooling and freezing techniques and their impact on nutrition in twentieth-century Germany', Hartog, A. den (ed.) *Food Technology, Science and Marketing: European Diet in the Twentieth Century*, East Linton, pp. 51-61.

Tilmann, G. (1933) *Die Entwicklung der Seefischerei im Zeichen moderner Verkehrswirtschaft*, Diss. agr., Berlin.

Timmermann, G. (1957) *Die nordeuropäischen Fischereifahrzeuge, ihre Entwicklung und Typen*, Stuttgart.

Tülsner, M. (1933) *Fischverarbeitung vol. 1: Rohstoffeigenschaften und Grundlagen der Verarbeitungsprozesse*, Hamburg.

Vereinigte Fischmärkte Altona und Hamburg (1937) *Von Fischerei und Fischmärkten in Hamburg und Altona zur 50-Jahrfeier*, Altona.

Winter, C. (1909) *Die deutsche Fischkonservenindustrie*, Diss. phil., Leipzig.

10 From crisis to cream.
The Scandinavian food system in the interwar period

Flemming JUST, University of Southern Denmark, Esbjerg

I. Introduction

Throughout Europe the history of farming and of the food industry in the interwar period is a bleak one, not only in economic terms but also political as the crisis years made many farmers throughout Europe an easy prey to ultra-right wing movements and parties (Tracy, 1989). In many countries agriculture was strongly divided along political, religious, linguistic and geographical lines.

In the three Scandinavian countries – Denmark, Norway and Sweden – the crisis lasted as long as elsewhere. Prices in 1931-1932 were only half of the 1927 prices. Many farmers went bankrupt, and in Denmark in the period 1931-1933 almost 5000 farm sales were forced sales out of a total number of 200,000 farms (Just and Oamholt, 1984: 49-50). Several movements were set up in response. One example was the LS movement in Denmark which was formed at the end of 1931 and within a year claimed to have 100,000 members (Brogård, 1969: 26-28). In Norway a similar movement, *Bygdefolkets Krisehjelp* (Rural Crisis Help), also gained widespread membership from the end of the 1920s, but lost most of this support as market conditions improved from the mid-1930s and the protest movements in both countries turned more and more to the extreme right.

However, democracy in the three Scandinavian countries was not threatened. Political parties on the extreme right received only a few per cent of the votes, as did the Communist parties. Another special feature in Scandinavia was that the existing agricultural organisations remained in the lead and resisted attempts to create new and more radical organisations. In this way agricultural organisations became stable democratic institutions in society, and contrary to some other countries, especially France, the countryside was not split into harsh ideological camps (Gueslin, 1990).

From the middle of the decade agriculture slowly recovered. Most remarkably – and an often neglected topic – the agribusiness sector as a whole came out of the decade more influential than it went into it. The two main arguments in this article are a) that the crisis regulations favoured a co-operative organisation of food processing and sales, which meant that the agricultural sector extended its influence in relation to other trades and in relation to society in general, and b) that as a result of the crisis negotiations the professional organisations became some of the leading government players in the post-war years. At the same time the organising of rural interests into formal organisations in itself became a democratic bulwark towards extreme political ideologies.

The article will also investigate how and why the strong cooperative organisation came about, and why Scandinavia followed this 'Sonderweg' (or special path). What would the alternatives have been? The article will conclude by showing how the decisions made in the 1930s still have importance for the food industry today.

Methodologically, the analysis is based on existing literature from all three countries and of course also the general literature. Theoretically, the point of departure is a historical institutionalistic approach where the emphasis is not so much on individual players as on the paths laid down by earlier developments which created specific values, areas of negotiation, and institutional arrangements. Such an approach at the same time indicates that it is important also to investigate the paths that led to the outcomes of the 1930s (see Steinmo, Thelen and Longstreth, 1992; Thelen, 1997).

In the following sections, some of the above-mentioned assumptions will be investigated by looking at how the interplay between agriculture and the state developed between 1880 and 1950 in the three Scandinavian countries: Norway with its heavily subsidised agriculture and import restrictions; Denmark with its large export trade and low subsidies; and Sweden in a middle position with a rather interventionist state where farmers were protected against world markets but not from internal competition.

II.   1880-1930: The formation of organisations

In all three Scandinavian countries the period after 1850 witnessed the creation of a huge number of local associations within all areas of civil society. As mentioned above it was a time of fundamental change in society when the new democracies and pre-industrial societies had to find new answers to the new challenges. Two major social movements became the most important players in the economy and society from the last quarter of the nineteenth century, namely the labour movement and the farmers' movement. Here I will only touch upon the latter.

Although the rural economy experienced a severe crisis all over Europe from the end of the 1870s until the middle of the 1890s, it was also a time of organisational activity in all local communities. Physically, it manifested itself in the erection of village halls and meetings houses, cooperative associations, cooperative dairies, and wholesale farm-supply associations. Partly as a reaction to urban culture and in order to strengthen rural civil society, an enormous number of associations were formed to stimulate the religious, cultural, political and social life in the countryside. These included the creation of local and regional farmers' unions in competition with the existing regional farming household societies that were dominated by the bigger landowners, priests and others from the ruling classes.

In all three countries agricultural ministries were established around the turn of the century in order to promote farming interests and to represent them inside the government and administration.

II.1. Norway 1880-1930

The farming household societies were the first to organise economic activities among farmers, though gradually the farmers themselves took over. The purpose of such joint companies or associations was to take as much of the middleman profits as possible. The strongest were the many parish-based cooperative dairies set up in the last quarter of the nineteenth century (Furre, 1971). Cooperative slaughterhouses were first created after the turn of the century. However, until 1930 the cooperative companies were primarily locally or regionally based, and were often in competition with each other. The few nationwide organisations, for instance The Norwegian Dairy Association from 1881, were rather weak and had little influence on national policy-making.

The economic crisis at the end of the nineteenth century and the Europe-wide demand for protection through imposing high customs duties on agricultural products, especially grain, became an important impetus for setting up professional organisations for farmers in Norway as elsewhere. *Norges Bondelag* (The Norwegian Farmers' Union), created in 1896, worked for the economic and social interests of farmers, which in its view would be furthered by the introduction of customs duties. However, such duties would in effect be detrimental to smallholders who were forced to buy grain. Consequently, the smallholders formed their own association in 1913, both with local units and the national Norwegian Farmers' and Smallholders' Union. As this union's views were in line with those of the liberal government, it was very quickly recognised by the government and obtained some state support for its activities (Just and Omholt, 1984: 83).

The numerous production and pricing regulations of the World War I period made the agricultural sector aware of the need to have adequate organisational resources if the sector itself and not the central administration was to take charge of the regulations. Thus, in 1919 the different farmers' unions and societies formed an agricultural council whose main purpose was to represent the united interests of Norwegian agriculture in dealings with the public authorities. The intention was that the Ministry of Agriculture would consult the council and allow it to draw up regulations on all relevant matters. In fact the initiative to form a council came from the state's civil servants and politicians who saw the advantage of having a strong organisation that would absorb the criticism that would inevitably arise from farmers being subjected to public regulations. However, the attempt failed due to growing antagonism between the different groups in the rural community in the 1920s where even the hitherto influential liberal party, Venstre, had to deal with the emergence of a new protectionist party, The Farmers' Party *(Bondepartiet)* (Jerbøvik, 1991).

To sum up the situation at the end of the 1920s in Norway: since the 1880s there had been a widespread local organisation of farmers' interests both in the unions and in cooperatives. There were only a few national organisations, and they were all weak vis-à-vis state and other organised interests. Furthermore, the cooperatives competed with each other and had a weak position in relation to distribution and sales.

II.2. Denmark 1880-1930

From the mid-nineteenth century Danish agriculture became part of an international division of labour. Although corn production increased, Denmark became a net importer of grain from the beginning of the 1880s due to a sharp rise in livestock production. From the turn of the century imports of feeding stuffs for cows, pigs and poultry were considerable as was the consumption of imported artificial fertilisers.

After the 1860s, price relations favoured animal production due to the larger quantities of overseas and Russian grain on the European markets. Living standards also gradually increased for the working class in the two neighbouring industrialised countries, the UK and Germany. This meant a growing demand for more expensive fats and proteins from animal products like butter, bacon and eggs. Milk, which was processed into butter, became the single most important product until the 1930s.

Gradually pig production also increased. In the 1880s exports of big live pigs and sows to the German market was still important. However, the closing of this market in order to protect domestic production, and above all increased demand in the UK for bacon and meat meant much more interest in producing and processing lean pig meat, and from the 1930s pig meat became the single most important agricultural product. Cattle and beef production was still important both as a by-product of milk production and as a specialised activity in regions with enough pasture.

Table 10.1   Various agricultural products and agricultural exports as share of total agricultural production value (%)

	Vegetable products	Milk and milk products	Cattle and beef	Pigs and meat	Others	Export's share of production value
1830-1839	39	13	15	12	21	22
1870-1879	29	22	19	14	16	37
1900-1919	6	42	19	21	12	53
1930-1939	6	36	11	36	11	67

Source: H. Chr. Johansen, 1985: p. 50.

All in all the development towards livestock production with an emphasis on milk, pigs and poultry was an advantage for the small and medium-sized farmers, first of all due to the strong cooperative organisation – and to some degree also the sales – in milk, pig and egg processing. Besides, imports of feeding stuffs and fertilisers were also organised in a cooperative way from around 1900. This meant that smallholders could make a living through intensive production on a relatively small acreage. Second, they could profit from the anti-protectionist outcome of the crisis management discussions in the 1890s. A dominating agricultural sector (with 90 per cent of exports still in 1914) with a need for large quantities of imports and exports made the free market choice relatively easy for this small, open economy. This meant that grain-buying smallholders could compensate for the lack of sufficient grain and feeding stuffs by buying imports at relatively cheap prices.

A third smallholder-friendly path was the legislation passed in favour of parcelling out and incorporating hitherto non-productive land (by cultivating heather, draining lakes and moors). Smallholders made up two-thirds of all agricultural producers but counted for only about 10-12 per cent of total production. Why then this relatively positive development for an otherwise resource-weak group which first began to organise independent smallholders' unions from the turn of the century? One explanation is that the cooperatives and the anti-protectionist approach were not invented to promote the interests of smallholders. They were just lucky that they could profit from initiatives taken by leading groups in society. The parcelling out was in the first phase initiated by the Conservative government in 1899 as the estate owners needed to attract a larger labour force to the countryside at a time when industrial workplaces in the towns offered better-paid jobs. But the parcelling out measure was also supported by the Liberal Party which received a majority of rural votes, including many smallholder votes. From 1905, however, the Liberal Party split in two and the Social Liberal Party was formed with strong support from smallholders. This small but very influential centre party played a significant political role throughout the twentieth century and was able to keep parcelling out and smallholder-friendly legislation on the agenda.

This development, however, must not obscure the fact that the key to understanding the successful development of Danish agriculture is to be found in the extremely strong position of the middle class farmers. They made up one-third of all cultivators and counted for around 70 per cent of all agricultural production.

Until the 1870s the landed proprietors were the dominant group in the countryside both politically and economically. They benefited from the strong grain sales period from 1830 to the mid-1870s and were the most important innovators, e.g. through the introduction of creameries. These landowners established regional professional unions in the 1840s and following decades.

The 1870s witnessed the beginning of a very strong organisational drive among middle class farmers. Many local farmers' unions were established in the 1870s and 1880s, but now with the farmers in the leading role. At the same time political tensions between the governing Conservatives and the Liberals were fierce and were in fact the parliamentary expression of the fight between the large landowners and middle class farmers about political and economic hegemony in the country. In 1893 there were 75 farmers' unions, in 1900 there were 102 unions with 60,000 members, and twenty years later 134 unions had 112,000 members. In comparison the smallholders' unions in 1920 included 1,102 local unions with 81,000 members (Just, 1990: 142).

Together the two unions organised almost all the farmers. The main incentive for paying the membership fee was that both unions created their own local advisory system with extension specialists mainly paid by the government. The smallholders' unions also created some joint buying associations, but in general the unions did not offer their members

217

any economic services (supply of feeding stuffs, capital, or other means of production). Such services were covered by other independent institutions, usually organised in a cooperative way. In 1919 the farmers' unions and the cooperative organisations formed the Agricultural Council. Based on experiences from World War I, the sector realised the importance of a strong, monolithic organisation. The smallholders chose not to be members of the Council as they feared being marginalised.

All in all Danish agriculture still played a significant role in society and in the economy at the end of the 1920s. The agrarian sector was very strongly organised with almost all farmers joined a union, and with both unions closely associated with two of the most influential political parties. Most of the principal products were processed in cooperative companies that were also in close collaboration with the farmers' unions and the Liberal Party. But agricultural production was in a vulnerable position as more than 95 per cent of exports went to only two markets, namely Britain and Germany.

II.3. Sweden 1880-1930

In many respects the agrarian development in Sweden was somewhere midway between that of Norway and Denmark. The geographical conditions in the northern two-thirds of the country were the same as in Norway with home-market oriented producers who also depended on extra income from other industries, primarily forestry. Growing conditions in the southern part were better and it could participate in the export markets, especially the export of butter to Great Britain and Germany and live pigs to the Danish slaughterhouses.

In 1925 there were 600 cooperative dairies with two-thirds of all milk producers as members. Still at the end of the nineteenth century butter accounted for a considerable part of the exports (c. 12 per cent). Due to the impressive industrialisation of Sweden from the turn of the century, agriculture's share of exports soon decreased. However, the state did not lose interest in the importance of the sector and supported the stationing of a dairy agent in Manchester and the introduction of a national trademark, a rune mark, in 1905, both initiatives strongly inspired by similar Danish decisions. Cooperative slaughterhouses were only established in the south and were meant for exports. However, they were never as successful as their Danish counterparts, and several of them went bankrupt during World War I due to the lack of raw materials (Rydén, 1998: 71-78).

The feeding stuff and fertiliser associations played an important role in the Swedish cooperative movement. They were seen as a means of improving the difficult situation of many smallholders, and for that reason the Ministry of Agriculture supported the establishing of regional associations. In 1905 a national organisation, SLR (*Svenska Lantmännens Riksförbund*) was formed by seven regional companies. In 1920 more than 20 regional and 1350 local associations existed with more than 85,000 members. Quite a lot of the members had joined the ranks during the World War as membership entitled them to receive some of the decreasing amounts of raw materials (Ryden, 1998: 67).

On the whole, the cooperative movement had some success in Sweden before 1930 and helped to integrate many smallholders into the market in some way. In general, however, a political and professional organisation of farmers and smallholders was not successful. With the Danish smallholders' unions as a model, several attempts were made in Sweden to organise the cottagers, both by the cottagers themselves and by the larger farmers and the state, but they proved abortive. The cottagers had insufficient resources to travel and participate in activities. But the most important hindrance was perhaps the lack of a collective identity that could have compensated for the lack of resources. The smallholders in the north and south of Sweden had different interests and many different outside jobs with which many of them identified (Rydén, 1998: 61).

Nor were the middle size farmers able to establish a strong and monolithic organisation. The north-south divide once more made this difficult, but more important was the fierce clash of interests with regard to customs duties. This fight also made it impossible to rally round a single political party. The only group to create a major interest organisation were the larger farmers who had had their regional farming household societies since the beginning of the nineteenth century. In 1917 they created SAL (*Sveriges Altmänna Landtbruktssälskap,* or The General Agricultural Society of Sweden) in response to the crisis regulations when the agricultural interests discovered that they lacked a central body like industry's. Despite wanting to play a key role in a prospective corporatist arrangement with the central administration and politicians, the Agricultural Society did not get a prominent position. One obvious reason was that it did represent only a small percentage of Swedish farmers. In 1920 only 3,400 single members, together with some regional societies and other associations were paid up members (Rydén, 1998: 53). Farmers and smallholders could not see the advantages of membership. The desire to have a 'free ride' was a major problem in forming these organisations. As a president of SAL stated in 1928: 'Our farmers are reluctant to pay membership fees to an organisation that, even if it is of the utmost importance for farming in general, does not give the individual farmer any direct economic advantages, and in any case these advantages can also be gained by farmers who do not directly belong to the organisation' (Rydén, 1998: 94).

Thus, the organisational situation in Sweden at the end of the 1920s was characterised by its ambition to have a 'Danish' model with very strong general organisations and cooperative organisations for each of the main agricultural products. But due to a low degree of membership and the absence of strong political alliances as a result of disagreement on the issue of customs duties, farmers and smallholders did not have a strong political voice. In that respect, the cooperative milk sector was in better shape, but even there it was difficult to speak with one voice because of divisions between north and south and between suburban dairies and more rural plants.

219

III. Crisis regulations and agribusiness in the 1930s

III.1. Norway

The difficult situation of Norwegian farming in the 1920s laid the basis for the regulations of the following decade. From the middle of the 1920s the Norwegian Farmers' Union succeeded in persuading the politicians of the need for customs duties on grain and later of subsidising the export of surplus production. In the egg sector a production surplus followed by export subsidies resulted in a nationwide union of egg producers in 1929 as a precondition for the distribution of subsidies. This was the first case where public regulations resulted in the organisational unification of a sector. With the crisis regulations in the 1930s this would be the model for all other branches in agriculture.

The milk sector became the driving force in the organising process and in the closer collaboration with the state. More than 40 per cent of the farmers' incomes derived from the sale of milk, and as one-third of the population lived by farming, the consequences were widely felt when prices fell dramatically at the end of the 1920s because of overproduction and sharp competition between local dairies.

In 1929 *Norges Bondelag* (the Norwegian Farmers' Union) took the initiative to form regional co-operative milk centres with a national association at the head that would regulate supplies and abolish competition. The farmers joined in great numbers, but still many 'free riders' did not participate, and the problem of how to finance the loss-making export of milk products was not resolved. So the farmer-friendly, social-liberal government passed a trade act in June 1930 with the support of all political parties. The purpose of the act was, with the help of co-operative organisations, to establish a central point for the sale of milk, butter, cheese and meat. The Labour Party especially found it important to strengthen the co-operative organisations vis-à-vis private traders and regarded co-operatives as a means of developing social equity (Just and Omholt, 1984: 91).

A Trade Board with the right to collect fees followed the implementation of the Trade Act. It consisted solely of agricultural and co-operative organisations and had only one representative from the businessmen's association. The Ministry of Agriculture did not want to be represented. With almost all other interests excluded from the regulation of the market, the agricultural organisations were given extensive possibilities for self-government. The co-operatives in particular became more powerful. On the administrative side the regulations required strong national organisations to take responsibility for local implementation. An example is the establishing in 1931 of the farmer-dominated *Norges Kjøtt- og Fleskecentral* (The Norwegian Central Office for Beef and Meat) as a direct consequence of the Trade Act. On the business side the co-operative enterprises profited from the requirement that local associations become members; an instance is the milk sector where co-operatives got a monopoly from 1936 on thanks to government policies.

III.2. Denmark

At the threshold of the new decade, the agribusiness sector in general was in a good shape after a troublesome post-war decade. Low prices on inputs made animal production profitable and many new cooperative slaughterhouses were established. Around 90 per cent of butter and bacon were processed at cooperative enterprises, and these two products comprised more than 60 per cent of the value of Denmark's exports.

Belief in the free market was very strong, and as agricultural products made up 80 per cent of exports, both the farmer-friendly Liberal Party (in government for most of the 1920s) and the Social Democratic/Social Liberal government coalition of 1929-1940 swept aside all Conservative desires for a customs policy to protect the weak industry sector.

When the world crisis hit Danish agriculture in late 1930, the sector once again reacted in classical market terms by expanding production and attempting to regulate a proportion of the exports on a voluntary organisational basis. However, extended protectionism such as quotas and differentiated levels of duty on the second most important market, Germany (20-25 per cent of the exports), made voluntary regulation inadequate. At the same time strong pro-Commonwealth feelings in the UK threatened Danish access to the main market.

In November 1932 the UK demanded a 'voluntary' and immediate reduction of 30 per cent from Danish bacon exporters. This demand, together with the huge problems with butter exports, brought the resistance to state intervention to a reluctant end, and a Bacon Board and a Butter Export Board were established under the Ministry of Agriculture. In the next three years similar export boards were formed for most other agricultural products. Their purpose was to regulate supplies, distribute export quotas among the export firms, and collect duties. The latter were introduced because of multiple pricing systems and bilateral trade agreements fixing very different prices for the same product in different markets at different times. Most of the duties were paid back to the producers so that all received the same price, but some of the money was kept in central funds and later became an important tool in the organisations' struggle for influence (Just, 1990).

Soon agriculture learnt to value state intervention. First of all the agricultural organisations discovered that state boards could be useful tools in pursuing organisational goals without losing their independence. The boards were totally dominated by the agricultural and cooperative organisations with very few seats for the private export sector, and there was no thought of giving seats to external representatives such as workers' and consumers' organisations. Nor did civil servants participate despite the boards' public law status. The Ministry just took cognisance of the boards' transactions.

Furthermore, the organisations had control over the authorities bestowed by the Agricultural Export Act of 1932 which empowered the Minister to establish export boards, stipulating that the Minister could not take any initiative without recommendations from the Agricultural Council (Just, 1993).

Another important factor was that the administration of the boards was located either in the cooperative organisations if they controlled most of the products concerned (butter, cheese and bacon), or else in the Agricultural Council (cattle, horses, eggs, potatoes). The organisations soon realised the possibilities this gave for increasing their influence on the export trade at the expense of the traditional private export sector that used to control a majority of export sales.

An important aspect of state intervention was the built-in right of the export boards to collect fees on exports. Gradually the organisations were allowed to use the pools of exports for all purposes provided they served common agricultural interests. One of those interests was to subsidise the secretariats in the organisations (Just, 1992).

The export boards were just one aspect of the regulation. In 1933 pig production itself was strongly regulated by every farmer being assigned the production of a fixed number of pigs. According to Michael Tracy, the so-called 'pig cards' were probably the first example in the world of a marketing quota system for an agricultural product (Tracy, 1988: 210). The different regulatory schemes were drawn up at the same as the Social Democratic-Social Liberal government made a major agreement with the Liberal Party. It was a kind of horse trading where town and countryside, workers and farmers agreed to collaborate on reducing the consequences of the crises. This agreement became important for the peaceful development of the labour market and trade for the rest of the decade, and together with much other legislation it contributed to relieving the crisis from the mid-1930s.

Last it should be noted that in general the government had a rather positive attitude towards the cooperative organisation of agribusiness as they considered this model to be fairer to producers than private companies. An example is the production of potato flour. When a new act on the promotion of Danish potato flour was passed in 1933, the Social Democrats made it a requirement that the planned seven new potato flour plants all should be cooperatives in order to prevent disagreement between producers and owners (Philip, 1939: 87).

III.3. Sweden

In Sweden agriculture was not organised to the same extent as in Denmark. When crisis hit the Swedish farmers and especially the dairy farmers, they did not have a strong co-operative organisation to regulate supplies and prices. The farmers' union, the SAL, therefore asked the social-liberal Minister of Agriculture to propose an act giving it a monopoly on regulating the market. Furthermore the cooperative dairies should be supported by having the right to collect fees even from non-members, and internal competition between dairies should be abolished by the establishing of a national dairy organisation (Rothstein, 1992: 112). The

Milk Regulating Act was carried through in 1932 and must be characterised as the initial step towards a very close collaboration between state and organisations in Sweden[1].

In 1933 this collaborative policy was cemented in the great Crisis Agreement between the new Social Democratic government and the non-socialist parties (in Swedish called *ko-handel* or cow trading). This agreement led to a strengthening of the farmers' and workers' organisations by giving them resources to monopolise respectively the supply of agricultural products and the work force (Micheletti, 1990: 62). It also gave them influence in the implementation and administration of state policies. The 'cost' for the organisations was an acceptance of state intervention. The experiences from World War I had shown that a solely bureaucratic regulatory system was not flexible enough and failed to gain legitimacy among both consumers and producers (Heckscher, 1946: 231). Thus, a broad organisation functioned both as an appropriate method of implementation and as an instrument to discipline farmers and their organisations into following the line in the national agricultural policy.

The result was a remarkable growth in membership. The proportion of organised industrial workers rose from 63 per cent in 1930 to 83 per cent ten years later. Within agriculture the development was more profound. A national organising drive was first initiated around 1930. At that time the producer cooperatives had 160,000 members. State policies helped them to expand to 721,000 members only ten years later. The RLF, which later became the leading farmers' union, was established in 1929. Its membership rose from 20,000 in 1931 to 75,000 in 1940 (Ryden, 1998: 137).

IV. Permanent institutionalisation in the 1940s

Previous sections have shown that while many farmers in Scandinavia were in deep crisis in the interwar years, their unions and cooperative organisations had considerable political and administrative influence behind the scenes. Although many of the schemes and interventions established in the 1930s were seen as temporary responses to a temporary crisis, no country dared to choose a pure market solution to solve the problems. The social and political consequences of letting market forces sweep over the countryside were among the reasons for the growing and continuing interventionism. The long cycle of production also made it difficult to adjust to strict market conditions without strong fluctuations. Another reason was that the war experiences had taught governments the need to pay for self-sufficiency. Last but not least was the changed attitude among politicians, civil servants, experts and organisational leaders towards regulations in general. Both supporters of Keynesianism and of a planned economy demanded more active state involvement in the economy and more targeted policies.

---

[1] Usually the collectivism and corporatism that afterwards became so significant for Swedish society are connected with the Social Democratic Party. But in fact the Social Democrats were opposed to the Milk Regulating Act. They found it harmed the consumers and that it was a very dangerous course to extend the authority of private organisations to non-members, forcing them to contribute to an organisation to which they did want to belong (Just, 1994).

IV.1. Norway

The crisis regulations of the 1930s laid the basis for a more permanent institutionalisation of the cooperative sector's stronghold in both policy-making and implementation and in the agribusiness market.

Many of the schemes and interventions of the 1930s were seen as temporary responses to a temporary crisis. The political consciousness of agricultural interventionism as a more permanent state first came after World War II, when politicians and organisational leaders realised that more permanent regulations were necessary given protectionism, unstable world markets and rural decline by contrast to the blossoming urban sector.

In Norway the two professional farmers' organisations, The Farmers' Union and the Smallholders' Union, were banned during the war, whereas the cooperative associations were allowed to continue their work. After the war the different national associations within the cooperative processing and sales of agricultural products formed the Joint Office (*Felleskontoret*) to take care of common interests, especially in dealing with the government administration. Norway became a relatively planned economy after the war. The country needed rebuilding after the occupation, and at the same the dominant Labour Party developed an integrated welfare state programme in which all sectors of society would participate. Agriculture's part in the deal was to provide the home market with food and by rationalising its workforce make manpower available for the more productive urban trades. The gains for agriculture would be more stable prices at levels comparable with other sectors of society. A first agreement was reached in 1947 and was cemented in the so-called Central Agreement in 1950. For the next four decades this agreement became the framework for negotiations between the agricultural sector and the government.

In Norway negotiations on agricultural prices were extremely segmented, which meant that non-farming interests were kept on the outside. An obvious reason was that one-quarter of the Norwegian MP's were farmers, and several of the leaders in the Farmers' Union and the cooperative organisations held seats for the non-socialist Farmers' Party. Furthermore, the Labour Party got 10 per cent of its votes from farmers. Among the agricultural organisations the Farmers' Union and the Smallholders' Union were given an exclusive right to negotiate with the Ministry of Finance and the government. The *Storting* (the parliament) and the other parties just functioned as a rubber stamp and did not change anything when agreements were reached. All of the implementation and administration were left to the cooperative organisations, which in this way acquired a monopoly in several sectors and at the same time acted as semi-state agencies.

There were several reasons for the unions' superior role. One reason was that the smallholders were close to the Labour government; another was that the Norwegian agenda was broad in an attempt to regulate production, regional distribution, prices and wages. For a labour government it was more natural to negotiate on such questions with farmers' unions than with cooperative organisations which were were instead delegated the task of administering the many regulations (Just and Omholt, 1984: 94-104).

IV.2. Denmark

By the outbreak of war in 1939 the government issued the export committees, which were controlled by the cooperative organisations and the Agricultural Council, with a monopoly on all agricultural exports. This did not mean that private trading firms could not participate in the market, but gradually cooperative companies took a larger bite of the cake. During the German occupation both professional and cooperative organisations had a central position in the negotiations with the Germans and in the supplying the agreed deliveries.

After the war powerful demands for abolishing the state export boards and returning to normal trade soon arose, but the agricultural organisations were not willing to let the state withdraw. They wanted to retain central direction of agricultural exports and considered control over the export sector to be as essential as control over the processing industry in the form of cooperative enterprises had been since the 1880s. A continuation of a more planned economy was not seriously on the agenda. The Social Democrats were attracted by the Norwegian and British approach, but had no aspiration to fix agricultural production and wages à la Norway since the Danish rural economy was much more intertwined with international markets.

Politically, the Social Democrats and the Liberal Party were united in supporting the state boards. However, the growing criticism from Conservatives and private business towards state regulation of the important agricultural export trade forced the Liberals to change course, and from Spring 1950 the export act was not prolonged. The agricultural organisations had to give in to the superiority of politicians, and they hurried to establish new export boards on an organisational basis in order to preserve most of the advantages of the state boards. The new boards were to be allowed to take over the huge capital collected in the former boards. In the succeeding decades the cooperative export organisations were able to out-compete private firms, and at present the cooperatives almost have a monopoly in both processing and sales.

IV.3. Sweden

Despite several similarities between Norway and Sweden, e.g. agriculture's share of the population and exports, it is also possible to point to fundamental differences. The large percentage of farmers in the Norwegian parliament has already been mentioned. In Sweden the farmers did not have the same possibilities of being heard. In 1965 only one per cent of the votes for the Social Democratic Party came from farmers. Neither could the Swedish farmers expect undivided support from the Liberal Party. In the 1950s and 1960s the Swedish Centre Party switched to being as much a party for white collars as for farmers, and the proportion of MP's with connections to agriculture dwindled to about 12 per cent (Steen, 1985: 59).

This situation had obvious consequences for the forms of political organisation. In Sweden political organisation was much more corporative. Negotiations took place in special committees with representatives from the farmers, workers and consumers. In this way the organisations were played off against each other, while the state kept its role as a 'neutral store-keeper'. The equalisation of incomes was the central objective from the outset, but increasing criticism from the Social Democrats and the labour movement meant more weight was given to consumer prices. So subsidies were given in exchange for greater efficiency in farming through structural rationalisation. Only farms of more than 10 hectares would be considered as possible rational units. The task of implementing and administering the price regulations was to be delegated to regional state bodies.

Organisationally, however, the regulatory crisis of the 1930s and during the Second World War meant a sharp increase in membership. Especially successful were the dairy and livestock sectors where the market shares of the cooperatives rose to 95 and 70 per cent respectively. Legislation helped the dairy organisation to control the milk producers fully. It was given the right to collect a milk duty from all producers, non-members as well as members. A bill demanding pasteurisation prohibited the sale of milk directly to consumers, and many small private dairies and slaughterhouses were forced to close down for sanitary reasons (Rydén, 1998: 236).

V.   Results

It has been shown above that the political economy with its crisis regulations in the 1930s played an important part in the future role and strength of cooperative organisations and enterprises in the three Scandinavian countries. In Norway and Sweden the regulations contributed to almost one hundred per cent support for the cooperatives and a much stronger position for the farmers' unions.

In Denmark both unions and cooperatives had reached such a point of maturity already before 1930. Here the regulations helped the cooperative sector to extend its influence from the processing sector predominantly to the sales of agricultural products as well. Furthermore, the collection of funds became an important instrument for the functioning of the unions and for the cooperative sector to take new initiatives (e.g. sales promotion, storing facilities).

It must be emphasised that this development was part of a special Scandinavian welfare model, which developed through the 1930s and 1940s. It was part of a deal between the working class and the farming community. The labour movement and the peasants' movement had been the two major social forces in all three countries since the end of the nineteenth century. With the deep crisis of the 1930s the clash of class interests was replaced by class cooperation. The regulatory schemes required the sectors to be well organised as the integrating organised interests is about bartering: in exchange for regulatory advantages the organisations promise to make the rank-and-file comply with the rules. It is also about conflict problem-solving. In making an agreement with the government, the organisations in question protect the government against strong criticism as the organisations have to take joint responsibility. Last, it is about the burden of administrative overload. In all three countries the state administration at that time was still rather undeveloped, and the experiences from the First World War made politicians realise that state administration was not the best way to solve very complex market problems.

In all three countries the cooperative organisations emerged at the end of the period as semi-state bodies. One reason was that both the Social Democratic and non-socialist parties had a lot of confidence in this type of organisation of economic life, the first-mentioned because they could see the cooperatives as non-profit organisations, the farmer-friendly parties because of the cooperatives' important role in the rural economy and the producers' influence on the processing and sale of agricultural products.

The integration of cooperatives in the crisis regulations was first of all a practical way of problem-solving. Intentional or not, it also became a profitable way of extending market shares in both processing and sales for the cooperative part of agribusiness. And more intentional: the integration of farmers' and smallholders' unions and of cooperative associations – which were all deeply rooted in democratic traditions, and especially in Norway and Denmark had played a keyrole in the fight for parliamentary democracy – helped to combat the extreme right in the crisis movements.

Bibliography

Brogård, P. (1969) *Landbrugernes Sammenslutning. LS's organisatoriske opbygning og virksomhed 1930-1936*, Aarhus.

Furre, B. (1971) *Mjølk, bønder og tingmenn [Milk, peasants and parliamentarians]*, Oslo.

Gueslin, A. (1990) 'Agricultural Co-operatives versus Farmer's Unions in France: From Coalescence to Conflictual Collaboration', Just, F. (ed.) *Co-operatives and Farmers' Unions in Western Europe*, Esbjerg, pp. 80-99.

Habermas, J. (1984) *The Theory of Communicative Action, Volume One: Reason and the Rationalization of Society*, Boston.

Heckscher, G. (1946) *Staten och organisationerna [The state and the organisations]*, Stockholm.

Hellström, G. (1976) *Jordbrukspolitik i industrisamhället med tyngdpunkt på 1920- och 30-talen [Agricultural policy in the industrial society in the 1920s and 1930s]*, Stockholm.

Helmer Pedersen, E. (1975) 'Landbruget', Dybdahl, V. (ed.) *Krise i Danmark. Strukturændringer og krisepolitik i 1930'erne [Crisis in Denmark. Structural changes and crisis policy in the 1930s]*, Copenhagen, pp. 74-122.

Johansen, H. Chr. (1985) *Dansk økonomisk statistik, 1814-1980* [Danish historical statistics 1814-1980], Copenhagen.

Just, F. and Omholt, K. (1984) 'En komparativ analyse af udviklingen i og årsagerne til det ændrede samspil mellem andelsorganisationer, generelle landbrugsorganisationer og staten i Danmark og Norge i periode ca. 1930-80', Just, F. et al. (eds.) *Samspillet mellem staten, landbrugsorganisationerne og landbrugskooperationen [The interplay between the state, the agricultural organisations and the cooperative movement]*, Esbjerg, pp. 1-194.

Just, F. (1990) 'Butter, bacon and organisational power in Danish agriculture', Just, F. (ed.) *Co-operatives and Farmers' Unions in Western Europe. Collaboration and Tensions*, Esbjerg, pp. 137-156.

Just, F. (1993) 'Agriculture and Controls in Denmark 1930-1950', *Scandinavian Economic History Review*, vol. XLl, 3, pp. 269-285.

Just, F. (1995) 'Administrative management in Denmark of the crisis in the 1930s', *Jahrbuch für europäische Verwaltungsgeschichte*, 7, pp. 101-118.

Just, F. (1994) 'Agriculture and Corporatism in Scandinavia', Lowe, Ph., Marsden, T. and Whatmore, S. (eds.) *Regulating Agriculture*, London, pp. 31-52.

Micheletti, M. (1990) *The Swedish Farmers' Movement and Government Agricultural Policy*, New York.

Nerbøvik, J. (1991) *Bønder i kamp. Bygdefolkets Krisehjelp 1925-35 [Peasants in conflict. The crisis relief organisation Bygdefolkets Krisehjelp 1925-35]*, Oslo.

Philip, K. (1939) *En Fremstilling og Analyse af den danske kriselovgivning 1931-38 [The Danish crises legislation 1931-38]*, Aarhus.

Rothstein, B. (1992) *Den korporative staten: intresseorganisationer och statsförvaltning i svensk politik [The corporatist state: interest organisations and public administration in Swedish politics]*, Stockholm.

Rydén, R. (1998) *At åka snålskjuts är icke hederligt. De svenska jordbrukarnas organisationsprocess 1880-1947 [The organisational process among Swedish farmers, 1880-1947]*, Gøteborg.

Steen, A. (1985) 'The Farmers, the State and the Social Democrats', *Scandinavian Political Studies*, Vol. 8, 1-2, pp. 45-63.

Steinmo, S., Thelen, K. and Longstreth, F., (eds.) (1992) *Structuring politics: historical institutionalism in comparative analysis*, Cambridge.

Thelen, K. (1997) 'Historical Institutionalism in Comparative Politics', *Annual Review of Political Science*, 2, pp. 369-404.

Tracy, M. (1989) *Government and Agriculture in Western Europe 1880-1988*, London.

Tullberg, P. (1977) *Bönder går sammen. En studie i Riksförbundet Landsbygdens Folk under världskrisen 1929-1933 [The National Association of Rural People and the world crisis 1929-1933]*, Stockholm.

11 Danish agriculture and economic collaboration during the German occupation

Mogens R. NISSEN, University of Southern Denmark, Esbjerg

I. Introduction

Five weeks after the German occupation of Denmark a crucial meeting took place in the *Reichsministerium für Ernährung und Landwirtschaft* (the German Ministry of Nutrition and Agriculture).[1] The topic for the meeting was in short *Dänemark*. High-ranking civil servants from the ministry participated and the *Staatssekretär* Herbert Backe chaired it. Backe and his subordinate Dr. Alex Walter, who was chief negotiator in relation to Denmark, explained the economic policy towards Denmark. They made it clear that it was not possible to manage the Danish economy and production by means of restrictions and control. Instead Danish production should be managed by differentiated prices on agricultural products and rations on food should be introduced only if the Danish population had a direct interest in this. This economic policy, which was implemented and continued through all five years of the German occupation, depended heavily on the voluntary collaboration of the Danish authorities and agricultural organisations.

Thus, from the beginning of the occupation the German authorities wanted to change Danish agricultural production by the use of differentiated prices on food. At the same time they managed to create and maintain the *Produktions- und Lieferfreudigkeit der dänischen Landwirtschaft*[2] as food prices after the occupation increased significantly. In this way the German authorities made it easy for Danish farmers and agricultural organisations to collaborate. At the same time the German authorities were aware of the impact that increasing food prices had on other classes in Danish society and that too-high food prices could cause social disturbances.[3] They were also aware that overly high food prices would make it difficult for the Danish government to collaborate.[4]

The aim of this article is to show the importance of the economic collaboration of the agricultural organisations under the Agricultural Council of Denmark (*Landbrugsraadet*) in the implementation of German economic policy in Denmark. This includes an evaluation of the degree of shared or divergent interests between the German occupying power,

---

[1] Bundesarchiv (henceforth BAarch), R16/1306. Denmark was occupied on April 9th 1940 and the meeting was on May 15th 1940.
[2] The expression means 'the Danish farmers' eagerness to produce and supply' and it was a mantra in many German reports for all five years. It is important to note that the expression was not just a phrase, as the Germans in practise did much to maintain the 'Lieferfreudigkeit' for all five years.
[3] In a report to van Scherpenberg in Auswärtiges Amt on February 18th 1941 Dr. Alex Walter was very pessimistic about Denmark. He was concerned that further price increases on food would cause social disturbances in Denmark. BAarch, R901/68311. See also (Salmon, 1997: 366)
[4] BAarch, R901/68311. The same report of February 18th 1941 as mentioned in note 3.

the Agricultural Council of Denmark and the Danish government.

II. The pre-war situation

Danish food production was of significant importance to the German occupying power during World War II as it ensured a substantial contribution of food to the German population and armed forces. Nevertheless, there are no indications that Danish food production played any role in the decision to occupy Denmark on April 9th 1940 (Giltner, 1997: 3; Brandt, 1953: 299; Poulsen, 2002: 11 and others). This has led to the conclusion that the leading German authorities knew little about the Danish economy and production before the occupation (Giltner, 1998: 171). However, though there was no German master-plan (Nissen, 1972: 52; Kirchhoff, 2002: 17; Andersen, 2003: 38-39), civil German authorities in the *Reichsministerium für Ernährung und Landwirtschaft*, the *Reichswirtschaftsministerium* (the Ministry of Economic Affairs), the commercial department of the *Auswärtiges Amt* (the Ministry of Foreign Affairs) and the economic department of *Oberkommando der Wehrmacht* (the Armed Forces) already had extensive knowledge about the Danish economy and production even before the occupation.

It is fair to describe the situation of the Danish economy in general and Danish agriculture in particular as problematic when Denmark was occupied. Denmark had Europe's most open economy with Britain as the single most important market for Danish goods, while Germany was the second most important export market.[5] Some 75 per cent of all Danish exports were agricultural products and 97 per cent of bacon exports and 75 per cent of butter exports were sold to the British market in 1938 and 1939.[6] At the same time almost all feed and most fertilizers were imported from overseas destinations, which were controlled by the British navy. This meant that when Denmark was occupied it had to stop all exports to England and send all goods to Germany or German-controlled areas instead. At the same time it was impossible to import feed and phosphorus fertilizers because the German occupying power did not supply Denmark with these goods; so the farmers had to reduce herds of animals and decrease the production of animal products. The significance of this was most uncertain just after the occupation, as no one could predict the length of the war and the dimensions of the Danish harvests, but both German and Danish experts agreed that herds of animals had to be drastically reduced.

The time between the outbreak of the war and the German occupation was difficult for Danish farmers. Even though the Danish Ministry of Foreign Affairs succeeded in making an arrangement, the so-called 'Maltese-Trade-Agreement', which allowed Denmark to maintain normal trade with both sides in the war, exports of agricultural products suffered (Giltner, 1997: 333-346; Nissen, 2003: 124-157). The British government introduced a war economy from the beginning of September 1939 with price controls, food rationing and currency devaluation; this meant a drop in both export quantities and export prices. Thus, in March 1940, after six months of hard negotiations between the Danish and Brit-

---

[5] Statistisk Aarbog 1940. The exports to England and Germany were in total some 75 per cent of the value of the total exports.
[6] Statistisk Aarbog 1940.

ish authorities, an agreement regarding the total amount of Danish exports in 1940 was reached. According to this agreement export quantities of butter and bacon would fall some 40 per cent compared to 1939 (Skade, 1940: 19). At the same time, in February and March 1940, the German negotiators indicated their willingness to pay higher prices for Danish agricultural products if Denmark could increase its exports.[7] But because of Danish neutrality it was not possible to transfer export quantities from the British to the German markets.

From that perspective one can argue that the German occupation was convenient for the Danish farmers. Suddenly it was possible to sell all their goods at higher prices. The Danish Minister of Foreign Affairs, Peter Munch, characterized the attitude of the Danish farmers in a draft of his memoirs: 'In parts of the rural population the British price policy caused significant aversion. In 1914-1918 there had been a strong feeling that even though the Germans were willing to pay higher prices than the British, Britain was still Denmark's best market. This time many [corrected from 'most', MRN] found the British price policy unfair. But this did not prevent the vast majority from remaining sympathetic to the British side'.[8]

This had significant importance. During the first years of the German occupation, the negotiators from the agricultural organisations tried to increase prices on food to the maximum because they felt entitled to the highest possible gains in order to rectify the imbalance in prices. The reason was partly that they felt the Danish government and parliament had not satisfactorily helped the hard-pressed farmers prior to the German occupation.

III. The negotiating positions

In 1934 the Danish-German Government Committees began to manage trade between the two countries and these negotiations continued during the occupation. The trade negotiations concerned quantities and prices at the top level, while detailed agreements about agricultural products were concluded between the different Reichsstellen and the Danish Export Committees. Normal practice was that the Government Committees met every three months and this did not change after the occupation. However, after the German occupation there was much more frequent contact between the chairmen of the committees.

---

[7] The archives of Landbrugsraadet: Meddelelse fra Præsidiet til Raadets Medlemmer (Reports from the Presidency to the Members of the Council. Henceforward written as PR., no.7/1940.
[8] Rigsarkivet (henceforth RA), The archives of P. Munch, no. 6663, package no. 109, s. 104. My translation of: 'I Dele af Landbefolkningen fremkaldte den engelske Prispolitik en ikke ringe Uvillie. 1914-18 havde der været en stærk Følelse af, at selv om Tyskerne betalte bedre end Englænderne, var og blev England dog Danmarks bedste Marked. Denne Gang fandt mange [rettet fra 'de fleste'] den engelske Prispolitik ubillig. Men dette hindrede ikke, at Sympatierne hos langt [tilføjet med blyant] de fleste blev paa engelsk Side.'

The key figure on the German side in negotiations with Danish authorities was the head of the commercial department in the *Reichsministerium für Ernährung und Landwirtschaft, Ministerialdirektor*, Dr. Alex Walter.[9] When Denmark was occupied he had been the chairman of the German Government Committee since 1936, and Danish negotiators in the Danish Government Committee and in the Export Committees knew him very well. Besides being chairman of the German Government Committee in negotiations with Denmark, Dr. Walter also was chairman for German negotiations with Sweden and Holland – that is until the German occupation of Holland. In addition he was a member of several other German negotiation committees in South-East Europe before the war.[10] His direct superior was *Staatssekretär* Herbert Backe, who in the *Vierjahresplan* (the Four-Year Plan) was head of the preparations for future agricultural production in the German *Grossraum* (New Order). From May 1942 he was acting Minister of Nutrition and Agriculture (Wistrich, 1995: 6-7), and he was the kingpin for the so called 'hunger plan' that in December 1940 calculated the number of Soviet citizens who would die of starvation when Germany occupied the Soviet Union. Dr. Walters' most important job was to negotiate imports from Denmark, while the deputy chairman in the German Government Committee, Waldemar Ludwig from the Reichswirtschaftsministerium (the Ministry of Economic Affairs), dealt with exports to Denmark. Other civil authorities participated in the German Government Committee as representatives, but in reality Walter and Ludwig were in charge.[11] Thus, decisions made at the aforementioned meeting in the *Reichsministerium für Ernährung und Landwirtschaft* on May 15th 1940 were of crucial importance for Danish production and exports, as it was this ministry that managed German imports from Denmark. Dr. Walter had a powerful supporter in Backe, and everything seems to indicate that Backe confirmed all the overall economic agreements concerning Danish production and exports.[12]

Dr. Walter and his superior in the *Reichsministerium für Ernährung und Landwirtschaft* not only controlled the German Government Committee, but also the *Reichsstellen*

---

[9] So it is very misleading, when Sjøqvist, 1971: pp. 326 mentions that Walter belonged to the Auswärtiges Amts commercial department. Walter participated frequently in the important intergovernmental committee for foreign trade, Handelspolitischer Ausschuss, as representative for the Reichsministerium für Ernährung und Landwirtschaft.

[10] BAarch, R901/67939 and R9I/949.

[11] Several times before and during the occupation Dr. Walter made reports to the central inter-ministerial committee Handelspolitischer Ausschuss. In this committee all the important economic authorities were represented by an assistant secretary or a civil servant of equivalent status. Many German decisions concerning foreign trade were concluded here. Representatives from the economic departments of the Armed Forces (OKW) and the Navy also participated in the committee meetings. BAarch, R901/68939 and R9I/949.

[12] In his memoirs the President of the Agricultural Council of Denmark, Henrik Hauch, made it clear that he met Herbert Backe many times before and during the occupation and regarded him as an 'energetic, active and competent' man, who was an old friend of the Agricultural Council. Hauch ascribed great importance to Backe with regard to German economic policy in Denmark before and during the occupation. (Hauch, 1978: 86-92).

(shown in figure 11.1), which was subordinated to this ministry.[13] The negotiators in the *Reichsstellen* received detailed instructions before negotiations with the Danes,[14] and they always had to report back to Dr. Walter since he and his superiors controlled all major decisions. In this way the German side was very hierarchical with *Staatssekretär* Herbert Backe as the top decision-maker and Dr. Walter as his subordinate, who was responsible for implementation.

On the Danish side senior officials from the Ministry of Foreign Affairs, with Assistant Secretary Matthias Wassard as chairman, were negotiating in the Government Committee (Mau, 2002: 25; Andersen, 2003: 33). Representatives from trade organisations did not normally participate in the Government Committee, but before the Danish-German meetings the members of the Danish Government Committee met with civil servants from the Ministry of Agriculture, the Ministry of Trade and other economic authorities as well as with representatives from the trade organisations to prepare proposals for the German negotiators.[15] As mentioned, the Danish-German Government Committees managed trade between the countries before and during the occupation at a superior level. The committees concluded almost normal international trade agreements under the most unusual circumstances, because Denmark was still officially a sovereign state after the occupation. Both sides saw an advantage in maintaining this illusion, and so the model was continued for all five years under German occupation.

Figure 11.1 Relations between different German and Danish trade committees

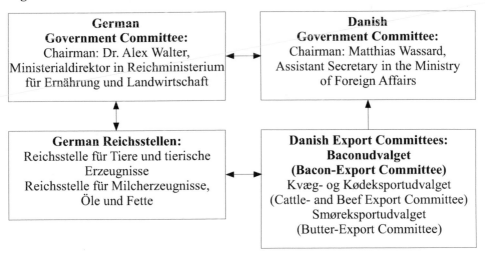

---

[13] The negotiators in the Reichsstellen were all senior civil servants in Reichsministerium für Ernährung und Landwirtschaft and all correspondence with Danish authorities was on the ministry's writing paper.

[14] An example is a meeting on June 19th 1940, where Dr. Walter instructed the German negotiators in different Reichsstellen about their authorisation in future negotiations. BAarch, R901/67771.

[15] The meetings in Berlin in May 1940 are an important exception, as representatives from the trade organisations participated in these negotiations.

As noted already, detailed trade agreements were concluded between the *Reichsstellen* and the Export Committees. The Danish Export Committees were officially subordinated to the Ministry of Agriculture, but in reality they were managed and controlled by the agricultural organisations placed under the Agricultural Council of Denmark (*Landbrugs-raadet*), as all the chairmen of the Export Committees were members of the presidency of the Agricultural Council. In this way the presidency coordinated and organised the policy in relation to the German negotiators and the Danish government. In November 1932 agricultural organisations under the Agricultural Council asked the Minister of Agriculture, Kristen Bording, to establish central export committees in order to control and regulate production and exports (Just, 1992: 121 and 138). The Minister agreed with the Agricultural Council on the purpose of these committees, and several times during the 1930s parliament passed export acts ensuring the committees' monopoly on the export of agricultural products. In September 1939 the Minister promised the Agricultural Council that the Export Committees in the future would have not only a monopoly on exporting agricultural products but also the exclusive right to negotiate price agreements with the *Reichsstellen* (Just, 1992: 138 and PR no. 26/1939). Thus, in the period between September 1939 and June 1941 the Export Committees, controlled and managed by Danish farmers, had the exclusive right to negotiate prices on Danish food exports with the Reichsstellen.[16] Needless to say it had a significant impact on the development of food prices in the time period between April 1940 and June 1941. The German negotiators wanted to promote specific types of production, primarily of milk and butter, by letting prices increase significantly. Danish farmers had a similar interest in getting higher prices to increase their income.

On June 14th 1940 Dr. Franz Ebner[17] sent a telegram to the *Auswärtiges Amt,* announcing that the President and Secretary General of the Agricultural Council, Henrik Hauch and Arne Høgsbro Holm, wanted to get in contact with Dr. Walter to negotiate prices on agricultural products.[18] Dr. Ebner explained that Høgsbro Holm had little confidence in the negotiators Wassard and Jacobsen[19] in the Danish Government Committee, and wanted to have direct contact with Dr. Walter without interference from the Ministry of Foreign Affairs. Dr. Ebner recommended that Dr. Walter contact Hauch and Høgsbro Holm via the Ministry of Agriculture, as they controlled all the Export Committees. On June 24th 1940 Dr. Walter replied to Dr. Ebner, agreeing that prices on food had to increase considerably and that he was aware of the different views of the Danish Government Committee and the Export Committees. He finished his letter by stating that future price negotiations were to be carried out between the *Reichsstellen* and the Export Committees.[20] So direct contact was established between Dr. Walter and the Agricultural Council.

---

[16] PR no. 2671939 and PR, no. 14/1941.

[17] Dr. Ebner was appointed Economic Commissioner at the German Embassy in Copenhagen. He was Ministerialrat (permanent undersecretary) in the commercial department in the Reichsministerium für Ernährung und Landwirtschaft and Dr. Walter was his direct superior before the occupation. Even though he was employed at the German Embassy, and thereby officially attached to the Auswärtiges Amt, he was in reality still connected to his former employer. BAarch, R3601/69.

[18] BAarch, R901/67771 and Jensen, 1971: 73.

[19] A. P. Jacobsen was Agricultural Attaché at the Danish Embassy in Berlin.

[20] BAarch, R901/67771.

At the same time the German top negotiator announced the decision that price negotiations concerning exports of Danish agricultural products in the future would take place directly with organisations under the Agricultural Council. Thus, important negotiating rights were taken away from the Danish Government Committee, which caused important disagreements between the Ministry of Foreign Affairs, the Ministry of Agriculture and the Agricultural Council.[21]

The Danish government more or less supported increased prices on agricultural products, mostly because it was concerned that the low prices in April 1940 would result in a significant fall in production and lead to a shortage of food supplies. At a meeting on March 6th 1941 the Minister of Agriculture, Kristen Bording from the Social Democratic Party, explained to the presidency of the Agricultural Council that initially the government wanted higher prices on food to secure maximum agricultural production, but now had to stop further price advances.[22] Another reason is that the important liberal agrarian party in the coalition government, Venstre, which had close connections to the Agricultural Council, was fighting against attempts to take away the exclusive rights from the Export Committees.

Increasing food prices had a strong impact on the distribution of wealth, especially in the first two years of the occupation period. This historical fact, which Hans Kirchhoff has described as a class struggle between farmers and workers, will not be dealt with here (Kirchhoff, 2002: 53). However, it is important to keep in mind that rising food prices meant worsening conditions for classes in society other than farmers as wages did not increase correspondingly.

IV. German price policy at the beginning of the occupation period

Even before the occupation of Denmark Dr. Walter indicated that he thought prices on Danish agricultural products were too low and that he feared a drop in production which could cause a stop in exports to Germany. But since Britain was the most important market for Danish goods the Germans would not pay more than Britain and there was nothing he could do.[23] The German negotiators used the same argument several times in the following months.[24] At a meeting in Berlin on March 28th 1940 the German negotiators declared that

---

[21] In an interview with Sigurd Jensen on September 2nd 1968, Knud Sthyr, who was Director and so-called 'German expert' in the Ministry of Foreign Affairs, distinctly gave voice to these disagreements. He also explained that the Minister of Agriculture and the president of the Agricultural Council in tandem were very powerful, so they could ensure the Export Committees' exclusive rights. RA, Håndskriftssamlingen: Sigurd Jensen IV.T.42: *Materiale vedr. besættelsestiden.*

[22] PR, no. 5/1941.

[23] Summary of a meeting November 21th 1939 in the Danish-German Government Committee. RA, UM 1909-45, H64-205.

[24] The last time was on March 27th 1940 when Dr. Walter in a letter to the Auswärtiges Amt explained that he did not want to pay higher prices for Danish goods, because the Danes could use it to reduce prices on the British market. But he was open to price increases if there were political reasons for this. BAarch, R901/67747.

it probably would be possible to pay the same butter prices as the British, plus a bonus equivalent to the costs of carriage to England, which were much higher than the costs of carriage to Germany.[25] This was the first time German negotiators expressed a willingness to pay more for Danish goods than the British authorities. Already on April 3rd 1940 the chairman of The Butter-Export Committee reported to the presidency of the Agricultural Council that the Germans had offered to pay some 18 per cent more for Danish butter than the British had offered to pay according to the trade agreement of March 1940, as long as the better prices were given to dairies exporting to Germany.[26]

Thus, the German authorities began their price policy even before the occupation of Denmark. They had agreed to pay more for Danish butter than the Danes could get on the British market as long as the higher prices benefited producers exporting to Germany. In this way one can conclude that the German price offer was made for political, not economic reasons, as Dr. Walter for economic reasons was against increased export prices.[27]

Still there was much uncertainty during the first days after the German occupation. Dr. Walter was in Sofia and it was not possible to get in touch with him until April 15th 1940 (Jensen, 1971: 20). Huge stocks of butter and bacon targeted for the British market were warehoused in Danish harbours and at meetings in the Ministry of Foreign Affairs on April 10th and 12th 1940 representatives from the Export Committees were anxious to find solutions so that the goods could be sold to Germany instead. Wassard was convinced that the Germans were interested in buying all the food stocks, and all future agricultural products, but the question was how much they were willing to pay.[28] When Dr. Ebner arrived in Copenhagen on April 12th 1940, Danish negotiators informed him of their desire to sell the stocks of food to Germany. On April 15th 1940 Dr. Walter notified the Danish Ministry of Foreign Affairs by phone and by telegraph that Germany intended to buy all stocks.[29] Still there was no sure indication of the prices the Germans were willing to pay, but it seems likely that Secretary General Høgsbro Holm had a clear feeling that it was possible to get higher prices, because on April 12th 1940 he recommended that the Ministry of Foreign Affairs should ensure only the sale of existing stocks while the Export Committees should still negotiate the prices. Wassard agreed to this.[30]

From 15th to 21st April 1940 meetings were held between Danish and German nego-tiators, at which agreements on the sale of the food stocks were confirmed and the prices were set at the pre-occupation level.[31]

---

[25] PR, no. 10/1940.
[26] PR, no. 12/1940. The offer was mentioned as 'Dr. Walters' offer', and on April 5th and 6th 1940 it was communicated in the Government Committee. See also BAarch, R901/67738 for detailed German calculations on March 29th and April 1st 1940.
[27] BAarch, R901/67747. See note 38.
[28] PR, no. 13/1940.
[29] RA, UM 1909-45, Dmk-Tyskland Handelstraktatforhold XXXIX, nr. H64-190. The German Em-bassy sent a matter-of-fact telex on April 12th 1940 to Berlin explaining that Wassard had offered 3,800 tons of butter, 187,000 slaughtered pigs and 2 million eggs. BAarch, R901/67741.
[30] PR, no. 13/1940.
[31] PR, no. 14/1940.

In the first days of the occupation agricultural experts attached to the Agricultural Council had already worked out a plan for future production based on the fact that it was not possible to import supplies of feed. The main conclusion was, as it was in a similar plan worked out by the Danish Agricultural Attaché A. P. Jacobsen at the Embassy in Berlin, that livestock numbers should be kept almost unchanged while the population of pigs had to be reduced by least 33 per cent.[32] Secretary General Høgsbro Holm informed the presidency of the Agricultural Council on April 17th 1940 that the Danish delegation at meetings with German negotiators '... had found it expedient to introduce this information to the German negotiators because it is to be anticipated that the Germans will be interested in future supplies from Denmark, and if Denmark does not have a plan for preparations of future production, there is a risk that the German negotiators will try to influence the preparations of future production.[33]

All indications are that the German agricultural experts employed by the *Reichsministerium für Ernährung und Landwirtschaft*, in the *Reichsstellen* or in the *Reichsnährstand* were already very familiar with the situation in Danish agriculture. A central idea in the Nazi agricultural policy was to create a self-sufficient food production in the German *Großraum* (see for instance Backe, 1942 and Walter, 1942/1943: 351-359), and several times before the occupation of Denmark German negotiators had recommended a change in Danish production so that the dependence on supplies of feed would be diminished. But the quotation shows that the Danish negotiators were interested in giving all information and estimations on future production to the German authorities in an attempt to prevent them from forcing through the preparations.

Danish negotiators succeeded in this and the German occupational power did not impose the proposed production structures. Instead the Germans used price mechanisms. The first sign of this was given at a meeting on May 3th 1940 between Danish agricultural experts, leading members of the presidency of the Agricultural Council and German experts from the Reichsnährstand. The German representatives told the Danes that there was no prospect in getting supplies of feed from Germany or German-controlled areas, so Denmark had to reduce the population of pigs, poultry and beef cattle. Instead the Germans wanted to maximise the production of milk-products. One German expert, Dr. von Hasselbach, stated, '...by the use of price-regulations it is possible to regulate the production on different markets.'[34] At a meeting on May 23th 1940 Secretary General

---

[32] PR no. 14/1940 and RA, UM 1909-45, Dmk-Tyskland Handelstraktatforhold XXXIX, nr. H64-190. At the same time the population of poultry had to be drastically reduced. The reason was that since no one could foresee how long the occupation would last, farmers had to reduce the population of pigs which could be regenerated again quickly, while it would take years to regenerate herds of cattle.

[33] PR, no. 14/1940. My translation of: '... havde man anset det formaalstjenligt at forelægge saadanne Oplysninger for de tyske Forhandlere, da man maatte forudse, at Tyskerne vilde være interesseret i Fremtidige Leverancer fra Danmark, og hvis Danmark ikke havde en Plan for den fremtidige Tilrettelæggelse af Produktionen rede, vilde man kunne risikere, at de tyske Forhandlere vilde prøve at faa Indflydelse paa den fremtidige Produktions Tilrettelægning.'

[34] RA, UM 1909-45, Dmk-Tyskland Handelstraktatforhold XXXIX, nr. H64-190. My translation of: '... ved hjælp af Prisreguleringer i ikke ringe grad kunde regulere Produktion paa forskellige Markeder.'

Høgsbro Holm reported to the members of the Agricultural Council that the representatives from the Reichsnährstand at meetings in the beginning of May 1940 had declared: 'Can we help you estimate relative price increases, so that the goods in which you – and thereby they mean we – have a special interest will receive a special price increase, and where there has to be a rapid cut that the price increase will not be strong? These relative prices should help Danish agriculture through – I will just mention it in the same breath, because it is something we will have 'the pleasure' of.'[35]

This shows that the Danish negotiators from the Agricultural Council at this early stage of the occupation sensed the overall objectives of the German price policy. The content of the policy is confirmed by a summary of the aforementioned meeting chaired by Backe in the Reichsministerium für Ernährung und Landwirtschaft on May 15th 1940, in which Dr. Walter, Dr. Ebner and Dr. von Hasselbach participated.[36] The Germans planned on how to change Danish agricultural production, in order to reduce the population of pigs and poultry dramatically and to keep the livestock almost unchanged during the summer. But they did not want to force changes through by means of control and rationing, because they very much doubted that this would bring about the desired results. Only if the Danish population had a direct interest in rationing would they voluntarily accept it, and only if this were the case did the German authorities see rations as a possibility. Instead they carried out a set of differentiated price changes, which made the Danish farmers change their agricultural production out of economic interests. Prices for poultry had to increase significantly while prices on eggs needed to fall. Prices on pork were already high and should remain unchanged while prices on milk-products had to increase considerably. In this way the German authorities managed to carry out the desired changes in Danish agricultural production.[37]

Several conclusions can be drawn from this meeting. 1) *Staatssekretär* Backe and his subordinates had from the beginning of the German occupation of Denmark a substantial knowledge about Danish agriculture, and very soon had a plan for carrying out the desired changes in production. 2) From the beginning of the occupation period the leading civil servants in the important *Reichsministerium für Ernährung und Landwirtschaft* realised the importance of maintaining the '*Produktions- und Lieferfreudigkeit der dänischen Landwirtschaft*' and they had a plan to ensure this. So it was of significant importance that the agricultural organisations cooperated voluntarily.

---

[35] PR, Raadsmøder 1938-40. Udførligt referat af Landbrugsraadets Møde den 23/5-1940 (detailed report of the meeting in the Agricultural Council on May 23th 1940). My translation of: 'Kan vi ikke hjælpe jer ved at ansætte relative Prisforhøjelser, saaledes at for de Varer, som I – og dermed mener de vi – er særlig interesseret i, sker der en særlig Prisstigning, og der hvor der skal skæres stærkt ned, bliver Prisstigningen ikke ret stor? Disse relative Priser, der skulde hjælpe dansk Landbrug med at komme igennem, vil jeg blot lige nævne i samme Aandedræt, for det er noget, vi kan faa 'Fornøjelse' af.' At the summaries the reporter made quick notes, and the quotation is also very odd in Danish.
[36] BAarch, R16/1306.
[37] BAarch, R16/1306.

At a meeting of the presidency of the Agricultural Council on May 30th 1940 Secretary General Høgsbro Holm recommended that the negotiators in the Export Committees start to discuss price increases with the different *Reichsstellen* in order to 'sense' their views on the subject.[38] At another meeting of the presidency of the Agricultural Council on June 27th 1940 the chairmen of the Export Committees reported back from the negotiations with the different *Reichsstellen*. The chairman of the Butter Export Committee reported that the negotiations went well, and that it would be possible to get better prices for the products. The chairman of the Cattle and Beef Export Committee reported that the German negotiators accepted a small price increase, and that negotiations were conducted in a 'businesslike, strictly business-oriented manner', while the chairman of the Bacon Export Committee reported that it was impossible to get price increases on pork. The last to report was the chairman of the Egg Export Committee. He reported that when he had asked for a price increase on eggs the Germans had turned him down: '*Reichsstelle* made it clear that Germany was more interested in butter and pork than eggs, and the chairman of the German negotiations delegation, Dr. Walter, had said that from the German side they would be more sympathetic to price requests on butter and pork, whereas egg prices had to be more or at the same level as in 1939. As long as this decision was in force, it was completely out of the hands of *Reichsstelle* to change the current prices.'[39]

This view is confirmed by a summary of a meeting on June 19th 1940, where Dr. Walter and *Regierungsrat* Meyer-Burckhardt from the *Reichsministerium für Ernährung und Landwirtschaft* instructed the negotiators of the *Reichsstellen* about their authority in the coming negotiations with the Export Committees. The instructions followed the above noted line of direction at the meeting on May 15th 1940 but were more detailed with regard to prices on different agricultural products, and they made it clear to all that they were directions worked out by *Staatssekretär* Backe.[40]

---

[38] PR, no. 19/1940.

[39] PR, no. 25/1940. My translation of: 'Heroverfor gjorde Reichsstelle gældende, at Tyskland i højere Grad var interesseret i Smør og Flæsk end i Æg, og Lederen af den tyske Forhandlingsdelegation, Dr. Walter, havde udtalt, at man fra tysk Side Skulde stille sig mere velvilligt overfor Prisønsker på Smør og Flæsk, hvorimod Priserne for Æg nogenlunde maatte holdes paa samme Niveau som i 1939. Saa længe en saadan Bestemmelse var gældende, var Reichsstelle ganske ude af Stand til at ændre de nuværende Priser.'

[40] BAarch, R901/67771. See also Patrick Salmon: *Scandinavia and the great powers 1890-1940*. Cambridge, 1997, 366.

Figure 11.2 Producer prices on various agricultural products (Kroner per 100 kg), 1940-1943

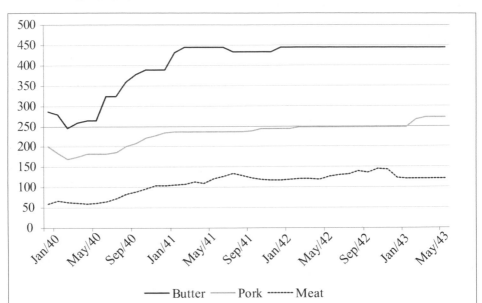

Source: Statistisk Aarbog 1940 – 1945.

To show that these were not just words it is important to look at the actual development in food prices during the German occupation. In figure 11.2 prices on the three most important agricultural products are listed showing a similar development as stated by *Staatssekretär* Backe, Dr. Walter and their subordinates. Prices on food increased rapidly in the first two years under German occupation. Butter prices increased quickly, while it took longer before prices on pork and beef peaked. From March 1940 to April 1941 producer prices on butter increased some 81 per cent, and stayed at this level for the remaining years of the German occupation. Prices on pork increased 39 per cent until April 1941 and 60 per cent until May 1943. For the rest of the occupation period prices on pork were constant. Prices on meat increased 70 per cent the first year and 91 per cent until February 1943, after a few months with increases at 125 per cent compared to the pre-occupation level.

What determined the producer prices were the export prices. The figure above does not show the full extent of the German price policy in the first years of the occupation, as export prices for butter and pork increased much more than the producer and consumer prices.

Figure 11.3 Various prices for butter (Kroner per kg), 1940-1941

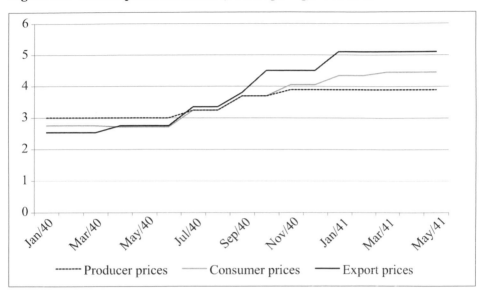

Source: Statistiske Meddelelser, 4. Række, 127. Bind, 3. Hæfte. *Landbrugsstatistik 1945. Herunder Havebrug, Skovbrug m.v.* Published by Det Statistiske Departement 1947.

Besides showing the rapid increase in export prices for butter figure 11.3 above also shows another interesting aspect about the development in the first year of the German occupation. While consumer prices were higher than producer and export prices until July 1940, export prices increased the most and were higher than producer and consumer prices after August 1940. In this way prices for butter exported to Germany after October 1940 were much higher than the prices Danish consumers had to pay.[41] This shows that the German occupying power did not want to introduce any kind of price rationing in Denmark, and it fits in with several statements made by Dr. Walter during the occupation. It was important for the German authorities to ensure that the Danish population had enough food at prices they could afford, as a way of pursuing their main objective; namely to keep law and order in Denmark.[42] This was the overall policy for all five years, and it is a fact that the domestic consumption of food in Denmark was fairly constant

---

[41] From January 1941 export prices on butter were 5.10 kr./kg while the domestic consumer prices were 3.89 kr/kg, so export prices were 31 per cent higher than consumer prices. In the first quarter of 1941 the Germans offered to pay 5 kr/kg for an export quantity of 8,000 million kg and 5.25 kr./kg. for export quantities above 8,000 million kg as an incentive to export more butter and to maintain the 'Lieferfreudigkeit'.

[42] BAarch, R901/68311. See note 3.

during all the years under German occupation, while the export of food varied in line with production.[43]

This does not mean that the German authorities did not want to reduce the Danish consumption of food. At meetings between the Government Committees in the beginning of November 1940 the Germans increased their demand for food supplies for the coming year to such a level that Danish consumption had to be reduced. But they left it to the Danish authorities to decide how the reduction should be implemented, as they were convinced that increased consumer prices on food would cause a strong reduction in Danish consumption.[44]

Figure 11.4 Various prices for pork (Kroner per kg), 1940-1943

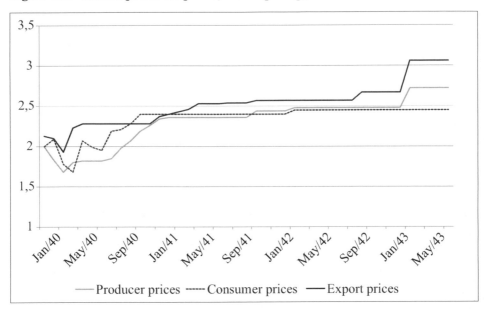

Source: Statistiske Meddelelser, 4. Række, 127. Bind, 3. Hæfte. *Landbrugsstatistik 1945. Herunder Havebrug, Skovbrug m.v.* Published by Det Statistiske Departement 1947.

The evolution in prices for pork is almost similar to that for butter prices, but it took much longer before price increases were completely carried out. After the summer of 1943 export prices were about 25 per cent higher than the prices Danish consumers had to pay. They were also higher than producer prices, but figure 11.4 above does not show

---

[43] At meetings between the Danish-German Government Committees in the beginning of July 1942 Dr. Walter was very clear in his statements. He declared that the domestic market for food had precedence over exports, and that this had been the policy from the beginning of the occupation, RA, UM-1909-45, F.02, H64-194.

[44] BAarch, R901/67771.

the full extent of the producer prices after October 1942, since Dr. Walter demanded that the producers should have 0.16 kr./kg more, if the deadweight of pigs were over 80 kg.[45] Originally the Germans wanted a bigger bonus for even heavier pigs, but they agreed to this model.[46] The traditional explanation for the German demand is that Germans preferred to eat fat schnitzels, not lean chops like the British, but this explanation is wrong. The German authorities based their demands on recommendations made by German nutritional experts, who believed that heavy pigs absorbed the feed better than smaller ones and that heavy pigs could be fed boiled potatoes successfully.[47] The preferences of the German consumers were not involved in the decision.

It was very important for the Danish government to ensure relatively low prices on butter, milk and pork for the Danish population so that all groups of society could afford to buy these products. Since producer costs were higher than consumer prices, the bonuses given to the producers were financed partly by high export prices and partly by subsidies. Again it is important to note that the German authorities accepted a policy in which Danish consumers gained from relatively low prices on food compared to export prices.[48]

---

[45] Statistiske Meddelelser, 4. Række, 127. Bind, 3. Hæfte. *Landbrugsstatistik 1945. Herunder Havebrug, Skovbrug m.v.* Published by Det Statistiske Departement 1947, p. 150. The dead weight of pigs in Denmark was normally between 60 and 80 kg.

[46] In a report from Dr. Ebner on February 24th 1942 to the Auswärtiges Amt about negotiations in the Government Committee he explained that the Germans wanted to give 0.24 kr./kg more for pigs with a deadweight between 85 and 100kg and 0.32 kr./kg for pigs with a deadweight between 101 and 130kg. BAarch, R901/ 67772.

[47] On April 22nd 1942 Dr. Walter sent a telex to Backe notifying him that the Danish negotiators had accepted the price arrangement for heavy pigs and that it was important to get some 300 potato-boiling plants before August 15th 1942. On April 24th 1942 Backe replied that it was possible to get the plants and that he would do everything possible to get them before the desired date. BAarch, R901, 67782.

[48] Even though Dr. Walter in a letter to the Auswärtiges Amt on July 17th 1940 noted that the domestic prices on food in the long term had to be at the same level as the export prices. BAarch, R901/68173.

Figure 11.5 Various prices for first-class beef (Kroner per 100 kg), 1940-1943

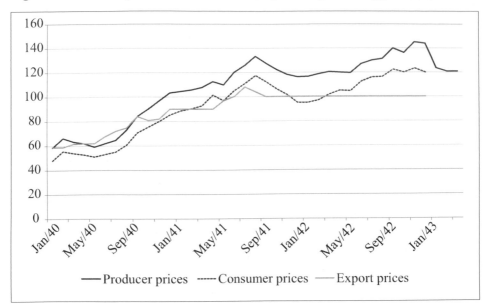

Source: Statistiske Meddelelser, 4. Række, 127. Bind, 3. Hæfte. *Landbrugsstatistik 1945. Herunder Havebrug, Skovbrug m.v.* Published by Det Statistiske Departement 1947 and Statistisk Årbog 1940 – 1943.

Beef prices developed quite differently from those for butter and pork. Until March 1943 there was a relatively free quotation on producer prices while consumer prices were restricted, leading to greater increases on producer than on consumer prices. In the Autumn of 1940 the German authorities wanted to reduce livestock and introduced short-term advances on export prices, so the prices in September 1940 temporarily increased by 10 per cent, in October 1940 by an additional 5 per cent, whereupon the temporary increases were annulled.[49] It was important that the temporary price increases be made available to the public so the farmers could make the best possible use of them. Though the temporary price increases became permanent, this shows again that the German authorities used prices to manage production and to regulate the dimensions of the Danish livestock population.

In the first half-year of 1941 prices for beef increased sharply, primarily caused by increased prices on pork. Following a price reduction in the second half of 1941 due to rising supplies of beef after a very poor harvest in 1941, price increases in 1942 were caused by failing supplies of pork, and prices for poor quality beef in particular increased significantly. This led the Danish government to reduce domestic prices for beef, and

---

[49] PR, no. 30/1940. At a meeting June 19th 1940 Dr. Walter informed the negotiators in the Reichs-stellen that it would be necessary to increase export prices on beef in the Autumn of 1940, as a means of reducing livestock numbers. BAarch, R901/67771.

from the beginning of 1943 fixed prices were carried through. This meant that producer and consumer prices from January 1943 and throughout the rest of the occupation period were at a constant 20 per cent higher than export prices.

The lower export prices for beef compared to domestic prices caused a periodic disappearance of beef for export. This annoyed the German negotiators on several occasions, but still they did not establish any means of changing the pricing system. Again it is clear that the Danish consumers had precedence over food exports.

V. German price policy during the second half of the occupation period

As mentioned in the beginning, it was only in the first year of the occupation period that the Export Committees had exclusive rights to negotiate prices. After July 1941 the Danish government took away these rights but had already in March 1941 warned the presidency of the Agricultural Council about this. Obviously the top executives of the Agricultural Council were very dissatisfied with this decision as they felt that food prices were still low. At a meeting on March 6th 1941 with the Minister of Agriculture, Kristen Bording, the presidency argued that prices had to increase more if production were not to fall. The basis of this argument was that import prices had increased more than export prices since the German occupation, so expenditures were rising more than revenues. Since the presidency doubted that it would be possible to stop further increases on import prices and on wages, they wanted to keep the right to negotiate prices on food.[50]

The Minister made it clear that the government had to stop further price increases on food, as other groups in society were suffering. Sections of the presidency of the Agricultural Council, including the President Henrik Hauch, and the representatives from the organisation of smallholders (*Husmandsforeningerne*) agreed that the government would control future price negotiations as long as the Minister of Agriculture was in charge; this was accepted by the presidency of the Agricultural Council as the model for the future. In this way the Agricultural Council tried to make the best of it as for eight years they had had a very close and confident cooperation with the Social Democratic minister.[51] They also agreed that it would be advantageous for Denmark and for the farmers, if both import and export prices were kept constant.[52] As mentioned there was little confidence that German export prices would fall or even be constant in the future, but the members of the presidency had reason to believe that the German authorities

[50] PR, no. 6/1941.
[51] The President of Landbrugsraadet, Henrik Hauch, and the Minister of Agriculture Kristen Bording especially had a very good relationship. Hauch joined the coalition government on April 10th 1940 as a supervisor for the Minister of Agriculture (Hauch was member of the Parliament during the German occupation elected by the liberal agrarian party Venstre), and apparently they had a very good understanding of one another. In an interview with the historian Johan Hvidtfeldt on September 13th 1962 Bording explained the very close and trustful relationship he had with Hauch. Bording made it clear that the close relations were established during the crises in the 1930s and were of crucial importance during the occupation. RA, Kristen Bordings privatarkiv nr. 5187, pakke 1.
[52] PR, no. 6/1941.

would let prices on Danish exports increase equivalent to import prices, no matter what the Danish government did.

During the Autumn and Winter of 1940 senior civil servants from the Ministry of Foreign Affairs, the Ministry of Trade, the Ministry of Agriculture, the Statistical Department and the National Bank discussed the German price policy in Denmark. There was a general feeling among the civil servants, especially in the Ministry of Foreign Affairs, that the Germans wanted to set prices in Denmark as in Germany, and the civil servants' objective was to present suggestions that would benefit Denmark the most (Andersen, 2003: 117-129). They feared the Germans might want to adjust Danish food prices to the German level, which would benefit the farmers but also cause inflation and harm other classes in Denmark. Wassard, Chairman of the Danish Government Committee, therefore asked the Agricultural Attaché A. P. Jacobsen at the Embassy in Berlin to calculate the price difference between domestic prices in Germany and Danish export prices for food, included freight costs. On February 15th 1941 Jacobsen sent the required calculations to Copenhagen, which indicated that German butter prices were some 25 per cent higher than the Danish export prices, egg prices were 51 per cent higher and beef 29 per cent higher. Export prices for pork were almost the same as the domestic prices in Germany.[53]

Apparently Jacobsen made a miscalculation on the butter prices, because on September 10th 1941 he dispatched new calculations to Copenhagen, where he noted that the difference between export prices and domestic German prices on butter had disappeared in the Spring of 1941. He also noted that there were no differences between the prices for pork or eggs, while there was still a minor difference for beef.[54] There is every reason to trust the fundamental conclusions in Jacobsen's calculations of September 1941, as he had extensive knowledge of the trade between Denmark and Germany. He had been stationed in Berlin since 1921 and had important connections to senior civil servants in the German ministries, especially in the Reichsministerium für Ernährung und Landwirtschaft. All together his evaluations should be ascribed great importance. Therefore the conclusion is that from the spring of 1941 export prices for butter and pork were at almost the same level as domestic German prices. The price gap for beef was reduced during the Summer of 1941 to less than 10 per cent, and apparently it disappeared in the Autumn of 1941.

This is confirmed by a summary of a meeting in the inter-departmental committee *Handelspolitischer Ausschuss* on April 7th 1941. The revaluation of the Danish currency against the Reichsmark was discussed, and Dr. Walter argued against a revaluation as he feared it would cause a reduction in agricultural production. To prevent this, Danish farmers had to be indemnified by higher prices on food. As the export prices already were at the same level as domestic German prices, this was not possible.[55]

---

[53] RA, Landbrugsministeriet 1948-. J04: Statskonsulenten i Berlin 1921-1945 (14/200/13). Pakke nr. 9-12.
[54] Ibid. In his letter of September 2nd 1941 Wassard asked Jacobsen to send a list of which currencies – Reichsmark or Kroner – different agricultural products were traded in. It is amazing that the Danish head negotiator did not know this before.
[55] BAarch, R9I/949. Dr. Walter points out the same conclusion in a letter on February 18th 1941 to Legationsrat van Scherpenberg in the Auswärtiges Amt; see BAarch, R901/68311.

So it seems as if the German authorities in the first year and a half of the occupation implemented a price policy in Denmark based on domestic German prices. In this way Danish agricultural production was an integral part of the Nazi regime's agricultural policy, with prices in general at the same level as German domestic prices. After this time food prices increased very little, and if export prices went up, they were either short-term increases to ensure rapid supplies or to stimulate the production of special goods, for instance heavy pigs. This does not necessarily mean that the objective was to engineer similar prices in Denmark and Germany, as there aren't any definite indications of this in the records. But when the German authorities implemented price increases on Danish food exports it was on the basis of the established German food pricing system. With repeated price increases during the first year and a half of occupation, Danish export prices eventually reached the level of German domestic prices. In this way price increases were more a consequence than an objective.

This article has not examined whether prices on goods imported from Germany were at the same level as domestic German prices. But it is reasonable to expect that this was the case from the Autumn of 1941, as import prices did not increase much after the turn of the year 1941-1942, indicating that the German authorities had agreed to adjust prices in Denmark to the German level but had no interest in causing inflation.[56] They shared the interest of the Danish Government, the Danish trade organisations and of the Agricultural Council in fighting inflation.

Figure 11.6 Wholesale price index on imports and exports (1935=100), 1940-1945

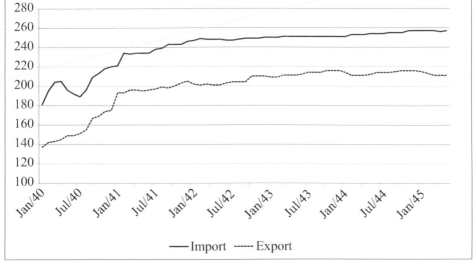

Source: Statistisk Aarbog 1940-1946.

---

[56] See also Walter (1942/43), p 355-356.

As shown in figure 11.6 above both export- and import prices were almost constant during the last three years of the occupation. It was not possible for the Export Committees to enter into agreements with the Germans that would result in higher prices on food, technically because the Danish government did not permit such negotiation, but in reality because the German authorities did not want higher prices. At the same time the presidency of the Agricultural Council changed its opinion. From the turn of the year 1941-1942, there are only a few instances of demands for price increases on food, because the members of the presidency, including the chairmen of the Export Committees, agreed that constant prices and low inflation favoured farmers. The most important exception happened at the beginning of March 1943, when the Chairman of the Bacon Export Committee Marius Byriel, prompted by the government's decision to raise wages, demanded higher prices for pork. In this Byriel was supported by the presidency.[57] The argument was that because of higher expenditures for farmers and for slaughterhouses caused by the increased wages the profitability of pork production was gone. All members of the presidency agreed that Danish consumer prices had to remain unchanged, while export prices and subsidies should ensure higher producer prices. All those involved were sure that the German negotiators would agree to pay higher prices for pork to ensure the production and export of pork. Following four days of difficult negotiation between the Bacon Export-Committee and the presidency of the Agricultural Council on one side and the Minister of Agriculture, Bording, on the other, an agreement was concluded on March 5th 1943. It was decided that producer prices on pork would increase some 10 per cent, and that the higher producer prices would be paid for by a 15 per cent increase on export prices.[58] Before the agreement was concluded Bording managed to get the government's support, as it was important, especially for the Social Democrats, that consumer prices remained unchanged. Thus, it is difficult to evaluate how opposed the presidency of the Agricultural Council in reality was to increased consumer prices on food, or how much this was a tactical move to get the government's acceptance for increased prices on pork.

From March 1943 until the German surrender in May 1945 the Agricultural Council was very defensive concerning prices on food, and in the last two years of the occupation the presidency did not demand any price increases. Instead the desire was to keep the current prices, as criticism of the farmers' high profits was growing among other social groups and within the resistance movement. Thus, the Agricultural Council had to be more and more defensive, and had to work harder to counter the argument that farmers were reaping profit at the expense of other groups in society. They succeeded in keeping food prices at a high level for the rest of the occupation period, but they lost the argument. It has been the general opinion since the war that the farmers made a fortune because of the market conditions during the occupation.

---

[57] PR, no. 4/1943 and no. 6/1943.
[58] PR, no. 6/1943 and Statistiske Meddelelser, 4. Række, 127. Bind, 3. Hæfte. *Landbrugsstatistik 1945. Herunder Havebrug, Skovbrug m.v.* Published by Det Statistiske Departement 1947.

VI. Conclusion

At the beginning of the German occupation of Denmark the German Ministry of Nutrition and Agriculture drew up a plan for managing agricultural production in Denmark. The objective of the plan was to maximise agricultural production and ensure a maximum of food supplies to Germany. *Staatssekretär* Backe and the chief negotiator Dr. Walter formulated the plan, which was communicated to all German negotiators in Denmark and to the German Embassy in Copenhagen, so they were very familiar with the German economic policy in Denmark and were aware of their authority in negotiations with the Danish authorities.

The central element of the plan was to maintain and strengthen the farmers *Produktions- und Lieferfreudigkeit*. By using differentiated prices on various agricultural products, which were at a much higher level than that prior to occupation, the German authorities secured the full cooperation of the Danish farmers and the farmers' organisations, which was of significant importance for the implementation of the plan. The German authorities behind the plan saw it as the most rational policy in Denmark, and doubted that restrictions, control or rationing would lead to the desired objectives. Seen in retrospective, given the quantity of food supplies to Germany and the few resources the Germans had to spend on management in Denmark, one has to agree with them.

The basis of the German economic policy was the voluntary collaboration by the Danish authorities, the farmers and the farmers' organisations, and the focus of the policy was precisely to ensure this. To the advantage of both Germany and Denmark, *Staatssekretär* Backe and the chief negotiator Dr. Walter established a very efficient system in which the farmers and Danish society in general maximised production and exports to Germany in their own interest. Danish farmers also gained from the German policy. It ensured most farmers a good profit every year during the German occupation, and it ensured that their productive capacity was almost intact when the war was over. Finally it meant that the production structures in Danish agriculture, both on the farms and in the processing industry, remained almost unchanged.

The Danish government in many ways enjoyed the same benefits from the German policy as had Danish farmers. The arrangement ensured the Danish population the highest food consumption per inhabitant in Europe during the war. At the same time the Danish government succeeded in its objective of keeping the German authorities away from the Danish population and institutions, and letting the Danish authorities carry out most of the administration in Denmark. But the policy also had a strong impact on the distribution of wealth in Denmark. The farmers benefited and the unemployed and pensioners especially lost out. Some social groups in Denmark had a hard time especially during the first years of the war when the cost of food increased significantly and the rate of unemployment was very high.

It is important to note that the German occupying power used the same economic policy for all five years in Denmark. In the winter of 1944-1945 when starvation was epidemic in Germany, the German authorities did not try to impose harsh rations and increase food supplies. They still doubted that it would lead to the desired results.

Bibliography

Primary sources

Bundesarchiv (henceforth BAarch):
BAarch, R16/1306
BAarch, R91/949
Baarch, R901/67738
BAarch, R901/67741
BAarch, R901/67747
BAarch, R901/67771
BAarch, R901/67772
BAarch, R901/67782
BAarch, R901/67939
BAarch, R901/68173
BAarch, R901/68311
BAarch, R901/68939
BAarch, R3601/69

The archives of Landbrugsraadet: Meddelelse fra Præsidiet til Raadets Medlemmer (Reports from the Presidency to the Members of the Council. Henceforward written as PR.):
PR, no. 4/1943
PR, no. 5/1941
PR, no. 6/1941
PR, no. 6/1943
PR, no.7/1940
PR, no. 10/1940
PR, no. 12/1940
PR, no. 13/1940
PR, no. 14/1940
PR, no. 16/1306
PR, no. 19/1940
PR, no. 25/1940
PR, no. 30/1940
PR, no. 261939 and PR, no. 14/1941

Rigsarkivet (henceforth RA):
RA, The archives of P. Munch, no. 6663, package no. 109, s. 104.
RA, Håndskriftssamlingen: Sigurd Jensen IV.T.42: *Materiale vedr. besættelsestiden.*
RA, UM 1909-45, H64-205.
RA, UM 1909-45, H64-190.
RA, UM 1909-45, Dmk-Tyskland Handelstraktatforhold XXXIX, nr. H64-190.
RA, UM 1909-45, F.02, H64-194
RA, Kristen Bordings privatarkiv nr. 5187, pakke 1.
RA, Landbrugsministeriet 1948-. J04: Statskonsulenten i Berlin 1921-45 (14/200/13). Pakken 1, 9-12.

Secondary sources

Andersen, S. (2003) *Danmark i det tyske Storrum. Dansk økonomisk tilpasning til Tysklands nyordning af Europa 1940-41*, Copenhagen.

Brandt, K. (1953) *Management of Agriculture and Food in German-Occupied and Other Areas of Fortress Europe. A Study in Military Government*, California.

Giltner, P. (1997) 'Trade in 'Phoney' Wartime: The Danish-German Maltese Trade Agreement of 9 October 1939', *International History Review*, pp. 333-346.

Giltner, P. (1998) *In the Friendliest Manner'. German-Danish Economic Cooperation During the Nazi Occupation of 1940-1949, (*Studies in Modern European History, Vol. 27), New York.

Just, F. (1992) *Landbruget, staten og eksporten 1930-1950*, Esbjerg.

Kirchhoff, H. (2002) *Samarbejde og modstand under besættelsen. En politisk historie*, Odense.

Lund, J. (1999) *Danmark og den europæiske nyordning. Det nazistiske regime og Danmarks plads i den tyske Grossraumwirtschaft 1940-42*, Unpublished dissertation, Copenhagen.

Mau, M. (2002) *'Business as usual'. De dansk-tyske handelsrelationer under besættelsen. En analyse af Udenrigsministeriets embedsmænds politik i regeringsudvalget*, Unpublished dissertation, Copenhagen.

Nissen, H. S. (1973) *1940. Studier i forhandlingspolitikken og samarbejdspolitikken*, Copenhagen.

Nissen, M. R. (2003) 'Landbruget og den danske neutralitet efter udbruddet af Anden Verdenskrig', *Historie*, 1, pp. 124-157.

Poulsen, H., (1985) 'Danmark i tysk krigsøkonomi. Myter og realiteter om den økonomiske udnyttelse af de besatte områder under 2. verdenskrig', *Den Jyske Historiker*, nr. 31-32, pp. 121-131.

Poulsen, H. (2002) *Besættelsesårene 1940-1945*, Aarhus Universitetsforlag, Aarhus.

Salmon, P. (1997) *Scandinavia and the great powers 1890-1940*, Cambridge.

Sjøqvist, E. (1971) *Danmarks udenrigspolitik 1933-1940*, 2. ed., Copenhagen.

Skade, R. (1940) 'Danmark under Krigen. Økonomiske Foranstaltninger i 1940', a supplement to *Økonomi og Politik. Politiske og økonomiske Kvartalsoversigter.*

12 Cooperative cellars and the regrouping of the supply in France in the twentieth century

Jean-Michel CHEVET, INRA-ALISS

When speaking in France about the international situation of viticulture, one often hears that production structures and commercialisation of our wines are factors responsible for a drop in market shares, in particular in Great Britain. French wines are accordingly suffering more and more from the competition of what are called the 'new world' countries, for example that from Australian viticulture, which is more efficient producing wines at a lower cost than ours. The advantages of the latter viticulture are due to the giant size of its production structures and commercialisation provided by large international groups, whilst in France, production is still in the hands of small vinification units, those of the viticulturists.

Such criticisms, voiced by the Press, are unfortunately echoed in certain professional circles and indeed by certain opinionmakers (Berthomeau, 2001). Let us take a few examples to illustrate our remarks. Referring to an issue of the review entitled *Capital*, one reads that 'in the year 2000 our wines for the first time lost their leadership in the British market, the most dynamic in the world', the market share of French wines having fallen from 30 per cent in 1970 to 26 per cent in 2000. This was partly due to the fact that 'our winegrowing industry responds inadequately to the requirements of new consumers' – the things that are put into the mouths of the poor consumer! – and also because 'Californian and Australian producers have dynamited the French tradition with their giant vineyards, their peak techniques and their efficient marketing policies'. Currently, when deregulation is one of the key words of certain systems of representation, the responsibility for it is certainly attributable to our 'unduly meticulous regulations', which 'increase production costs'. To this must be added a volume of produce, which is 'too fragmented (of its 400,000 players only a handful have an annual turnover of more than 1,000 million francs)' and our lack of business efficiency, which results in a virtual absence of marketing. The journal *Le Monde* is not far behind. In its issue of 10th July 2002, it states that: 'In the United States and in Great Britain, wines from the southern hemisphere are in the process of supplanting claret, burgundy and other Languedoc wines'. We are also told (but without reference to French exports) that in the United Kingdom 'the market share of French wines is continually falling back'. Since here, too, there is a need to find scapegoats, the blame is put anew on our *appellations d'origine contrôlée* which add to the cost of production, whilst 'Australian winegrowers are authorised to blend grapes from the whole country, in order to achieve a constant quality'! To crown it all, a report by the permanent secretary of the Ministry of Agriculture confirms the analyses which appear in the Press. These state, in particular, that 'our market share is inexorably crumbling' and we are told about our 'export setbacks' and our 'market loss in England'! There is no appeal from that diagnosis; 'absence of brands' and 'too fragmented a volume' are the key phrases used in the analysis, France having no business companies able to 'provide the means for developing sales policies that bear comparison with those of the competition'. Of course there, too, the standards of reference are companies such as Gallo and Southcorp. After reading all of this, one would think that our vini-viticulture was antiquated!

A doubt nevertheless arises to the veracity of this diagnosis, which admittedly is more accurate than has been demonstrated, seeing that the references made to the brand, groups and to structures of viticulture (which are the true *leitmotifs* of all these analyses) are not really relevant to the performance of French vini-viticulture during recent years. Moreover, the origin of these references, whose efficiency is not really analysed, is to be found among the 'mythologies of our time', rather than in any real demonstration of knowledge. This, moreover, is not surprising, since the report by the permanent secretary of the Ministry of Agriculture relies on the dicta of 'consultants', who are naturally unwilling to saw off the branch on which they are sitting. It should also be added that the said dicta stem from collective beliefs and that in that respect, they are normative and should rather reflect marketing-related and group positions than those of the entire French viticulture.

Poor French viticulture! Fairy godmothers are no longer leaning over its cradle. It has nevertheless recorded marked export successes over the past thirty years. In fact, when we are told that our market share is slackening (in particular in Great Britain where our export percentage has fallen from 50 to 26 per cent) it has to be remembered that during the same period, the total of French exports has almost trebled. In terms of value, since they represent 34.2 per cent of the world total, they reached a maximum of 50.4 per cent in 1985-1989 and still accounted for 41.6 per cent during the 1995-1999 period - a performance that must surely be placed to the credit of our *appellations d'origine contrôlée*. Even in Great Britain, French performance is far from negligible. Certainly, our market share has fallen from 38-39.3 per cent to 24.5-29.6 per cent, compared with the 1978-1982 period. However, during the same period, Great Britain imported 2.5 times as much VQPRD and 3.3 times as much of 'vins de Pays' and of table wines! Whilst the latter accounted for 43 per cent in 1990, they accounted for 51.7 per cent in 2002.[1] If one does not remember that market share is an indicator, more relevant to marketing than to the macro-economy, one commits several errors of assessment of the true place of France in the world market. During discussions, I was able to realise that some people were equating loss of market share to a drop in exports. Nevertheless, a drop in market share does not mean that French exports have dropped. In fact, they only dropped in terms of a global volume, which had grown more rapidly than French exports. Could not an attempt at hegemony stirred up by a certain chauvinism, be the source of this blindness? Instead of blaming the hypothetical effects of marketing and the presence of groups, should one not question the results of a somewhat too Malthusian a policy, the competition only having filled the place which we were unable to fill? We will certainly not go as far as this but, to put it simply, such an explanation is no more mistaken than others (Chevet, Giraud-Hérault and Green, 2005).

---

[1] For the data see Office National interprofessionnel des Vins, 1994, 1999, 2003, 2004.

The context of this communication having been defined, it is necessary to set more precise limits, seeing that all the problems which have been mentioned cannot be dealt with in such a small available space. Here, the question will arise of the 'regrouping of the supply' and more particularly through a study of the part played by cooperative winecellars. Though not very modern, they do not play a key par t in the said regrouping of the supply? In certain regions, was it not the cooperative winecellars who in the final analysis could compete with the large Australian groups and who made it possible (in particular for table wines and the *vins de Pays*) to achieve the export successes which France has recorded over recent years as mentioned in the introduction?

After very briefly summarising the evolution of the production of French viniculture, we shall study the coming into existence and development of cooperative cellars on the national level, more particularly in the region of Languedoc-Roussillon. The latter shall serve as a point of comparison with Australian viticulture. An attempt will then be made to see whether the past thirty years, cooperative cellars have not experienced a movement towards concentration. The part played by groups of producers and by amalgamations, which took place between cooperative cellars, will be dealt with next. In conclusion, an attempt will be made to draw a comparison with the Australian groups. This study will show that France is (at least as far as vinification structures are concerned and at least in respect of one market sector) far from exhibiting the disadvantages which are commonly attributed to her. This is all the more true if one is mistaken about the gigantism of the Australian groups and their economic nature.

I. Evolution of the wine production in France during the 20th century

I.1. Evolution of the structure of vineyards and of wine production during the 20th century

Let us begin by briefly tracing the evolution of the French wine production during the twentieth century. For that purpose, we have assembled the principal data in the following table.

Table 12.1 Evolution of the surface area of vineyards, production and number of declarants in France, 1900-2002[2]

	Area	Index	Production	Index	Hl/ha	Index	Declarants	Index
1900-1909	1,699,387	100	55,832,695	100	32.9	100	1,776,700	100
1910-1919	1,547,046	91	43,191,696	77	27.9	85	1,536,512	86
1920-1929	1,524,713	90	59,976,571	107	39.3	119	1,482,216	83
1930-1939	1.531,146	90	58,759,058	105	38.4	117	1,507,839	85
1940-1949	1,443,900	85	42,206.992	76	29.2	89	1,493,448	84
1950-1959	1,352,871	80	52,935,171	95	39.1	119	1,498,500	84
1960-1969	1,261,324	74	60,507,613	108	48.0	146	1,243,450	70
1970-1979	1,182,763	69	68,515,681	123	57.9	176	894,950	50
1980-1989	1,038,744	61	68,346,903	122	65.8	200	634,600	36
1990-1999	807,895	47	47,724,636	85	59.1	180	333,310	21
2000-2002	787,628	46	47,577,667	85	60.4	184	212,700	13

Source: Annuaire national officiel des caves coopératives de France, 1949-1954; Annuaire national officiel des caves coopératives de France, 1955-1956; Guide des caves coopératives de France; Annuaire national des caves coopératives; Annuaire national des caves coopératives.

After three quarters of a century of growth during the nineteenth century, the area of vineyards reached its apogee with 2.465 million hectares in the middle of the 1870s. After that date and under the influence of the phylloxera crisis and the growth of the Algerian vineyards, French vineyards passed through a long period of decrease and as early as the turn of the century, only 1.700 million hectares of vineyards were under cultivation. Initially slow, the rate of decline increased after the Second World War to the extent that, after a hundred years, the area of vineyards had fallen by half. An examination of production in respect to the first half of the twentieth century, shows that due to the rise in yields, it remained relatively stable. It then rose between 1950 and 1990, breaking through all ceilings and resulting (at least at the beginning of this period) in a need to replace the

---

[2] All data given in figures and tables have been updated by Mr. Dhuisme and the members of the federations of the Languedoc-Roussillon cooperative cellars.

absence of Algerian wines. Finally, from 1990 onwards, it dropped appreciably below the most productive periods of the beginning of the century.

Certainly, although the number of declarants is not exactly the same as the number of viticulturists, we are nevertheless going to use it as an indicator. At the beginning of the twentieth century, France had some 1.78 million viticulturists and it is not of importance to know whether or not they only cultivated vineyards. As a result, every declarant cultivated on average less than one hectare of vineyard and produced 32.5 hectolitres of wine. Gradually the number of declarants dropped, initially in parallel to the size of the vineyard, with the average area remaining relatively stable until the 1970s. From this period onwards, the rate with which declarants disappeared increased, their number dropping by half during the decades from 1900 to 1970 and again by half that number between 1970 and the present. Since the total area of vineyards decreased gradually, the area per declarant rose from 1.32 hectares during the 1970s, to 1.64 hectares during the following decade and to 2.19 hectares at the beginning of the twenty-first century. Production per declarant, initially relatively stable until the eve of the Second World War, quadrupled during the second half of the twentieth century.

I.2. A brief look at the conditions of wine commercialisation during the twentieth century

Production was in the hands of a large number of small winegrowers and in certain regions, in Languedoc-Roussillon for example, of medium-sized and indeed by large growers. At the start of the twentieth century, merchants were commercialising a considerable part of the production, the remainder being consumed at home. The merchants bought the wine or the grapes, which they vinified and matured in large storehouses. It was the merchants who sent the wine to the major centres of consumption, for example, towns.

It is difficult to say how much was consumed at home prior to 1964. The quantity was certainly larger than that estimated by the agricultural land registry in 1964. At that time, 50.71 per cent of growers (739,500 out of 1,458,672) did not sell their wine. They cultivated a total of 197,684 hectares out of the 1,181,015 hectares of vineyards, thus 14,34 per cent. In 1988, 105,851 viticulturists were still not selling the produce of their vineyards. According to the agricultural inquiry in 2000, the rate of home consumption (which between 1964 and 1988 had already lost 85 per cent of its ratio) dropped considerably further still. At that date there were 33,878 producers who consumed their produce at home, i. e. a further decrease of 68 per cent. In 1988, they were only cultivating 30,396 hectares and in 2000 only 9,958 hectares or 1.1 per cent of the area of French vineyards. The sales quotient of viticulturists therefore fell less rapidly than shown in Table I. Henceforth, there will be no mention of viticulturists other than those who commercialise their harvest directly, or send it to cooperative cellars.

At the beginning of the twentieth century, the purchase and sale of wines was handled by brokers, introducing vendors and purchasers to one another. Dependant on the state of the market, the brokers who were established in towns and who possessed the power of merchants, exerted a more or less fair pressure on the small proprietors, weakened by the effects of the phylloxera crisis, securing their harvests at the lowest possible price.

These market dysfunctions (coupled with fraud and imports) induced certain representatives of viticulture to establish the first cooperative cellars in order to counter the power of the merchants[3]. The wine crisis and the conditions of wine commercialisation became the breeding grounds for cooperative cellars.

II. Creation of cooperative cellars in France during the 20th century

Figure 12.1 traces the creation and the disappearance of cooperative cellars in France and in the region of Languedoc-Roussillon. It is clear that before the First World War, only a small number of cooperative cellars had been established and that this primarily took place in the region of Languedoc-Roussillon.

Figure 12.1 Creation and disappearance of cooperative cellars in France and in the Languedoc-Roussillon region from 1900 to 2001

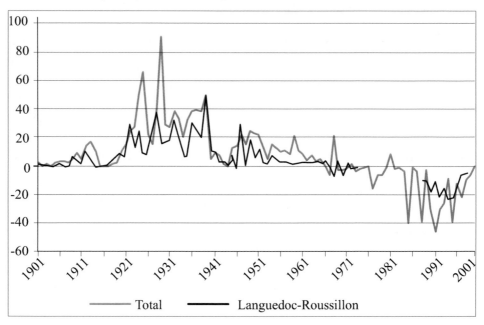

Source: see Table 12.1.

---

[3] There were also other reasons, arising in particular from the globalising society, responsible for the creation of cooperative cellars. It was also in order to counter the power of the merchants, that vineyard owners sold increasing quantities of their wines. In Burgundy and the Bordeaux region, this method of sale, which represented 5 per cent of the volume at the beginning of the 1960s, has presently reached 25-30 per cent.

The first cooperative cellar was established in 1901 at Maraussan near Béziers. Its name, *Les Vignerons Libres* (The Free Viticulturists), clearly demonstrates the climate, in which these first cellars had been founded. Initially, the cellar only accepted the grapes of small viticulturists. The wine was sold to labor cooperatives of the Paris region on provision that one quarter of the income would be distributed to socialist organisations and another quarter to labor cooperatives. It was not until 1907 that a second cellar opened at Bompas, in the Pyrénées-Orientales. Although vinification was collective, the sale of wine took place through the traditional commercial channels. In 1909, two new cellars came into existence, one at Lézignan in the Aude and the other at Siran in the Hérault. There, no discriminatory threshold was fixed for membership of the cooperative. Moreover it left it's members the option of selling their wine as they saw fit (Gavignaud-Fontaine, 1987). Generally, the first cooperative cellars came into existence in socialist 'fiefs', initially in irregular forms of cooperation that in time became ordered.[4]

With the exception of the department of Var, it was the region of Languedoc-Roussillon that showed rapid growth of the cooperative cellars (85 per cent of them had come into existence around 1910). Nevertheless, the movement became noticeable in other regions, where these cellars multiplied as other southern vineyards, resulting in 35 in the department of Var in 1914. They also appeared in the Côte-d'Or (8 in 1914) where the winegrowers had a reputation for being fiercely independent. Out of the 73 cellars which came into existence before the war, merely 33 were in the region of Languedoc-Roussillon (Ramade-Beaujour, 1975).

After a slight levelling off, (figure 12.2), the movement resumed its forward march from the nineteen-twenties onwards. It was between 1925 and the eve of the Second World War that the largest number of established cooperatives was recorded, the modalities varying from vineyard to vineyard. The Languedoc-Roussillon region accounted for 62 per cent of the established cooperatives between 1920 and 1925, then for 47 per cent between 1925 and 1930, and for 72 per cent between 1930 and 1935. After that, the movement ran out of breath and it accounted for only 9.5 per cent between 1935 and 1940. Moreover, the rhythm of expansion differed in the four departments.

---

[4] The movement grew just as successfully under the influence of republican and lay unionism, as under that of clerical and even monarchist trends. The establishment of cellars was affected by the political divisions of society at the time. Gavignaud-Fontaine, 1987: 11-22.

Figure 12.2 Evolution of the number of cooperative cellars in France and in the region of Languedoc-Roussillon, from 1900 to 2001

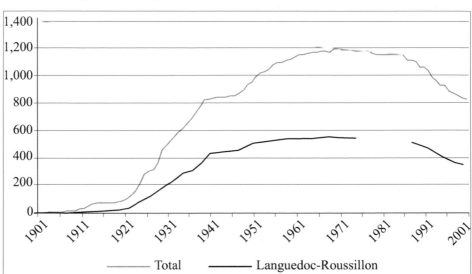

Source: see Table 12.1.

The Champagne region was particularly affected by the revolt of viticulturists at the beginning of the century. The small viticultural property predominated and the first cooperative cellar did not come into existence until 1920. This tardy development was due to a favourable economic climate, enhanced by the advantages of luxury production. Thanks to the merchant houses of the Champagne region, the product was sold at a price well above that dictated by the mass-producing vineyards. From 1920 onwards, growth was regular, interspersed with some levelling off. There were 7 cellars in 1925, 14 in 1930, 29 in 1940 and 52 in 1952. In the Pays de la Loire, the chronology was similar, but additions were much less numerous.

Finally, during the decade which preceded the Second World War, cellars where established in places where previously none existed. Between 1929 and 1934, 7 cellars were founded in the Beaujolais, 5 of which were in the vintage zone, the zone of large properties and crop-sharing. In the Bordeaux region, their coming into existence and growth were equally slow, the first cellar dating only from 1930. Their introduction then become rapid, resulting in 35 cellars in 1935 and 56 in 1940. Generally speaking, the southwest followed the same chronology as in the Bordeaux region, but less rapidly.

As seen in figure 12.2, approximately 80 per cent of cellars were created in 1940, compared with the maximum number in 1969. The Languedoc-Roussillon region slightly exceeded the national average whilst the departments of Rhône, the Côte-d'Or and the Gironde reached their apogee. Certain departments stayed behind. The departments of Gers and the Hautes-Pyrénées only reached 25 per cent of their potential, the region of

Toulouse only 37.5 per cent, whilst the central western region and the Charente reached only 55 per cent. This discrepancy was corrected after the war. New cellars were initially numerous, becoming infrequent from 1960 onwards.

From 1969 onwards, the movement went into reverse. Largely as a result of amalgamations, cellars had started to disappear. This deceleration proceeded gradually until 1985, when the 1960 level was reverted to and the movement accelerated before decelerating again from 1992 onwards. In 2000, the number of cellars had reverted to the pre-war level. This concentration of cellars was not homogeneous. It was very pronounced in the Côte-d'Or, the Dordogne and the Pays de Loire, where the number of cellars had fallen by over 80 per cent, followed by the regions of Toulouse and of the Tarn. The Mediterranean region lost some 40 per cent of its cellars, the losses being more numerous in the Bouches-du-Rhône (62 per cent) than in the other departments, the minimum occurring in Gard and in Vaucluse. The losses were small in the Drôme (12.5 per cent), in the Gironde (9.7 per cent) and in the Champagne region (4.4 per cent). During this period, only Lot-et-Garonne increased its number of cellars by two.

At least two reasons can be mentioned to explain the movement of cellar creation. Although some creations occurred during periods of relative prosperity – 1920-1929 and 1946-1950 –, it was generally speaking during periods of crisis – 1901-1914 and 1930-1939 – that most cellars were established. This appears to be all the more true, since even during the period of 1920-1929, the fall in prices which began in 1919 was accompanied by a rise in establishments, the peak of 1925 corresponding to a resumption of the rise in prices. On the other hand, the passing of the *Statut Viticole* (Wine Charter) increased the creation movement. In fact, in order to counter the slump, the law of 4 July 1931 restricted (to the vineyard and not to the trade) that portion of the harvest, which exceeded 400 hectolitres. The new legislation of 30 July 1935 lowered this limit to 185 hectolitres and imposed a rigorous spacing out of sales. The proprietors who no longer had the means of purchasing additional cellar equipment had every interest in becoming a member of the cooperative. Moreover, the legislation imposed charges on the large estates in respect of high yields and on compulsory distillation, which led certain proprietors to opt for the cooperative. The *Statut Viticole* and the fall in prices therefore restarted the movement of the establishment of new cellars and triggered it in those departments where cooperatives had not existed before.

Almost from their first introduction, cooperative cellars received considerable assistance from the authorities, which did not fail to affect their growth. That assistance was initially felt in the area of taxation, since the cellars enjoyed a certain number of exemptions. They were exempt from land tax on real estate and they did not pay tax on fixtures and fittings. They were no longer subject to licensing, or to tax on agricultural profits and to industrial or business taxes. They also had advantages in the form of subsidies and loans. The State provided 20 per cent of the preliminary flotation funding regarding the creation of cellars and 10 per cent of the development costs. For its part, the *Crédit Agricole Mutuel* granted to cooperatives maximum loans over thirty years at 3 per cent interest, with a limit of three times their company capital. The cooperative was therefore funded 20 per cent by the State, 60 per cent by loans and the remaining 20 per cent by the members of the cooperative. In the wake of the 'associationistic' movement, cooperative cellars were managed democratically on the principle of 'one man, one vote'. They were moreover private companies and partnerships. The company shares were registered and individual. The cost of subscription was based on expected services. The interest due on the capital was not always paid. Created by the initiative of the viticulturists, they were nevertheless subject to State approval. Their management was subject to the dispositions of the general assembly, which consisted of shareholders possessing equal rights. These shareholders had an overriding right to make decisions. The board of directors was a permanent collective body, responsible for the management of the cooperative. Salaried personnel, including an auditor, a managing director and a director filled the technical functions. The authorities exercised ongoing control of the cooperatives activity through the Ministry of Agriculture, the Senior Treasury Official and the regional funds of the Crédit Agricole. Operating costs were covered by the sale of a part of the harvest (wines and by-products) benefitting the cellar, in other words the 'cellar's share'. The balance was divided among the members of the cooperative, according to the weight and the mustmetric degree of each member's grape harvest.

II.1. Evolution of the area controlled by cellars and of the percentage of viticulturist-members of a cooperative cellar

Figure 12.3 traces the evolution of the total surface area of vineyards controlled by cooperative cellars. It likewise shows the evolution of the percentage of members, calculated in relation to the number of declarants. From the end of the Second World War until the middle 1980s, the surface area controlled by cellars rose continually, from some 18 per cent of vineyards to almost 50 per cent. Despite a slight regression since 1985, the findings are in no way subjected to doubt.

Figure 12.3 Evolution of the percentage of cooperative viticulturists and of the percentage of the area of their vineyards from 1944 to 2000

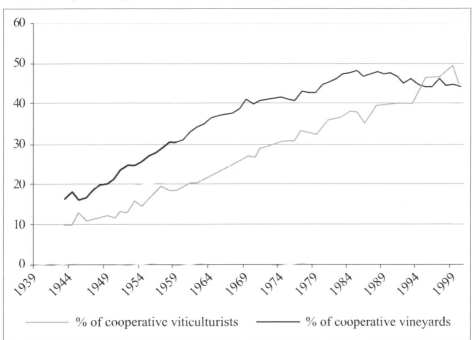

Source: see Table 12.1.

The figure likewise shows that some 10 per cent of declarants, on the eve of the Second World War, sent their grapes to a cooperative. This proportion has been rapidly on the upgrade, presently reaching almost 50 per cent. The cessation of the movement of cellar creation, which occurred in 1969, did not therefore prevent the growth of interest shown by vine growers in cooperative cellars. In 1945, there were on average 180 members per cellar, 241 in 1969 and 120 in 2000. A regression was found in the number of members but an even greater one in the number of declarants. On the other hand, the average area exploited by a member increased during the same period. It was 1.48 hectares in 1945, 1.71 hectares in 1969 and 3.84 hectares in 2000. The average size of a vineyard controlled by a cellar followed the same rising movement. In 1945, one cellar controlled 267 hectares. In 1969 this figure had gone up to 412 ha and to 460 ha in 2000.

II.2. Importance assumed by cooperative cellars according to region

The introduction of cooperative cellars did not experience uniform development in all regions. Let us for example look at their numbers at the end of the period being studied, with the help of the tables 12.2 and 12.3 below.

Table 12.2   Regional importance of cooperative cellars in 2000

	Cellars	Members	Declarants	Area of members' vineyards	Area of declarants' vineyards
Alsace	17	2,548	7,786	4,937	14,735
Aquitaine	76	7,977	24,783	35,619	145,839
Bourgogne	16	2,070	8,729	6,270	29,534
Centre	16	1,154	11,134	3,212	23,396
Champagne	131	13,161	13,797	9,998	29,344
Languedoc-Roussillon	352	29,333	43,044	212,110	296,468
Midi-Pyrénées	21	11,601	16,484	8,673	41,285
Pays de la Loire	4	680	9,751	4,291	37,710
Poitou-Charente	7	1,318	6,170	3,055	80,489
Provence	119	18,718	16,690	62,156	97,841
Rhône-Alpes	66	11,214	21,731	31,323	57,195
Other	15	785	43,789	954	17,947
Total	840	100,559	223,888	386,598	871,783

Source: see Table 12.1.

Not all regions had the same system of cooperative cellars. They were more numerous in the Mediterranean area than anywhere else, except in Champagne and to a lesser extent in Aquitaine. The picture is the same regarding the number of members. An examination of the sizes of members' vineyards largely confirms what previously has been said. Since it is difficult to judge the rightfull place occupied by the cooperative using the total figures, the indices of table 12.3 will put things in a clear light.

An examination of the first column of the table below shows that the average number of members per cellar varied from region to region. They were most numerous in the region of Midi-Pyrénées. A long way behind came a group of regions where the number of members per cellar was between 150 and 180. It was found that there were relatively few in Languedoc-Roussillon, which led to a pronounced intermeshing of cellars in the region.

Table 12.3   Regional importance of cooperative cellars in 2000

	Members per cellar	% of declarants	% of the area of members' vineyards	Area per declarant (ha)	Area per cellar (ha)
Alsace	150.0	32.7	33.5	1.94	290
Aquitaine	105.0	32.2	24.4	4.47	469
Bourgogne	129.5	23.7	21.2	3.03	392
Centre	72.0	10.4	13.7	2.78	201
Champagne	100.5	95.4	34.1	0.76	76
Languedoc-Roussillon	83.5	68.2	71.6	7.23	603
Midi-Pyrénées	552.4	70.4	21.0	0.75	413
Pays de la Loire	170.0	7.0	11.4	6.31	1,073
Poitou-Charente	188.5	21.4	3.8	2.32	436
Provence	157.5	112.2	63.5	3.32	522
Rhône-Alpes	170.0	51.6	54.8	2.79	475
Other	52.5	1.8	5.3	1.22	64
Total	119.5	45.0	44.4	3.84	460

Source: Agricultural census 2000.

The proportion of cooperative vine growers in relation to the total number also varied from region to region. We have already seen that the Mediterranean area had the largest proportion of viticulturists who sent their grapes to cooperatives, with the Champagne region not far behind. Aquitaine and Alsace came next. Among the large vineyards, the low degree of participation in the cooperative movement was noticeable in Burgundy, despite its early beginning. It was also noticeable in Poitou-Charente and first and foremost in the Pays de la Loire. The proportion of vineyard area owned by cooperative viticulturists did not reflect the number of members, explained by the drawing up of the statistics themselves.

Mediterranean vineyards distinguished themselves in their importance to the cooperatives. With about a third of their produce vinified by cooperatives, Alsace and Aquitaine represented the scope of the cooperative system. The area per member varied a great deal from region to region. It was largest in Languedoc-Roussillon with 7.23 hectares, followed by the Pays de la Loire with 6.31 hectares and by Aquitaine with 4.47 hectares. A group of regions followed with between 2 and 3 hectares. Finally, it was particularly low in the Midi-Pyrénées and in the Champagne region, resulting in a high percentage of member-viticulturists.

The average size of vineyards controlled by a cellar is very important because this dictates the size of the economic entity. The Pays de la Loire with an area of 1,073 hectares per cellar had the highest average. The Languedoc-Roussillon region followed with a much smaller area. The size of the vineyard area in Aquitaine, Poitou-Charente, Provence and Rhône-Alpes was half of that in the Pays de la Loire; in Burgundy and in

Alsace the vineyard area was once again halved. The Champagne region came last and a long way behind.

II.3. Is it possible to explain the incongruity of cellar creation?

Historically, vineyards were not only planted where ease of transport allowed the commercialisation of their produce in the large centres of consumption. According to the availability or otherwise of these facilities, the number of vineyards producing wine for their own consumption varied considerably, while this is not the sole explanatory variable, it cannot be disregarded. It is moreover reasonable to suppose, that in regions with a high degree of home consumption, the vineyard plays only a marginal part in the majority of enterprises. It is even possible that the grapes for producing wine destined to be consumed by personnel, were cultivated by a large enterprise. It is quite diffirent in regions in which winegrowing represents the sole agricultural activity. There, it is the principal source of income, even for the small vineyard. Its production and sales are imperatives, which cooperative cellars have greatly facilitated.

The statistics recording home consumption are recent, dating from the establishment of the agricultural land registry in about 1960. For lack of anything better, the information which can be obtained from them, can nevertheless be used, particularly since the evolution of viticulture in the first half of the twentieth century was slow. The registry shows that the proportion of vineyard area cultivated to produce wine destined solely for home consumption, is very low in the Mediterranean departments, namely 0.45 per cent in Pyrénées-Orientales, 2.05 per cent in Gard, and 4.56 per cent in Bouches-du-Rhône. Comparable figures elsewhere are 2,02 per cent in the Gironde and 1.03 per cent in the Marne. These mostly relate to departments with large sales and to those where cooperatives are particularly well established. Everywhere else, the proportion of home consumption is more than 10 per cent, as in the Dordogne with 34.13 per cent, which nevertheless is a department with a large production.

For purposes of illustration, let us take the example of two departments, the Hérault and the Indre. In the first of these, where the exclusive cultivation of the vine predominates, over 98 per cent of the vineyards produce for the market, involving 99.8 per cent of the surface area. In the Indre, only 14.25 per cent of the vineyards sell their wine produced on 29.7 per cent of the surface area. In the Hérault, 91.6 per cent of the vineyards with less than 0.25 hectare sell their wine. When this threshold is passed, the proportion exceeds 98.5 per cent. The situation is quite different in the Indre, where only 5.8 per cent of the vineyards with an area of less than 0.25 hectares sell their wine. For the category owning between 0.25 and 1 hectare, this proportion is 21.6 per cent; for the category, which owns from 1 to 3 hectares, the proportion amounts to 74.52 per cent. Since the number of viticulturists there is lower, only 3,950 compared with 63,955 in the Hérault, it is understandable that under these circumstances the conditions for the growth of cooperative cellars were not really favourable.

III. Assessment of the number of vinification units in France at the end of the twentieth century

In France, the statistics concerning viti-viniculture are most frequently accumulated from the angle of viticulture, rather than that of viniculture. Vineyards, areas under vines and other important details are counted. These statistics tell us little about vinification units. Although it is possible to ascertain the number of vineyards that carry out the vinification of their own grapes, it is very difficult to ascertain the hierarchy of the various units. In Australia on the other hand, statistics tell more about the aspect of vinification (the wine-industry) than about the aspects of viticulture. Because of this, a comparison of the respective viti-vinicultures of the two countries proves sensitive. In order to assess the differences which exist between the two countries, it is therefore necessary to compare that which can be compared and not to contrast the world of French viticulturists with the large Australian groups. It is for that reason that we are going to assess the number of vinification units existing in France.

To do this, let us start from the most recent agricultural census. Of the 108,604 wine-growing companies which sell their harvests, there are 37,875 or 34.9 per cent, which vinify their own harvests in cooperative cellars. On the other hand, 64,651 viticulturists or 60 per cent, send them to the cooperative cellars. As a result, these companies cannot be put on a par with vinification units which, in contrast to the aforementioned, vinify in private cellars. The 19,604 companies which sell their harvests in the form of grapes are not members of vinification units either. Unfortunately, we cannot at present count those units, which are most frequently in the hands of merchants. Nevertheless, because of the importance of this type of vinification (with the exception of the Champagne region) the results which we have obtained, cannot be far from the truth. According to the table 12.4, vinified volumes make private cellars appear more prominant (48.6 per cent) than their number.

Table 12.4 Division of vineyards and vinified volumes, according to the method of vinification adopted, 2000

	No of vineyards	%	Volume (hectolitres)	%
Vinification in private cellars	37,875	31.39	28,147,157	48.6
Vinification in cooperative cellars	65,166	59.53	25,972,068	44.8
Sale of new and other harvests	19,604	9.09	3,795,810	6.5
Total	108,604	100	57,915,035	100

Source: see Table 12.1.

The statistics of cooperative cellars make it possible to ascertain the approximate number of vinification units. They tell us that in 2000, there were 877 cooperative cellars, which, as we have seen, had a membership of 65,166 viticulturists. These 877 cooperative cellars together with the private cellars, were vinification units. The same calculation was carried out for 1980 and 1988 (cf. table below).

Table 12.5 Estimate of the number of vinification units in France at the end of the 20th century (in round figures), 1980-2000

	1980	1988	2000
No of vineyards selling their harvests	236,000	166,000	108,600
No of vinification units	109,600	57,900	33,100
Average area controlled by a private cellar (hectares)	4,9	8.1	15.2
Average area controlled by a cooperative cellar (hectares)	433.0	413.0	460
Average area controlled by one vinification unit (hectares)	9,4	15.9	26.5

Source: see Table 12.1.

As table 12.5 above shows, in twenty years the number of vinification units fell by over two thirds, whilst a little more than half of the vineyards disappeared. This can be attributed to the fact that the number of vineyards vinifying their own grapes had fallen much more rapidly than that of vineyards sending their grapes to cooperatives, the fall in the number of cooperatives playing only a small part in this phenomenon. In 1980 the number of vinification units had dropped to less than half of the number of vineyards. This ratio rose to 2.9 in 1988 and to 3.3 in 2000, pointing to the existence of a concentration of vinification during this period. This smaller number of vinification units in relation to the number of vineyards, had a significant effect on the areas relying on them. In 1980, a vineyard which was independent from the point of view of vinification, heading an area of 4.9 hectares or 0.7 hectares more than the average area of all vineyards, had a vinification unit measuring 9.4 hectares; whilst a cooperative controlled on average 433 hectares. Within the space of 20 years, the importance of concentration was apparent in growth of the average area of an independent vineyard. These became 3 times larger while the area of vinification units increased 2.8 times. Only the average area controlled by the cooperatives remained relatively stable, with an increase of 6 per cent per cooperative.

Without taking into account the sales of new vintages, the 103,041 viticulturists produce an average of 525 hectolitres. Whilst a vineyard vinifying in a private cellar produces an average of 743 hectolitres, one which sends its harvest to a cooperative cellar only produces 398 hectolitres, in the form of grapes. On the other hand, the corresponding vinification units, namely the cooperative cellars, regroup the supply by very markedly limiting the explosion of production from half of the vineyards, since each of them vinifies an average of 29,614 hectolitres of wine.

Different methods of vinification produce different results, depending on the category of the vines and hence of the produce obtained. In fact, the most recent census shows that AOC wines are more frequently produced in private cellars than the vins de Pays and table wines. The harvests from 51 per cent of the 'appellation d'origine controlée' (AOC) vineyard area are vinified in private cellars, compared with 46.3 per cent in cooperative cellars, with the remaining 2.70 per cent coming from new vintages. Only 26.9 per cent of the 'vins de Pays' and 32.9 per cent of table wines are vinified in private cellars. Of these latter two categories of wine, cooperative cellars vinify grapes respectively from 71.6 per cent and 65 per cent of the area of the vineyards. This state of affairs is largely the outcome of the geographical distribution of cooperative cellars, which are more numerous in the vineyards of the vins de Pays and of table wines (Languedoc-Roussillon) than in AOC vineyards.

We can see that on national level, cooperative cellars play an important role in the regrouping of the supply. As the number of vineyards controlled by cooperative cellars varies widely from region to region, a reproduction the national analysis concerning regions where cooperative cellars are of great importance (as is the case in Languedoc-Roussillon) would be of interest. The study of that region is all the more essential, because it produces large quantities of table wines and vins de Pays, making it comparable to Australian regions. We refer in this case to the census made in 2000.

Table 12.6   Distribution of vineyards and of volumes vinified according to the method of vinification adopted in Languedoc-Roussillon, 2000

	No of vineyards	%	Quantity (hectolitres)	%
Vinification in private cellars	3,027	9.24	4,693,877	24.26
Vinification in cooperative cellars	28,756	87.75	14,263,255	73.74
Sale of new vintages and others	979	3.00	386,751	2.00
Total	32,762	100.00	19,343,883	100.00

Source: see Table 12.1.

It can be seen from the table 12.6 that the region of Languedoc-Roussillon shows somewhat anachronistic behaviour within the French framework, since 73.74 per cent of the harvest is vinified in cooperative cellars. When one adds the 2 per cent of new vintages sold to merchants, only 24.26 per cent of wines vinified in private cellars remain. This is half of the national figure. Considering the number of vineyards, the situation is further emphasized by only 9.24 per cent of vineyards having a vinification unit available to them.

As before, private cellars correspond to vinification units. The 352 cooperative cellars of the Languedoc-Roussillon region reflect as many vinification units. The sum of these two entities, appearing in the table below, reflects less than the true total, the number of units owned by merchants being unknown.

Due to the importance of cooperative cellars, the number of vinification units is almost ten times less. Whilst private cellars account for 9.24 per cent of total number of vineyards, they represent almost 90 per cent of the number of vinification units (compared to over 97 per cent nationally). Private cellars vinify an average of 1,551 hectolitres of wine, or twice the national figure. As far as the cooperatives are concerned, they vinify an average of 40,521 hectolitres or 36 per cent more than the national figure. On average, a vinification unit vinifies 5,610 hectolitres compared with the national figure of 1,396 hectolitres. This shows that cooperative cellars play their part in the regrouping of the supply and that three quarters of the wine production comes from the large vinification structures.

Table 12.7 Distribution of vineyards and of volumes vinified according to the method of vinification adopted, 2000

	Units	%	Vinified volume (hectolitres)	%	Average volume vinified per unit (hl)
Private cellars	3,027	89.6	4,693,877	24.76	1,551
Cooperative cellars	352	10.4	14,263,255	75.24	40,521
Total	3,379	100	18,957,132	100	5,610

Source: see Table 12.1.

Roughly, these volumes correspond to the following areas: the vintage of a private cellar comes from a vineyard with an average area of 28.5 hectares, a cooperative cellar takes grapes from approximately 600 hectares. A vinification unit in the Languedoc-Roussillon region vinifies a harvest from an average area of 89 hectares.

It was said earlier on that most of the Languedoc-Roussillon wine production came from large vinification units. However, since the size of such a unit can only be established by comparison, we are going to make a comparison with the Australian wine-industry.

Australian vineyards have passed and are still passing through a phase of enormous growth, which makes France tremble. In 1995, the area of production was 62,454 hectares, reaching 100,000 hectares in 2000 and 143,400 hectares in 2002 (a sixth of French vineyards and half of the vineyards of Languedoc-Roussillon). The number of Australian vinification units known as 'wineries', have also increased. There were 845 in 1995 and 1,625 in 2002. A slight increase in concentration of the grapes taken to every winery, between these two dates also took place. In 1995, a winery vinified an average of 74 hectares of vineyard, rising to 88-89 hectares in 2000-2003.

If one compares the average area controlled by a winery (88-89 hectares) with the area controlled by a vinification unit in France (26.5 hectares), it becomes undeniably clear that Australia has vinification units almost three times larger than France. However, on national level, it can be seen that the presence of cooperative cellars (which vinify almost half of French production) plays an important part, especially since it is difficult to compare all French vineyards with Australian ones. On the other hand, as previously mentioned, the comparison with Languedoc-Roussillon appears to be fairer, since the average area of a vinification unit, 89 hectares, is about the same as in Australia.

The abovementioned averages may reflect a different distribution of respective vinification units and wineries. In order to verify and to refine what has been stated, we will now study the distribution by category of size of vinification units in Languedoc-Roussillon and in Australia. In Languedoc-Roussillon, that distribution will be used to analyse the concentration of vinification units between 1980 and 2000.

III.1. Concentration of volumes in the cooperative cellars of Languedoc-Roussillon between 1980-2000

We now have to see whether amalgamations of cooperative cellars led to a concentration of their production, and whether the said concentration was not hindered by the fall in global production, although a slight increase in the vinified quantity was noticed. This work has been carried out solely for the region of Languedoc-Roussillon, since we currently lack the data for other regions. The results have been assembled in the tables 12.8 and 12.9.

Table 12.8   Distribution by size category of the number of cooperative cellars and of the volumes vinified in 1980

Volume in hectolitres	No	%	Cumul. %	Volume	%	Cumul. %	Average volume
Over 200,000	3	0.57	0.57	655,007	2.97	2.97	218,336
100,000 - 200,000	20	3.77	4.34	2,521,936	11.45	14.43	126,097
75,000 - 100,000	45	8.49	12.83	3,959,088	17.98	32.41	87,980
50,000 - 75,000	94	17.74	30.57	5,626,992	25.56	57.96	59,862
25,000 - 50,000	194	36.60	67.17	6,930,029	31.47	89.44	35,722
10,000 - 25,000	114	21.51	88.68	1,988,588	9.03	98.47	17,444
Less than 10,000	60	11.32	100	336,978	1.53	100	5,616
Total	530	100		22,018,618	100		41,545

Source: see Table 12.1.

In 1980, only 30.57 per cent of cellars vinified over 50,000 hectolitres a year, a figure which corresponded roughly to an area of approximately 720 hectares. These cellars vinified about a quarter of the total production of all cellars. More than half of them had more modest capacities, from 10,000 to 50,000 hectolitres. These were responsible for approximately 40 per cent of the production.

Table 12.9 Distribution by category of size of the number of cooperative cellars and of the volumes vinified in 2000

Volume in hectolitres	No	%	Cumul. %	Volume	%	Cumul. %	Average volume
Over 200,000	3	0.85	0.85	872,769	6.17	6.17	290,923
100,000 - 200,000	24	6.78	7.63	3,097,192	21.89	28.06	129,050
75,000 - 100,000	12	3.39	11.02	976,185	6.90	34.96	81,349
50,000 - 75,000	54	15.25	26.27	3,065,017	21.66	56.62	56,760
25,000 - 50,000	131	37.01	63.28	4,318,163	30.52	87.13	32,963
10,000 - 25,000	97	27.40	90.68	1,650,643	11.67	98.80	17,017
Less than 10,000	33	9.32	100	170,022	1.20	100	5,152
Total	354	100		14,149,991	100		39,972

Source: see Table 12.1.

Twenty years later, when the number of cellars had fallen by 33 per cent, the situation hardly changed. It was found that the number of cellars vinifying over 50,000 hectolitres, have dropped slightly and that these cellars are currently vinifying a slightly smaller part of the production. Only the share of the larger cellars, those of over 100,000 hectolitres, has slightly increased. In 2000, the number of structures producing 10,000 to 50,000 hectolitres have increased, vinifying a smaller quantity than in 1980.

An examination of turnovers, available to us for 2000 only, shows that these turnovers were slightly less concentrated than the volumes. This could be attributed to the fact that the smaller cellars had to vinify the high quality (AOC) wines, whilst the larger ones vinified table wines and the vins de Pays.

It would of course be necessary to refine the calculations, the adjustments which we were obliged to make, perhaps having produced distortions on the level of concentration. It could, however be possible that the drop in number of cellars did not lead to a concentration of production. If this had been true, there would not have been an accentuated regrouping of the supply, as desired by the authorities over the past twenty years. An even more remarkable fact (one which is in opposition to the diagnosis of the permanent secretary of the Ministry of Agriculture) is that the absence of the concentration of cooperative cellars would not have impeded on exports - quite to the contrary.[5]

---

[5] It would be necessary to extend the analysis to the entire territory of France and to take the year 1970 as reference.

III.2. Comparison of the distribution of the respective vinification structures in Australia and in Languedoc-Roussillon at the end of the 20th century

The present condition of Australian data available to us, has made it necessary to prepare estimates, which will subsequently need to be verified. The thresholds adopted are not always identical and have introduced some approximate comparisons, which nevertheless will not spoil our conclusions.

Australia has a large number of small vinification units, corresponding particularly to our independent viticulturists whose harvests come from an area of less than 4 hectares. In the larger category, that of 4 to 36 hectares, a part of these units belong to independent vineyards, whilst others are the property of groups. Due to the importance of these two categories, the proportion of the larger units is, without a doubt, small in comparison with France. In respect of this, it has not been possible for us to split up the viticulturists who vinify for their own account. Since one may assume that the situation has hardly changed since the census of 1988, the majority of these units were split up in these two categories.

Table 12.10 Distribution by category of size of wineries, number and volume vinified in Australia in 2002

Category in hectares	Number	%	Volume		% Volume/unit
Over 1,800	9	0.68	5,381,810	57.26	597,979
900 to 1,800	11	0.83	1,163,813	12.38	105,801
36 to 900	96	7.28	1,280,96	13.63	13,343
4 to 36	160	12.14	190,078	2.02	1,188
Less than 36	1,042	79.06	1,383,032	14.71	1,327
Total	1,318	100	9,399,698	100	7,132

Source: *Australian Wine Industry Overview*, 2000 and 2002; *Vintage. The Australian Wine Industry Yearbook 2002.*

Australia and the Languedoc-Roussillon region are almost equal in the distribution of the number of vinification units by category of size. Let us examine the most important units, on which the superiority of the performances of the Australian wine-industry is based. Whilst in Australia there are 9 units (0.68 per cent), which control over 1,800 hectares of vineyards, there are 27 (0.8 per cent) in Languedoc-Roussillon. The following categories do not really contribute to this comparison, since 8.8 per cent of wineries take the harvests from more than 36 per cent of the area of vineyards, while the percentage in Languedoc-Roussillon is about 10.8 per cent.

Table 12.11 Distribution by category of size of vinification units, number and volume vinified in Languedoc-Roussillon in 2000

Number	Category in ha	%	Volume	%	Volume/unit
27	Over 1,800	0.80	3,969,961	21.07	147,036
66	900 to 1,800	1.95	4,041,202	21.45	61,230
261	36 to 900	7.72	6,138,828	32.58	23,520
3,027	Less than 36	89.53	4,693,877	24.91	1,550
3,381	Total	100	18,843,868	100	5,573

Source: see Table 12.1.

The examination of distribution of volume produced by entities belonging to different categories of vinification, leads to somewhat different conclusions. The largest wineries, without being proportionally more numerous than in Languedoc-Roussillon, produce almost double the quantities produced by their peers in Languedoc-Roussillon. On average, they vinify four times more wine than a unit in Languedoc-Roussillon. The lower category reverses the situation by producing almost double the quantity of wine produced by an Australian winery. By limiting this analysis to these two categories, it is found that Languedoc-Roussillon produces 42.5 per cent of its wines in units, controlling over 900 hectares, compared to 69.6 per cent produced in Australia. It does not appear credible from an economic point of view that these differences account for the international importance accredited to the Australian wine-industry. This seems to be all the more true as the costs of grape production are higher there than in Languedoc-Roussillon. A certain number of Australian wineries belong to groups , a phenomenon which does not exist in France. Could this be effecting a greater economic efficiency in Australian viniculture?

III.3. Amalgamations, groups and economic conditions in Australia and in Languedoc-Roussillon

We will be getting complaints concerning the foregoing comparison, to the effect that we did not take into account the existence of the large Australian groups such as BRL Hardy, Southcorp, Orlando and a number of others. They control the bulk of the Australian wine market. BRL Hardy, which has been acquired by Constellation Wines, controls about 20 per cent of Australian production. BRL Hardy, Southcorp, McGuigan and Orlando control 62 per cent of the Australian wine-industry. The ten leading groups account for over 80 per cent of production. French viniculture, such as that of Languedoc-Roussillon, appears to be far from achieving such a level of concentration.

Is it this formidable concentration which has enabled the Australian wine-industry to achieve the rapid success which it has scored in the British market? Many commentators seem to think so. This is doubtful, to begin with, because it is with structures, now thought to be archaic, that French viti-viniculture was able to win the place which it occupies in the international wine-trade. Secondly, it is not necessary (contrary to what is said on the subject) to imagine that the Australian groups are giant centralised factories - vinifying the grapes from their vineyards dispersed over the Australian territory, resulting in the lower production costs shown on the level of wineries.[6] If that were the case, there would be certain irrationality in not having regrouped or planted the vineyards around the vinification centres.

The reality is quite different. The groups were formed through the gradual purchase of vineyards, which often existed already. Let us take the example of BRL Hardy which came into being in 1992 by the merger of Berri Renamano and Thomas Hardy Wines. In 1976, Thomas Hardy had acquired several estates, namely, Emu Wine Co, Hougton Wines and others. Since the merger, Yarra Burn Winery, Hunter Ridge and others also have been acquired. In 2002, the group owned, as far as is known, 17 estates to which must be added 3 New Zealand vineyards and 1 French vineyard in Languedoc-Roussillon. It vinifies 313,500 tons of grapes harvested on 27,261 hectares. It owns only 2,400 hectares or 8.8 per cent. A large proportion of the grapes used are bought from many small producers (around a thousand of them). The other groups, such as BRL Hardy, have several estates, explaining the number of vinification units and their smallness compared with groups, which were mentioned earlier.

Moreover, the groups are far from being monolithic entities. Let us continue the analysis, using the example of BRL Hardy. Whilst 'Berri Estates', without possessing any vineyards, vinified some 70,000 tons of grapes, the same model does not apply to the other vinification centres of the group. At 'Yarra Burn' for example, production is only 200 tons (the equivalent of some 200,000 bottles), 70 tons of which are produced on 9.5 hectares of vineyard. One might assume that the wines produced in the different vinification units are combined in order to equalise the quality of a restricted number of brands spread over millions of specimens. To think so would be a mistake, since most of the group estates appear to have largely retained the autonomy of management of the individual brands. The company itself can own numerous brands, 25 having been determined for BRL Hardy. Owing to this, the quantity-related importance of each brand is reduced.

The example of 'Deakin Estates', belonging to the Wingara Wine Group (the 19th largest Australian group) demonstrates the autonomy of an estate belonging to a group.[7] Nothing on the labels of it's wines shows to whom it belongs. An examination of the counter label refers to the website of the estate and not to that of the group. Only at the bottom of the label does the name of the group appear.[8]

The Australian groups hardly differ from the French LVMH group, except in that LVMH not only owns wine and spirit companies, but also a whole series of luxury products. In the area of wines, LVMH is a majority shareholder in the estate of château d'Yquem, and

---

[6] The price of grapes paid by Australian vinification units to the producers, is higher than the price the cooperative cellars in the Languedoc-Roussillon region pay to their members.

[7] In the United Kingdom, publicity for wines produced by Lindemans or by Penfold, who belong to the Southcorp Wines group, is published under their name. It is the same in the case of Jacob's Creek, which belongs to Orlando Wyndham, owned in turn by the Pernod-Ricard group.

[8] See the website of these compagnies: www.hardywines.com.au; www.wingara.com.au and www.lvmh.com.

owns two of the most prestigious Champagne region groups, Moët et Chandon and Veuve Clicquot. The purchase of these two groups does not appear to have greatly modified their autonomy. When Moët et Chandon launches publicity in the British market, there may perhaps be repercussions on champagne in general, but certainly not more on Veuve Clicquot than on estates not belonging to LVMH. In fact, Veuve Clicquot, Moët et Chandon and the estates belonging to these two entities, function with their own brands. This would imply that in business, two and two does not make four. In other words, when companies are bought, the old brands or structures usually remain. This form of concentration does not bring about any economy of scale, any more than it multiplies the financial means of the different companies. It is the same in Australia. is the merits of gigantism of the Australian groups are not enough to boast about. It is also necessary to discover what their means of existence is rooted in. This is what we shall now examine.

Apart from the veneration with which some people regard groups, the logic of group amalgamations and acquisitions and the economic analysis of their real nature deserves to be questioned. Here, we will confine ourselves to mentioning one track. A group is defined as *a number of companies, which are legally independent but which form a single economic unit because of their close financial links*. However, we have seen that companies belonging to such an unit appear to have a real day-to-day autonomy. Thus, Veuve Clicquot brand wines are vinified in the company cellars, sold under its brand name and the company arranges its own publicity. If a group has real economic unity, that unity only comes to light during the consolidation of accounts. In that case, the production scale is not affected. Profits only grow at the rate of sales.

According to certain institutionalists, owners and shareholders do not have the same interests as the directors of a company. This is one of the fundamental characteristics of a modern company. According to these economists, the objective of the management team is not one of maximising profits, but one of maximising the volume of company sales. The management will favour that objective, because their own incomes or prestige depend more on the level of sales, than on that of the profits earned. According to the theory of W. Baumol, 'when a level of profit is being considered insufficient to provide a minimum level of income demanded by the shareholders, the practical objective of the company will be one of maximising its sales, which corresponds to the objective of maintaining and increasing its market share' (Baumol, 1959). The purchase of companies and the formation of a group are an excellent way of achieving these objectives, since this also makes it possible to get rid of less profitable companies, should that prove necessary.

Since the 1960s, the authorities have favoured the grouping of viti-vinicultural structures, despite this not necessarily achieving the desired effect. Mergers of cooperatives appeared, particularly in Languedoc-Roussillon, where around 145 occurred in 2000. According to our estimate, their average volume of sales was lower than the average produced by cooperative cellars. In view of the regrouping of supply, mergers could not have been very successful.[9]

---

[9] Regarding this issue, certain investigations remain to be carried out.

Some merged cooperatives nevertheless became as huge as the Australian groups. In Languedoc-Roussillon, the two largest amalgamations, UCCOAR and Val-d'Orbieu (which functions as bottlers) follow just behind the two largest Australian groups, BRL Hardy and Southcorp, in terms of quantities of wines sold. Being rather less like conglomerates than the latter, these two amalgamations possibly derive economic advantages. Moreover, an examination of the profit and loss of UCCOAR shows that this amalgamation only exports 10 per cent of its sales. It can be deduced from this that there is no close link between the size of groups and their success in exporting.[10]

IV. Conclusion

The analysis proposed in the introduction and partly proven here is to be pursued, particularly focussing on the effects of marketing on sales, the nature of Australian viti-viniculture, the explanation of English infatuation with Australian wines and so on. Nevertheless, certain conclusions, which contrast with what can be read in the press, have come to light in the course of this study.

We have seen that cooperative cellars came into existence in response to the difficulties, encountered by producers in the vinification processes and in the commercialisation of their wines. In the course of half a century, the cooperative cellars succeeded in gaining control of half of French wine production, admittedly with large variations from region to region. Apart from the importance which can be ascribed to them and the hopes which are placed on them, the cooperative cellars, and it is this which appears to be of the essence, have played and still do play a leading role in the processes of regrouping the supply. They have succeeded so well, that they can claim to play a part equal to that of the Australian wineries.

If one considers that some amalgamations accentuate the regrouping of the supply and that the groups do not possess the economic efficiency which their spokesmen ascribe to them, one may ask oneself whether French viti-viniculture does not hold all the trumps needed to face the challenge of a competition which, to use the parlance of war, is called ferocious.

---

[10] See www.uccoar.fr.

Bibliography

Annuaire national officiel des caves coopératives de France, 1949-1956.

Australian Bureau of Statistics (2002) *Australian Wine and Grape Industry*.

Baumol, W.J. (1959) *Business behaviour, value and growth*, New York.

Beeston, J. (1999) *The wine regions of Australia*, Saint-Leonards.

Berthomeau, J. (2001) *Comment mieux positionner les vins français sur le marché de l'exportation*, Paris.

Carrière, P. (1979) 'Les coopératives vinicoles en Languedoc-Roussillon', *Bulletin de la Société Languedocienne de Géographie*, April-June, pp. I-VIII.

Chevet, J.-M., Giraud-Hérault E. and R. Green (2005) « El sector vitivinicola francés ante los retos del mercado », Capitulo XII, S. Mili y S. Gatti (ed.) *Mercados agroalimentarios y globalizacion*. Madrid.

CFVF (2000) *Guide des caves coopératives de France*, Paris.

CNCV (1976) *Annuaire national des caves coopératives*, 4th ed.

CNCV (1982) *Annuaire national des caves coopératives*, 5th ed.

Delbos Ch. and Furestier M. (1970) 'Caves coopératives et commercialisation du vin en Sommiérois et Vaunage', *Bulletin de la Société Languedocienne de Géographie*, April-June pp. 167-181.

Gavignaud-Fontaine, G. (1987) 'Les caves coopératives, bastions de la viticulture populaire en Roussillon au XXe siècle', *Revue de l'économie sociale*, pp. 11-22.

Gavignaud-Fontaine, G. (2002) *Les caves coopératives dans le vignoble du Languedoc et du Roussillon*, Montpellier.

Gavignaud-Fontaine, G. (2000) *Le Languedoc viticole, la Méditerranée et l'Europe au siècle dernier (XXe)*, Montpellier.

Lachiver M. (1997) *Vins, vignes et vignerons. Histoire du vignoble français*, Paris.

Ministère de l'Agriculture, Agricultural census 2000. *Agreste - Cahiers*, no 3.

Montaine, E. (1999) 'Normes, qualités des vins et mutations de la viticulture du Languedoc', workshop, *Market, Right and Equity: Rethinking Food and Agricultural Grades and Standards in a Shrinking World*, October 31th - November 3th, Michigan State University.

Montaigne, E. (1997) 'Les mutations de la viticulture languedocienne mises en perspectives par deux siècles d'histoire', *Canadian Conference in Economic History*, May 1997, Niagara-on-the-Lake.

Office National interprofessionnel des Vins (1994) *Statistiques sur la filière viti-vinicole, t. I, Série rétrospective jusqu'au milieu de la décennie 1980*, Paris.

Office National interprofessionnel des Vins (1999) *Statistiques sur la filière viti-vinicole, t. II, Données récentes à partir du milieu de la décennie 1980,* Paris.

Office National interprofessionnel des Vins (2003) *Onivinstats, Données chiffrées sur la filière viti-vinicole 1993-2002*, Paris.

Office National interprofessionnel des Vins (2005) *Onivinstats, Données chiffrées sur la filière viti-vinicole 1995-2004*, Paris.

Pech, R. (1980) 'L'organisation du marché du vin en Languedoc et en Roussillon aux XIXe et XXe siècles', *Etudes rurales*, April-December, pp. 99-111.

Ramade-Beaujour, Ch. (1975) 'La coopération vinicole en Vaucluse de 1920 à 1940', *Etudes Vauclusiennes*, January-June, pp. 13-22.

Rinaudo, Y. (1985) 'La naissance de la coopérative viticole: les caves du Midi au début du XXème siècle', *Revue de l'économie sociale*, January-March, pp. 17-30.

Roudié, Ph. (1994) *Vignobles et vignerons du Bordelais, (1850-1980)*, Bordeaux.

Roudié, Ph. and Hinnewinkel J.-Cl. (2001) *Une empreinte dans le vignoble. XXe siècle: naissance des vins d'Aquitaine d'origine coopérative*, Bordeaux.

Simpson J. (2000) 'Cooperation and Cooperatives in Southern European Wine Production', *Advances in Agricultural Economic History*, vol. 1, pp. 95-1026.

Touzard, J.-M. (2002) 'Recensement 2001 des caves coopératives: diversité des stratégies et des résultats économiques', *Agreste - Languedoc-Roussillon*, October.

Villaret, J. (2002) 'Influence de la concentration des distributeurs européens sur la filière viticole', *Bulletin de l'O.I.V.*, vol. 75, no 853-854, pp. 195-207.

Vintage. The Australian Wine Industry Statistical Yearbook 2002-2003.

Vintage. The Australian Wine Industry Statistical Yearbook 2002.

13 Food production and food processing in western Europe, 1850-1990 Some conclusions

Paul BRASSLEY, University of Plymouth

I. Introduction

The historical literature on food production, insofar as it is synonymous with agriculture, is extensive; that on food processing is much less so. Merely making this statement raises two questions: where should the line be drawn between the two, and is not drawing the line at the other end, between processing and retailing, equally difficult? However, ignoring these problems for the moment, it is worth noticing that while most of the general histories of the European economy discuss agriculture at length, few of them consider the activities of millers, bakers, dairymaids, cheesemakers, drovers, or butchers, and various industrial food processors, to say nothing of wholesalers, innkeepers, and grocers, important as all these trades might be in feeding the European population. Some more specialist works have devoted a little space to their activities: for example, Cleary (1989) discusses agricultural organisations, including co-operatives in twentieth-century France, and Collins (2000) devotes a chapter to milling, malting, and food manufacturing. Montague (2000) includes some material on corn merchants in a work otherwise largely devoted to the agricultural input trades. There is certainly a growing specialist literature on food processing, distribution, retailing, and diet, nutrition, food safety and cooking, stimulated in recent years by the establishment of research groups such as The International Commission for Research into European Food History (ICREFH) and the European Institute for Food History (Teuteberg, 1992; Burnett and Oddy, 1994; den Hartog, 1995; Schärer and Fenton, 1998; Fenton, 2000; Hietala and Vahtikari, 2003; Jacobs and Scholliers, 2003). However, this work rarely connects with the agricultural literature, or histories of the wider economy and society. Moreover, much of it deals with individual countries rather than with Europe as a whole, with Braudel (1981) and Davidson (2002), in their different ways, standing out amid the honourable exceptions to these strictures.

The papers in this collection therefore represent a welcome and important excursion into hitherto little-explored territory. Whether or not they represent the only possible approach is another question, and one that cultural, ethnological, or social historians might answer differently from economic historians (and since economic historians are prone to specialisation, it might be interesting to examine the different approaches of rural and urban specialists). There is perhaps an interesting debate to be had on what should be seen as historically significant in the changing linkages between the producers of farm products and the consumers of food; the following remarks, however, do not range as widely as that, but attempt to deal with the questions raised by the previous chapters in this book.

II. Common themes

Considering the detailed nature of their contents, an adequate summary of the papers contained in this volume would require almost as much space as they already occupy. But it is possible to discern several common themes which appear in many of them: the effect of technical change; the impact of producer co-operatives; the move from small- to large-scale operation and from family to company ownership; the changes resulting from increasing international trade and growing urban markets; and the effects of changing attitudes to risk and food safety. The comprehensive survey of the Netherlands meat industry over the whole period from 1850 to 1990 by Koolmees, for example, emphasises the significance of many of these. Others give greater prominence to just one or two. Technical change is a prominent feature of the two papers concerned with fish products. As Drouard points out, the whole sardine canning industry arose from the application, in the 1820s, of food preservation techniques developed in the preceding two decades. In the case of the German deep-sea fishing industry, as Teuteberg shows, the critical developments were steam propulsion and preservation by freezing, both of which came to dominate the trade between 1885 and 1930. Steam power was also introduced to dairy processing in the two decades before the First World War, as Bieleman demonstrates, and American techniques in canning, freezing and the manufacture of breakfast cereals had a big impact, according to Collins, on the British market between the wars.

However, technical change rarely worked on its own; it was usually accompanied by economic or social change of some kind. In some cases, social factors delayed the adoption of new technology, as in Belgium, where Lefebvre and Segers demonstrate that the perception of dairying as women's work hindered capital investment in it by men. In the case of the German fishing industry, the new technology, with its bigger, more expensive ships, required investments far beyond the resources of independent fishermen, who became instead the employees of public limited companies. Similarly, Lummel's paper on the food industry in Berlin shows how the introduction of new technology in cereal milling and livestock slaughtering also produced economic change. The combination of steam-powered mills and factory baking from the later 1850s kept down the price of bread, although it did not drive small bakers out of business. On the other hand, the establishment of the central slaughterhouse in 1881 led to a transformation in the activities of the city's butchers. Whereas previously they had been vertically integrated, carrying out all stages of meat processing from the initial purchase of cattle, through slaughtering, to the final sale to the consumer, they now bought carcases from the slaughterers and concentrated on the retail end of the butchery trade. Similar specialisation followed the opening of a new abattoir in Ghent in 1857, as De Waele demonstrates. This relationship between technology, labour specialisation, capital requirements, and the size structure of firms in an industry can also be seen in the dairying trade. In Belgium (see Lefebvre and Segers) and the Netherlands (Bieleman), producers responded by establishing co-operatives, which came to dominate milk processing in the years before the First World War. After the war the same process occurred in the French wine trade, with the peak of the creation of 'caves co-operatives' occurring in the early 1930s, according to Chevet. The increasing economic power of the co-operatives could lead to increasing political power, as Nissen and Just demonstrate in their two papers on Denmark and the rest of Scandinavia. While such cartelisation might have been seen as a source of supernormal profits in non-agricultural industries, in agriculture it seems to have been perceived more as an instrument of social policy. And it is perhaps worth noting that even where co-operation remained insignificant, as in Great Britain, the interwar years also saw producer organisations increasingly influencing the market, in this case through the development of producer-controlled marketing boards (Tracy, 1982: 161-162). In this half century between about the middle of the 1930s and the middle of the 1980s farmers' organisations of one kind or another dominated the politics of the food chain.

In the late nineteenth and early twentieth centuries producers and processors were also coming to terms with increasing urban populations and the greater availability of food supplies from outside Europe. For some this created an opportunity: Israelsson shows how one of the bigger Swedish farm estates expanded its dairy output to cater for a nearby urban market, although neighbouring small farms continued to operate a policy of domestic self-sufficiency during the same period. On a wider scale, the penetration of the British market by producers in Denmark, the Netherlands and Ireland may be seen as part of the same process, although of course it might also be seen as part of the internationalisation of the food industry that began in the late nineteenth century as producers from North and South America, and from the southern hemisphere, began to trade with European markets, with varying results in different countries (O'Rourke, 1997). Historians such as Michael Tracy (1982) have made us familiar with this as a commodity trade, but by the interwar years, as Collins shows, it was a more complex story, with firms from the United States developing an expanding demand for branded products and introducing American business practices and promotional techniques to the European market. This raises some interesting questions for those interested in the development of the global economy. Is the process described by Collins part of the expansion of international trade in food which had been going on since at least the 1870s, if no earlier, or is it something more, perhaps the one of the first signs of globalisation? The difference between the two is said to be that international trade simply requires the physical movement of goods, whereas in globalisation labour and technology transfer, branding, and foreign direct investment play a much greater role (Woods, 2000: 3). It would seem that the processes described by Collins involved much more than the simple importation of food products.

Maria De Waele identifies the 1960s and 1970s as the decades in which consumers became increasingly sensitive to food safety issues, and indeed it was so as far as their self-education and organisation, supported by the mass media, were concerned. Much earlier a small professional middle class were pushing in the same direction. As Koolmees points out, meat hygiene inspections were being carried out in the Netherlands from the 1860s onwards, and from the 1880s were increasingly based on scientific investigations. Similar policies were instituted in Belgium, France and Britain (Atkins, 2004). Indeed, food safety was an issue in Britain by 1850, and became even more so as the use of chemical preservatives increased after 1890 (Collins and Oddy, 1998). Health issues were affecting the dairy trade by the 1930s as consumers became increasingly aware that milk could transmit tuberculosis, and as the bigger retailers increased their market share the legal implications of food-borne disease had a greater and greater impact on the services they demanded from the food processing firms.

These are the common themes, at least in terms of subject matter, which emerge from a reading of the papers in this collection. There is also a methodological theme: much of the work reported here is more descriptive than analytical, which is perhaps inevitable, given the underdeveloped state of historical work on the food chain between the primary producer and the final consumer. There are terminological questions too: at what point does food processing cease to be a part of farm work and instead become part of manufacturing industry; how should we describe those who work part-time or seasonally in two or more industries; and precisely which terms, in what circumstances, are appropriate for women involved in farm, factory, or commercial activities?

As an alternative to this thematic approach, we might also attempt to summarise this mass of material by constructing a dualistic model, comparing 1850 with 1990. This requires much neglect of the nuances and the temporal and spatial details contained in these studies, but it may be useful as a basis for discussion and analysis. Thus we might compare the food production and processing industries at two different points in time as in figure 13.1.

Figure 13.1 General characteristics of the food production and processing industries in 1850 and 1990

1850	*1990*
Many small producers,	Many small but some big producers,
Many traditional processors,	Some traditional, and a few big high-tech often the same people as processors, selling, to
the large number of traditional retailers	some traditional and a few very big retailers,
selling commodities,	mostly branded goods,
to poor consumers	for rich consumers
with high price and income elasticities,	with low price and income elasticities
in an essentially local food chain,	in an essentially global food chain,
little influenced by politics	much influenced by national and international politics

The papers in this collection certainly provide evidence to confirm the 1850 model: the small herds described by Israelsson sold little milk or butter but used their production at home, and the people using the Gent municipal slaughterhouse in 1850 were largely the small butchers. Even as late as 1900, according to Chevet, 1.7 million hectares of vines were shared between 1.78 million growers, and in 1850 the Elbe fishermen were still operating in small sailing boats. On the other hand, there is also conflicting evidence: Koolmees dates the beginnings of the international meat trade to the 1840s, and by 1850 the butter producers in the Netherlands were already dependent upon the British market; sardine canning began in the 1820s, and, as Lummel shows, there were indeed small suppliers in farmers' markets in the Berlin of 1850, but there were also food factories. On the whole, however, this model does not seem too unrealistic. It might usefully be complicated by inserting some intermediate points, such as 1914 and 1950. In 1914 many of the features of the 1850 model might remain in place, although the food chain would no longer be so local in the free-trading countries, given increasing imports of cereals, meat, and dairy products. By 1950 the price and income elasticities would be falling, and the proportion of branded goods would be increasing as the modern food industry became established, but the dominance of the market by the big producers and retailers that is so much in evidence in 1990 would not be so apparent. But precisely how, when, and why the changes took place could still usefully be the subject of further elucidation.

III. The issues still to be investigated

It is clearly an impossible task to cover the history of the whole of food production and processing in all the North Sea countries over the last 150 years in a mere dozen papers. At this point therefore it is useful to identify the more obvious omissions from this volume.

Simply in terms of major commodities, the papers in this collection have little to say about cereals, sugar, potatoes, fruit and vegetables, and it would certainly be interesting to compare the development of meat and dairy markets, often the principal concerns of small farmers, with the major arable commodities, often produced by the bigger farms. Neither do they deal with imported, often tropical products, such as tea, coffee, oilseeds, or sugar, even though these were the products upon which the fortunes of several of the more influential processors (e.g.Unilever) and distributors were founded, and oilseeds and sugar would become major farm products in the North Sea area in the twentieth century (see, for example, Fine et al, 1996: chap. 5). What they do make clear, however, is that to see farm products as undifferentiated commodities is an oversimplification. As the marketing chain grew more complex over this period so buyers came to specify their requirements more precisely. Wheat varieties appropriate to the animal feed trade would not satisfy the requirements of the flour millers; feeding barley varieties do not make good malting samples; the potato varieties grown by Breton or Cornish producers aiming at early markets will not be the same as those used by maincrop producers. At the next link in the marketing chain, as Lefebvre and Segers demonstrate, the location of a co-operative in Belgium could have a major impact on the kind of product it could sell. And in any case, as Collins shows in the case of the English market, retailers and their consumers, by the interwar period, were, to a large extent, no longer buying commodities, but branded products. What we do not know, however, is how unusual the British market was in this respect, and comparative studies from other countries would be useful. They would also be useful on the question of vertical integration. Medieval farm women, taking their butter and eggs to town to sell on market days, were as vertically integrated as those producers who invest in farm shops in the 21st century. However, the main story of the link between farmer and food consumer is one of specialisation as the food processing and retailing industries developed. But to what extent did farmers remain, through co-operative ownership or management, in control of food processing and distribution? Equally, how much did retailers attempt to become processors, or food processors buy their own farms?

Most of the papers in this collection place their emphasis on describing what was happening, and given the current extent of our knowledge of the history of the food industry and the entire food chain on a European scale it is probably right that they should do so. Before we can proceed to analysis and explanation there are several further aspects of the story that merit investigation. One of these is surely transport cost changes, both their extent and the reasons for them. For perishable goods, such as milk, milk products, meat, and fish, the development of rail transport and refrigerated sea transport in the nineteenth century, and of road transport in the twentieth century, clearly had a major impact. Price (1983) has examined this process in the context of nineteenth-century France, but there appear to be few comparable studies. How and when these changes came about clearly needs to be investigated to a much greater degree than it has been hitherto. Transactions costs might be seen as related to transport costs, and it would be interesting to have a study of the impact of the telephone on the transmission of orders, or of the trade press and the broadcast media on the transmission of market information. It is impossible to imagine the late twentieth century food chain without the use of computers for stock control, but again little appears to be known about the origins and development of such systems. Another aspect is retailing. It is clear that the concentration and consequently increased market power of the food retailers in the late twentieth century has had a major impact on primary producers and processors, but how widespread this development has been across Europe, and when and why it began, are by no means clear. None of the papers in this collection deals specifically with this topic, although some work has been done elsewhere on individual countries (Freeman, 1989; Scola, 1992; Collins, 2000; Otterloo, 2000) and there are some relevant recent statistics for several European countries in Hughes (2002).

The position is similar with the next and final stage of the process: the consumer, and changing influences on consumer behaviour. Again, none of the papers here specifically examine consumers, although Koolmees and Lummel mention the impact of increasing disposable incomes on the demand for meat. In the United Kingdom the National Food Survey (MAFF, 1991) has now been in existence for more than 50 years and has produced an enormous amount of information on consumption patterns and demand changes in the second half of the twentieth century, but it appears to have been little used by historians. On the other hand, it has been extensively employed, along with other data sources, in recent studies of late-twentieth-century food consumption (Fine and Leopold, 1993; Fine et al, 1996). For a slightly earlier period, and on a Europe-wide scale, the papers collected in Just and Trentmann (2006) are also mostly concerned with aspects of consumption, as are those in Jacobs and Scholliers (2003) for the period from 1750.

Finally, it is interesting to note that few of these papers pay much attention to gender aspects and related changes. In the Netherlands, in the late nineteenth century, as Margreet van der Burg (2002: 346) points out, 'dairy processing was a woman specific labour domain'; by the middle of the twentieth century it had been 'externalised'. Mary Bouquet (1985) has noticed a similar process in south west England, and Joanna Bourke (1990), discussing it in an Irish context, points out its relationship with education and professionalization (on which see also Valenze, 1991, and Sommestad, 1992). Food processing and retailing are probably among the major employers of female labour, and it would be interesting to know much more about the changes that have undoubtedly occurred and their social and economic impacts.

IV. What should be done next?

There is still plenty of room for descriptive studies of food distribution, as the lacunae identified above suggest. Just as farmers have come to realise the significance of increasing concentration in the market chain as it affects their ability to negotiate prices, so agricultural historians are beginning to identify food processing and distribution as topics worthy of study. Hitherto such work has been in danger of falling into the gap between rural and urban history, but studies like those here, or, for an earlier period, those of Campbell et al (1993) on the food supply of medieval London, or Kaplan (1984), show the potential benefits of linking the disciplines. There is also clearly a place for more comparative studies in which the differing responses of different regions or countries to the same pressures can suggest fruitful lines of enquiry. The discussion above also suggests several explanatory variables which could usefully be investigated further: changing transport costs, the concentration of retailing, the evolution of consumer demand, and the various factors affecting transactions costs, for example.

Most Europeans now relate to food production at the supermarket shelf rather than the farm gate. Although food now takes only a small proportion of their disposable income – typically less than 20 per cent – it is still among the most regularly purchased and widely advertised of the products they buy. When it goes wrong – as in the case of BSE or Salmonella – it is among the most politically sensitive. The story of Camembert, as told by Boisard (2003), has been identified by Steven Shapin as a 'metonym for modernity' (Shapin, 2003: 15). This book reveals the enormous potential for further historical work on the impact of the developing modern (and post-modern?) economy on some of the most traditional of economic activities.

V. Acknowledgements

The author would like to extend enormous thanks to Margreet van der Burg, Yves Segers, Derek Shepherd, and Ted Collins for their very valuable comments and suggestions on an earlier draft of this chapter.

Bibliography

Atkins, P. (2004) 'The Glasgow case: meat, disease and regulation, 1889-1924', *Agricultural History Review*, 52, pp. 161-182.

Boisard, P. (2003) *Camembert: a national myth*, California.

Bouquet, M. (1985) *Family, Servants and Visitors: the farm household in nineteenth- and twentieth-century Devon*, Norwich.

Bourke, J. (1990) 'Dairywomen and Affectionate Wives: Women in the Irish Dairy Industry 1890-1914', *Agricultural History Review*, 38, pp. 149-164.

Braudel, F. (1981) *Civilization and Capitalism, 15th - 18th Century, vol I: The Structures of Everyday Life*, London.

Burg, M. van der (2002) *'Geen Tweede Boer': gender, landbouwmodernisering en onderwijs aan plattelandsvrouwen in Nederland, 1863-1986*, Wageningen (A.A.G.Bijdragen 41).

Burnett, J. and Oddy, D. (eds.) (2004) *The origins and development of food policies in Europe*, London.

Campbell, B.M.S., et al. (1993) *A Medieval Capital and its Grain Supply: Agrarian Production and Distribution in the London Region c.1300*, Institute of British Geographers, Historical Geography Research Series No.30.

Cleary, M.C. (1989) *Peasants, Politicians and Producers: the organisation of agriculture in France since 1918*, Cambridge.

Collins, E.J.T. (ed.) (2000) *The Agrarian History of England and Wales*, vol. VII, 1850-1914, Cambridge.

Collins, E.J.T. and Oddy, D.J. (1998) 'The centenary issue of the *British Food Journal*, 1899-1999 - changing issues in food safety regulation and nutrition', *British Food Journal*, 100 (10/11), pp. 434-460.

Davidson, A. (2002) *The Penguin Companion to Food*, London.

Den Hartog, A.P. (ed.) (1995) *Food technology, science and marketing: European diet in the twentieth century*, East Linton.

Fenton, A. (ed.) (2000) *Order and disorder: the health implications of eating and drinking in the nineteenth and twentieth centuries*, East Linton.

Fine, B. and Leopold, E. (1993) *The World of Consumption*, London.

Fine, B., Heasman, M. and Wright, J. (1996) *Consumption in the age of affluence: the world of food*, London.

Freeman, S. (1989) *Mutton and Oysters: the Victorians and their Food*, London.

French, M.J. and Phillips, J. (2003) 'Sophisticates or dupes? Attitudes towards food consumers in Edwardian Britain', *Enterprise and Society*, 4, pp. 442-470.

Mamilton, S. (2003) 'The economics and conveniences of modern-day living: frozen foods and mass marketing 1945-65', *Business History Review*, 77, pp. 33-60.

Hietala M. and Vahtikari, T. (eds.) (2003) *The landscape of food: the food relationship of town and country in modern times*, Helsinki.

Hughes, D. (2002) 'Grocery Retailing in Europe', *EuroChoices*, 1(3), pp. 12-16.

Jacobs, M. and Scholliers, P. (eds.) (2003) *Eating out in Europe. Picnics, gourmet dining and snacks since the late eighteenth century*, Oxford.

Just, F. and Trentmann, F. (eds.) (2006) *Food and conflict in Europe in the age of the two World Wars*, London.

Kaplan, S.L. (1984) *Provisioning Paris: merchants and millers in the grain and flour trade during the 18th century*, Ithaca.

MAFF [Ministry of Agriculture, Fisheries and Food] (1991) *Household Food Consumption and Expenditure 1990*, London.

Montague, D. (2000) *Farming, Food and Politics: the merchant's tale*, Dublin.

O'Rourke, K.H. (1997) 'The European Grain Invasion 1870-1913', *Journal of Economic History*, 57(4), pp. 775-801.

Otterloo, A.van (eds.) (2000) 'Voeding', Schot, J.e.a (eds.) *Techniek in Nederland in de twintigste eeuw. III. Landbouw. Voeding*, Zutphen, pp. 235-375.

Price, R. (1983) *The Modernization of Rural France*, London.

Schärer, M.R. and Fenton, A. (eds.) (1998) *Food and Material Culture*, East Linton.

Scola, R. (1992) *Feeding the Victorian city: the food supply of Manchester 1770-1870*, Manchester.

Shapin, S. (2003) 'Cheese and Late Modernity', *London Review of Books*, 25 part 22 (20 November 2003), pp. 11-15.

Sommestad, L. (1992) 'Able dairymaids and proficient dairymen: education and de-feminization in the Swedish dairy industry', *Gender and History*, 4, pp. 34-49.

Teuteberg, H.J. (ed.) (1992) *European Food History. A research overview*, London.

Tracy, M. (1982) *Agriculture in Western Europe: challenge and response 1880-1980*, London.

Valenze, D. (1991) 'The art of women and the business of men. Women's work and the dairy industry c.1740-1840, *Past and Present*, 130, pp. 142-169.

Woods, N. (ed.) (2000) *The Political Economy of Globalization*, London.